ELLIPTIC PROBLEM SOLVERS

ACADEMIC PRESS RAPID MANUSCRIPT REPRODUCTION

Proceedings of the Elliptic Problem Solvers Conference
Held in Santa Fe, New Mexico
From June 30–July 2, 1980
Sponsored by the Los Alomos Scientific Laboratory

Elliptic Problem Solvers

Edited by

Martin H. Schultz

Department of Computer Science
Yale University
New Haven, Connecticut

ACADEMIC PRESS 1981
A Subsidiary of Harcourt Brace Jovanovich, Publishers

New York London Toronto Sydney San Francisco

ACADEMIC PRESS, INC.
111 Fifth Avenue, New York, New York 10003

United Kingdom Edition published by
ACADEMIC PRESS, INC. (LONDON) LTD.
24/28 Oval Road, London NW1 7DX

Library of Congress Cataloging in Publication Data
Main entry under title:

Elliptic problem solvers.

Proceedings of the Elliptic Problem Solver Conference held in Santa Fe, N.M., June 30-July 2, 1980, which was sponsored by the Los Alamos Scientific Laboratory.
Includes index.
1. Differential equations, Elliptic--Numerical solutions--Congresses. I. Schultz, Martin H. II. Elliptic Problem Solver Conference (1980 : Santa Fe, N.M.) III. Los Alamos Scientific Laboratory.
QA377.E53 515.3'53 81-3611
ISBN 0-12-632620-7 AACR2

PRINTED IN THE UNITED STATES OF AMERICA

81 82 83 84 9 8 7 6 5 4 3 2 1

Contents

Contributors

Numbers in parentheses indicate the pages on which the authors' contributions begin.

John C. Adams (187), National Center for Atmospheric Research, Boulder, Colorado

Owe Axelsson (191), University of Nijemegan, The Netherlands

Petter Bjørstad (203), Department of Computer Science, Stanford University, Stanford, California 94305

Randolph E. Bank (1), Department of Mathematics, University of Texas at Austin, Austin, Texas 78712

Th. Beelen (383), N.V. Philips' Gloeilampenfarieken, Eindhoven, The Netherlands

Marsha Berger (197), Department of Computer Science, Stanford University, Stanford, California 94305

Garrett Birkhoff (17), Department of Mathematics, Harvard University, Cambridge, Massachusetts 02138

J. L. Blue (435), Scientific Computing Division, National Bureau of Standards, U.S. Department of Commerce, Washington, D. C. 20234

Ronald F. Boisvert (219), Center for Applied Mathematics, National Bureau of Standards, Washington, D. C.

Alain Bossavit (225), Electricite de France, Direction des Etudes et Recherches, 1, Avenue du General de Gaulle, 92140 Clamart, France

Achi Brandt (39), Weizmann Institute of Science, Rehovot, Israel

John J. Buoni (231), Department of Mathematics, Youngstown State University, Youngstown, Ohio 44555

B. L. Buzbee (85), Computer Science and Services Division, Los Alamos Scientific Laboratory, Los Alamos, New Mexico 87545

D. A. Calahan (241), Department of Electrical and Computer Engineering, University of Michigan, Ann Arbor, Michigan 48109

J. E. Dendy, Jr. (247), Los Alamos Scientific Laboratory, P.O. Box 1663, MS 233, Los Alamos, New Mexico 87545

Louis W. Ehrlich (255), Applied Physics Laboratory, The Johns Hopkins University, Laurel, Maryland 21218

Samuel C. Eisenstat (99), Department of Computer Science, Yale University, New Haven, Connecticut 06520

David J. Evans (269), Department of Computer Studies, Loughbourgh University of Technology, Loughborough, Leicestershire, United Kingdom

Vance Faber (271), Los Alamos Scientific Laboratory, P.O. Box 1163, MS 265, Los Alamos, New Mexico 87545

Wolfgang Fichtner (277), 2A-325, Bell Laboratories, Murray Hill, New Jersey 07974

Hartmut Foerster (285) , Institut für Mathematik, GMD, Postfach 12 40, Schloss Birlinghoven, D-5205 St. Augustin 1, Federal Republic of Germany

Patrick W. Gaffney (301), Computer Science Division, Oak Ridge National Laboratory, Oak Ridge, Tennesee

Charles I. Goldstein (307), Applied Mathematics Department, Brookhaven National Laboratory, Upton, New York 11973

William J. Gordon (315), Department of Mathematical Sciences, Drexel University, Philadelphia, Pennsylvania 19104

P. W. Hemker (383), Mathematisch Centrum, Amsterdam, The Netherlands

David J. Hunt (339), International Computers Limited, Computer House, 322 Euston Road, London NW1 3BD, England

J. M. Hyman (247), Los Alamos Scientific Laboratory, University of California, MS 233, Los Alamos, New Mexico 87545

Olin G. Johnson (345), System Technology Department, IBM Thomas J. Watson Research Center, Yorktown Heights, New York 10598

Sandra J. Kelly (315), Department of Mathematical Sciences, Drexel University, Philadelphia, Pennsylvania 19104

David R. Kincaid (163, 353), Center for Numerical Analysis, RLM 13.150, University of Texas at Austin, Austin, Texas 78712

Lois Mansfield (361), Department of Applied Mathematics and Computer Science, University of Virginia, Thornton Hall, Charlottesville, Virginia 22901

C. Wayne Mastin (369), Department of Mathematics, Mississippi State University, P.O. Drawer MA, Mississippi State, Mississippi 39762

R. A. Nicolaides (115), Department of Mathematics, Carnegie-Mellon University, Pittsburgh, Pennsylvania 15213

Joe Oliger (197), Department of Computer Science, Stanford University, Stanford, California 94305

Seymour V. Parter (375), Department of Mathematics, University of Wisconsin-Madison, Madison, Wisconsin 53706

George Paul (345), System Technology Department, IBM Thomas J. Watson Research Center, Yorktown Heights, New York 10598

S. J. Polak (383), N.V. Philips' Gloeilampenfabrieken, Eindhoven, The Netherlands

Wlodzimierz Proskurowski (391), Department of Mathematics, University of Southern California, University Park, Los Angeles, California 90007

John R. Rice (135), Mathematical Sciences, Purdue University, West Lafayette, Indiana 47907

Patrick J. Roache (399), Ecodynamics Research Associates, Inc., Albuquerque, New Mexico

Garry Rodrique (197), Computer Research, Lawrence Livermore National Laboratory, Livermore, California

Donald J. Rose (277), Bell Laboratories, Murray Hill, New Jersey 07974

Y. Saad (405), Department of Computer Science, University of Illinois at Urbana-Champaign, Urbana, Illinois 61801

Ahmed Sameh (405), Department of Computer Science, University of Illinois at Urbana-Champaign, Urbana, Illinois 61801

N. L. Schryer (413), Bell Laboratories, Murray Hill, New Jersey 07974

Martin H. Schultz (99), Computer Science Department, Yale University, New Haven, Connecticut 06520

L. Ridgway Scott (419), Department of Mathematics, 347 West Engineering Building, University of Michigan, Ann Arbor, Michigan 48109

Andrew H. Sherman (429), Department of Computer Sciences, University of Texas at Austin, Austin, Texas 78712

Vincent G. Sigillito (423), Applied Physics Laboratory, The Johns Hopkins University, Laurel, Maryland 20810

Klaus Stüben (285), Institut fur Mathematik, GMD, Postfach 12 40, Schloss Birlinghoven, D-5205 St. Augustin 1, Federal Republic of Germany

Paul N. Swartztrauber (187), National Center for Atmospheric Research, Boulder, Colorado

Roland A. Sweet (187), National Center for Atmospheric Research, Boulder, Colorado

J. F. Thompson (369) , Department of Mathematics, Mississippi State University, P.O. Drawer MA, Mississippi State, Mississippi 39762

Ulrich Trottenberg (285), Institut für Angewandte Mathematik, and Sonderforschungsbereich 72, Bonn University, Bonn, West Germany

Richard S. Varga (231), Department of Mathematics, Kent State University, Kent, Ohio 44242

A. Wachters (383), N.V. Philips' Gloeilampenfabrieken, Eindhoven, The Netherlands

Stephen J. Webb (339), International Computers Limited, Computer House, 322 Euston Road, London NW1 2BD, England

Alan Wilson (339), International Computers Limited, Computer House, 322 Euston Road, London NW1 2BD, England

C. L. Wilson (435), Electron Devices Division, National Bureau of Standards, U.S. Department of Commerce, Washington, DC 20234

David M. Young (163, 353), Center for Numerical Analysis, University of Texas, Austin, Texas 78712

Preface

This volume contains the proceedings of the Elliptic Problem Solver Conference held in Santa Fe, New Mexico, from June 30 to July 2, 1980. The conference was sponsored by the Los Alamos Scientific Laboratory with financial support provided by the Applied Mathematical Science Program, Office of Energy Research, U.S. Department of Energy. The program committee consisted of Garrett Birkhoff, Bill Buzbee, John Rice, Martin Schultz, and David Young.

All of the papers presented at the conference dealt with some aspect of the numerical solution of elliptic partial differential equations, a subject central to much of large-scale scientific computation. As discussed in the paper presented by Garrett Birkhoff, this subject has a long, rich history dating back to the first investigations of numerical methods for partial differential equations. The conference marked the 30th anniversary of David Young's landmark Ph.D. dissertation, *Iterative Methods for Solving Partial Differential Equations of Elliptic Type*.

The papers vividly demonstrated current advances in developing elliptic problem solvers and analyzing their performance. Interest in scientific computing is rapidly increasing, algorithms and the mathematical analysis of their performance are improving, and both hardware and software technologies are developing. These advances will undoubtedly continue for the foreseeable future. The organizing committee hopes that these proceedings will serve not only to summarize the current state of the art but also to help chart the course for the future.

It is with utmost gratitude that I thank Laurie Fusco and Mary-Claire van Leunen for their assistance in preparing these proceedings.

A MULTI-LEVEL ITERATIVE METHOD FOR NONLINEAR ELLIPTIC EQUATIONS

Randolph E. Bank[1]

Department of Mathematics
University of Texas at Austin
Austin, Texas

I. INTRODUCTION

In this work we consider the solution of the quasi-linear elliptic boundary value problem

$$-\Delta u + f(u, u_x, u_y) = 0 \quad \text{in} \quad \Omega \subset R^2 . \tag{1.1}$$
$$\frac{\partial u}{\partial n} = 0 \quad \text{on } \partial\Omega$$

We shall assume that f is such that a unique solution of (1.1) exists [15].

We will discuss the approximate solution of (1.1) using a standard Galerkin method employing piecewise linear triangular finite elements [1,17,20], a damped Newton iteration [7,8,10,11,12,18] using a multi-level iterative method for the resulting linear systems, [4,5,6,9,13,14,16] and an adaptive mesh refinement procedure [2,3]. The presentation here will be mainly of a nontechnical nature; the reader is referred to the references cited above for details.

In Section 2, we begin with a discussion of a multi-level method for linear problems. In Section 3, we discuss adaptive mesh refinement, while in Section 4 we describe a multi-

[1]Supported by ONR grant N00014-80-C-0645

ISBN 0-12-632620-7

level method for nonlinear problems. In Section 5, we give a numerical illustration.

II. A MULTI-LEVEL ITERATIVE METHOD FOR LINEAR PROBLEMS

Consider the linear elliptic boundary value problem

$$-\nabla\cdot(a\nabla u)+b\cdot\nabla u+cu = f \quad \text{in} \quad \Omega \subset R^2; \tag{2.1}$$
$$\frac{\partial u}{\partial n} = 0 \quad \text{on} \quad \partial\Omega.$$

We assume a is positive and all the coefficient functions are smooth. For convenience, we shall also assume Ω is polygonal. A weak formulation of (2.1) is: find $u \in H^1(\Omega)$ such that

$$a(u,v) = (f,v) \quad \text{for all} \quad v \in H^1(\Omega) \tag{2.2}$$

where

$$a(u,v) = \int_\Omega a\nabla u\cdot\nabla v+b\cdot\nabla uv+cuvdx; \tag{2.3}$$
$$(f,v) = \int_\Omega fvdx$$

Here $H^k(\Omega)$ denotes the Sobolev space equipped with the norm

$$\|u\|_k^2 = \sum_{|\alpha|\leq k} \|D^\alpha u\|_{L_2(\Omega)}^2 \tag{2.4}$$

For non-integral and non-positive k, $H^k(\Omega)$ is defined in a standard fashion [17]. We assume that there exists a constant $\alpha = \alpha(a,b,c,\Omega)$, $0 < \alpha \leq 1$, such that $u \in H^{\alpha+1}(\Omega)$ provided $f \in H^{\alpha-1}(\Omega)$ and

$$\|u\|_{1+\alpha} \leq C(a,b,c,\Omega)\,\|f\|_{\alpha-1}. \tag{2.5}$$

Let T_1 be a fixed triangulation of Ω satisfying quasi-uniformity and shape regularity assumptions [17,20]. We construct a sequence of quasi-uniform, shape regular triangulations T_j, $j > 1$, as follows: for each triangle $t \in T_{j-1}$, refine t into 4 triangles in T_j by pairwise connecting the midpoints. Let h_j denote the diameter of the largest triangle in T_j; then $h_j = h_1 2^{1-j}$.

We associate with each triangulation T_j the N_j-dimensional space of continuous piecewise linear polynomials M_j [1,17,20]. Because of our construction of T_j, $M_{j-1} \subset M_j$, and $N_j \doteq 4N_{j-1}$, $j > 1$.

We consider the following sequence of finite dimensional problems: find $u_j \in M_j$ such that

$$a(u_j,v) = (f,v) \quad \text{for all} \quad v \in M_j. \tag{2.6}$$

For h_1 sufficiently small, it can be shown [1,17,19,20] that the problems (2.6) have unique solutions and that

$$\| u-u_j \|_1 \leq K(a,b,c,\Omega,T_1,u)h_j^\alpha \tag{2.7}$$

In the multi-level scheme we shall use [4,5], we compute approximations $\tilde{u}_j \in M_j$ which satisfy error bounds of the form

$$\| u-\tilde{u}_j \|_1 \leq \tilde{K}h_j^\alpha. \tag{2.8}$$

The computational effort required to produce such solutions is asymptotically (as $j\to\infty$) proportional to the number of unknowns N_j.

Our multi-level iterative method consists of two distinct strategies:

(A) sequentially computing approximate solutions $\tilde{u}_j$ to (2.6), using $\tilde{u}_{j-1} \in M_{j-1} \subset M_j$ as the initial guess for $\tilde{u}_j$.

(B) using a method, called the j-level iteration, which has the property of reducing the initial error by a fixed factor $0 \leq \gamma < 1$ (γ is independent of j) in work which is proportional to N_j as $j\to\infty$.

First, assuming we have an iterative method with the properties described in (B) above, consider the following algorithm for solving (2.6).

Algorithm I

(i) for $j = 1$ solve (2.6) exactly ($\tilde{u}_1 = u_1$).

(ii) for $j > 1$, starting from $\tilde{u}_{j-1}$, use r iterations of the iterative method in (B) to compute $\tilde{u}_j$; r is chosen such that $\gamma^r = \varepsilon < 2^{-\alpha}$.

In view of (ii) above, we have

$$\| \tilde{u}_j - u_j \|_1 \leq \gamma^r \| \tilde{u}_{j-1} - u_j \|_1 < 2^{-\alpha} \| \tilde{u}_{j-1} - u_j \|_1 . \tag{2.9}$$

If u_j and $\tilde{u}_j$ satisfy (2.7) and (2.9), a simple induction argument can be used to show $u - \tilde{u}_j$ satisfies (2.8) with

$$\tilde{K} = K \cdot \left(1 + \frac{\gamma^r (1+2^\alpha)}{1-\gamma^r 2^\alpha}\right). \tag{2.10}$$

Let the work required for a single iteration of the j-level iteration (B) be bounded by CN_j. Then the work involved in Algorithm I is (using $N_j \doteq 4N_{j-1}$)

$$r\, C \sum_{i=1}^{j} N_i \leq r\, C' N_j \tag{2.11}$$

By producing good initial guesses for each problem after the first, Algorithm I avoids the "extra" $\log N_j$ factor commonly associated with work estimates for iterative methods. [5]

The j-level iteration is an iterative method which uses the spaces M_k, $k \leq j$, to solve a problem corresponding to M_j. To define this scheme, we must consider slightly more general problems than (2.6). Let $z \in M_j$ satisfy

$$a(z,v) = G(v) \quad \text{for all} \quad v \in M_j \tag{2.12}$$

where G is a linear functional ($G(v) = (f,v)$ in (2.6)). The j-level iteration is now defined inductively as follows:

Algorithm II

(i) if $j = 1$ solve (2.12) exactly.

(ii) if $j > 1$, one iteration of the j-level scheme takes $z_0 \in M_j$ to $z_{m+1} \in M_j$ as follows:

$$z_\ell = S(z_{\ell-1}), \quad 1 \leq \ell \leq m. \tag{2.13}$$

Compute $\tilde{\delta} \in M_{j-1}$, an approximation of $\delta \in M_{j-1}$ where

$$a(\delta,v) = G(v) - a(z_m,v) \equiv \bar{G}(v) \text{ for all } v \in M_{j-1}. \tag{2.14}$$

Use two iterations of the (j-1)-level scheme with initial guess zero. Set

$$z_{m+1} = z_m + \tilde{\delta}. \tag{2.15}$$

Equation (2.13) is called a smoothing iteration. It represents the use of some iterative method which can rapidly damp components of the error which oscillate on the order of h_j (Under-relaxed Jacobi or Gauss-Seidel iterations are often used.) [6,9]. Typically m is on the range $1 \leq m \leq 4$.

Once the error has been "smoothed" it can be well approximated by an element in a space of smaller dimension, i.e., M_{j-1}. In (2.14), δ is such an approximation of the error. In [4,5], it is proved that under reasonable hypotheses Algorithm II has the properties claimed in (B) above.

III. ADAPTIVE REFINEMENT

To implement Algorithms I and II, no information about the level-j grid T_j is required until $\tilde{u}_{j-1}$ has been computed. Thus we can use $\tilde{u}_{j-1}$ to adaptively determine T_j, rather than specify all grids a-priori. Our adaptive refinement scheme is based on the ideas of Babuška and Rheinboldt [2,3].

Let $t \in T_{j-1}$ have diameter h_t and let

$$\| z \|_t^2 = \int_t |\nabla z|^2 + z^2 dx \tag{3.1}$$

denote the $H^1(t)$ norm. We estimate the error in z using

$$\| \tilde{u}_{j-1} - u \|_t \cong C_1(a,t) h_t^2 \int_t R^2 dx + C_2(a,t) h_t \int_{\partial t} J^2 ds \tag{3.2}$$

where $R = -\nabla \cdot a \nabla \tilde{u}_{j-1} + b \cdot \nabla \tilde{u}_{j-1} + c\tilde{u}_{j-1} - f$, J is the jump in normal derivative of $\tilde{u}_{j-1}$ across ∂t, and C_1 and C_2 are computable constants which can depend on the geometry of t but not its size. The factors of h_t are required to make the homogeneity of the right and left hand sides of (3.2) consistent. For triangles with edges lying on $\partial\Omega$, (3.2) is modified in an appropriate fashion.

In our adaptive scheme, we compute the error estimate, e_t, in each triangle $t \in T_{j-1}$ using (3.2). Let e_{max} denote the largest of these errors and let $t_{max} \in T_{j-1}$ denote a triangle with estimated error e_{max}. From e_{max} we compute a threshold value, which is the estimated value of the error in the triangles resulting from the refinement of t_{max}. We then refine any triangles in T_{j-1} whose current error estimate is larger than the threshold.

Note that if the problem has severe singularities, the procedure described above may result in the refinement of relatively few triangles. Since we want the dimension of the spaces T_j to increase geometrically, this adaptive procedure may be invoked several times to compute a single triangulation [6].

Also, the adaptive proceudre is likely to produce triangulations which are highly non-uniform. For such triangulations, the convergence of the multi-level iterative method can still be proven, but our proof of the asymptotic optimal order work estimate may no longer hold.

IV. A MULTI-LEVEL ITERATIVE METHOD FOR NONLINEAR PROBLEMS

A weak form of (1.1) is: find $u \in H^1(\Omega)$ such that

$$a(u,v) = 0 \quad \text{for all} \quad v \in H^1(\Omega) \tag{4.1}$$

where

$$a(u,v) = \int_\Omega \nabla u \nabla v + f(u,u_x,u_y)\, v dx. \tag{4.2}$$

With T_j and M_j as in Section 2, we consider the analogue of (2.6): find $u_j \in M_j$ such that

$$a(u_j,v) = 0 \quad \text{for all} \quad v \in M_j. \tag{4.3}$$

We assume (4.3) has a unique solution, and that error bounds of the form (2.7) hold.

We define the form $b(u;v,w)$, bilinear in its last two arguments, by

$$b(u;v,w) = \int_\Omega \nabla v \nabla w + b\cdot\nabla v w + cvw dx;$$

$$b_1 = \frac{\partial f}{\partial u_x}(u,u_x,u_y); \; b_2 = \frac{\partial f}{\partial u_y}(u,u_x,u_y); \tag{4.4}$$

$$c = \frac{\partial f}{\partial u}(u,u_x,u_y).$$

For a fixed $u \in H^1(\Omega)$, $b(u;\cdot,\cdot)$ is of the form (2.3).

Our damped Newton iteration for (3.3) can be defined as follows: Let $u_j^{(0)} \in M_j$ be given. Then we find $\chi_j^{(k)} \in M_j$ such that

$$b(u_j^{(k)}; \chi_j^{(k)}, v) = -a(u_j^{(k)},v) \quad \text{for all} \quad v \in M_j \tag{4.5}$$

and set

$$u_j^{(k+1)} = u_j^{(k)} + t_k\chi_j^{(k)}. \tag{4.6}$$

($t_k \equiv 1$ corresponds to Newton's method). In this scheme the parameters t_k are chosen to guarantee convergence for any $u_j^{(0)} \in M_j$ and $\lim_{k\to\infty} t_k = 1$ in order to guarantee the ultimate superlinear convergence of the method. The elliptic problems (4.5) are approximately solved using several iterations of the j-level iteration defined in Section 2.

The overall process for computing approximate solutions $\tilde{u}_j$ of (4.3) is summarized in

Algorithm III

(i) if $j = 1$, start from some (arbitrary) initial guess and use s_1 damped Newton iterations to obtain $\tilde{u}_1$.

(ii) if $j > 1$, start from $\tilde{u}_{j-1} \in M_j$, and use s_j damped Newton iterations to obtain $\tilde{u}_j$.

The integer s_1 is chosen such that the resulting approximate solution $\tilde{u}_1$ will satisfy an error bound of the form (2.8). For $j > 1$, s_j is chosen such that the initial error, $\| \tilde{u}_{j-1}-u_j \|_1$, is reduced by a fixed factor $\varepsilon < 2^{-\alpha}$. Then, it can be shown, using arguments analogous to those given in Section 2, that the solutions $\tilde{u}_j$ will satisfy bounds of the form (2.8).

It is important to realize that for $j > 1$, we want to reduce the initial error by only a constant factor. One can show that eventually $\tilde{u}_{j-1}$ is in the region of super-linear convergence for u_j, and taking $s_j = 1$ will cause a decrease in error much larger than ε, if (4.5) were solved exactly. (In fact, we will use r j-level iterations to solve (4.5), where $\gamma^r < 2^{-\alpha}$. These are exactly the same computations which are required for a linear problem.) Thus the asymptotic complexity of the linear and nonlinear problems are identical.

V. NUMERICAL EXAMPLE

We consider the problem

$$-\Delta u + e^{u-w}-e^{v-u} + f(x,y) = 0 \quad \text{in} \quad \Omega \tag{5.1}$$

which has application in the field of semi-conductor modeling. The region Ω and boundary conditions are shown in Figure 1.

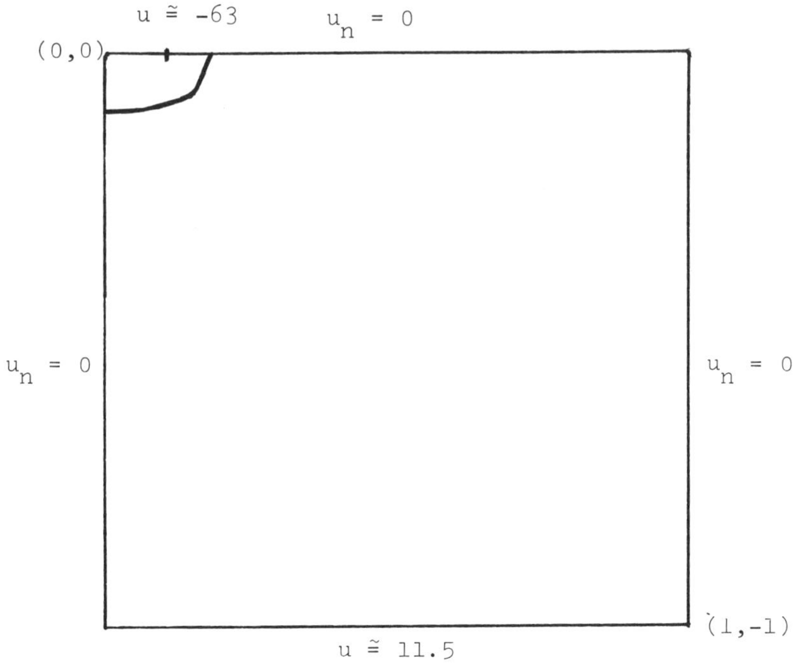

FIGURE 1.

The functions v and w satisfy $-40 \le v, w \le 0$; w is nonzero only near the "plate" (Dirichlet portion of the top boundary). $v \equiv -40$ in the P region. Both decay smoothly from -40 to zero. The doping profile f(x,y) is given by

$$\begin{aligned} f(x,y) &= c_1 \exp(-Ay)^2)\{\operatorname{erfc}(A(x-\bar{x})) - \operatorname{erfc}(A(x+\bar{x}))\} \\ c_1 &\cong -5.\times 10^9 \\ A &\cong 33.93 \\ \bar{x} &\cong .1 \quad . \end{aligned} \tag{5.2}$$

The function $f(x,y) = 0$ along the P - N junction, is positive in the N-region, and is negative in the P-region.

The problem was solved using 3 triangulations and the nonlinear multi-grid method described in Section 4.

Triangulations T_2 and T_3 were computed using adaptive refinement. The triangulations are shown in Figure 2.

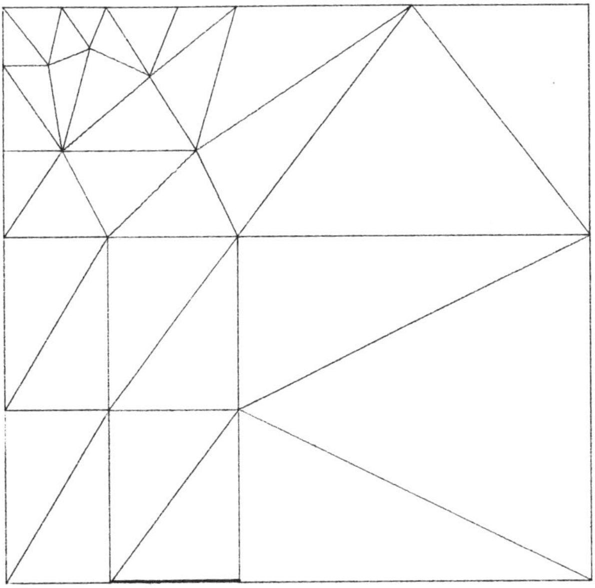

T_1

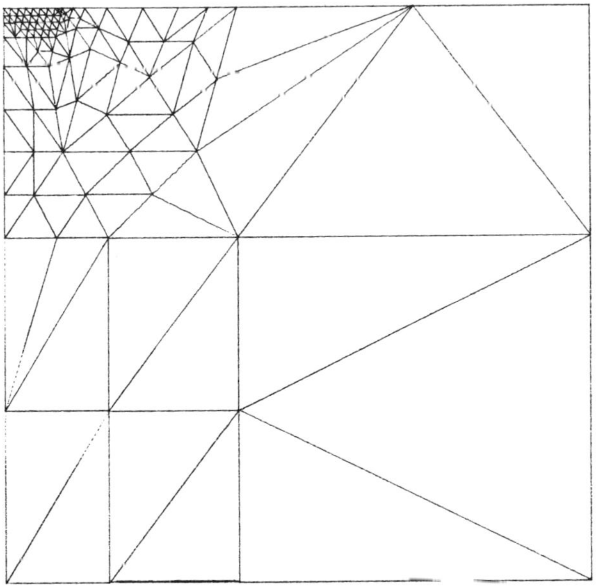

T_2

FIGURE 2.

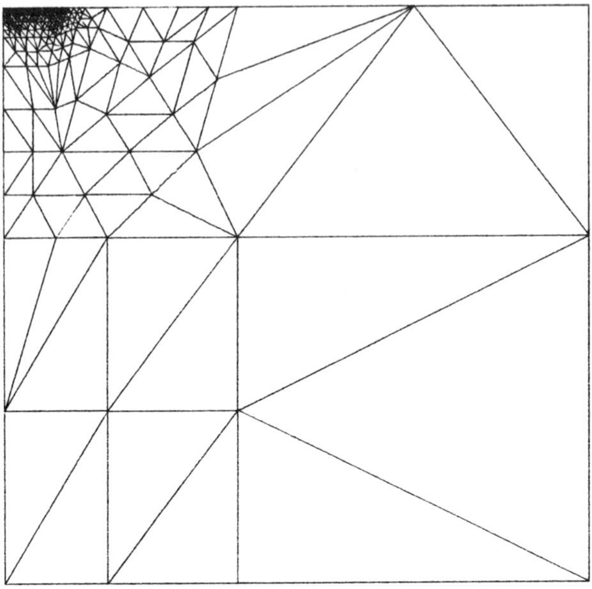

$\mathcal{T}_3$

FIGURE 2.

The computed solution is shown in Figure 3. (In Figure 3,

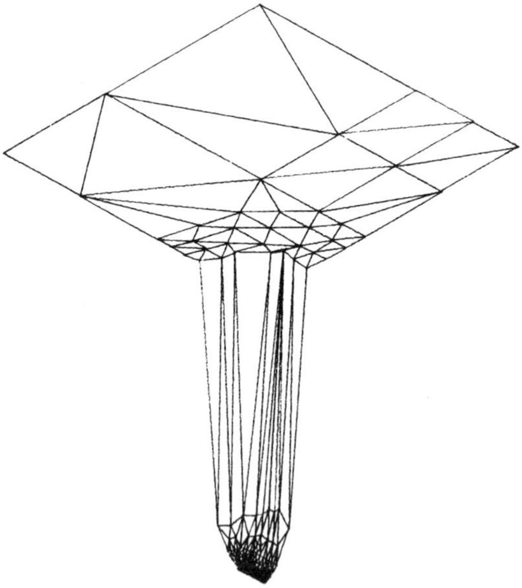

FIGURE 3.

the solution is shifted and scaled, $\hat{u} = \left(\frac{u-u_B}{u_T-u_B}\right)$, in order to make the three axes of comparable length.) In tables I and II we record the progress of the damped Newton iteration for N = 25 and N = 299. For illustrative purposes, we stopped the iteration when the estimated error $\| u_j^{(k-1)} - u_j^{(k)} \|_1 / \| u_j^{(k)} \|_1$ was less than 10^{-6}. The quantity ANORM is given by

$$(\text{ANORM})^2 = \sum_{i=1}^{N_i} a\left(u_j^{(k)}, \phi_i\right)^2 \tag{5.3}$$

where the ϕ_i are the standard nodal basis functions for $\mathcal{M}_j$. The initial guess for N = 25 was taken as zero for illustrative purposes (it is relatively easy to construct much better initial guesses).

TABLE I

		$N_1 = 25$	
k	t_k	ANORM	$\| u_j^{(k-1)} - u_j^{(k)} \|_1 / \| u_j^{(k)} \|_1$
0	-	4.74 (6)	-
1	1.1 (-4)	4.73 (6)	9.28 (3)
2	7.3 (-4)	4.70 (6)	8.11 (2)
3	7.3 (-3)	4.56 (6)	9.16 (1)
4	7.0 (-2)	3.41 (6)	1.12 (1)
5	2.8 (-2)	3.31 (6)	1.13 (1)
6	2.3 (-1)	2.21 (6)	6.26 (-1)
7	2.4 (-1)	5.27 (5)	1.32
8	9.3 (-1)	5.88 (4)	4.37 (-1)
9	6.1 (-2)	3.64 (4)	2.33
10	1.5 (-1)	3.11 (4)	1.56
11	6.7 (-1)	1.29 (4)	1.41 (-1)
12	9.8 (-1)	5.73 (3)	1.39 (-1)
13	9.9 (-1)	4.08 (3)	1.25 (-1)
14	7.4 (-2)	3.79 (3)	4.39 (-1)
15	3.5 (-1)	2.99 (3)	2.67 (-1)
16	8.7 (-1)	1.74 (3)	1.36 (-1)
17	3.2 (-1)	9.51 (2)	2.83 (-1)
18	8.9 (-1)	5.19 (2)	3.49 (-2)
19	7.7 (-1)	3.58 (2)	1.98 (-2)
20	7.7 (-1)	2.03 (2)	1.41 (-2)
21	9.8 (-1)	1.37 (1)	3.26 (-3)
22	9.9	1.44	2.23 (-4)
23	1.0	1.57 (-1)	2.36 (-5)
24	1.0	7.67 (-6)	2.90 (-6)
25	1.0	4.27 (-7)	1.40 (-10)

TABLE II

		$N_3 = 299$	
k	t_k	ANORM	$\| u_j^{(k-1)} - u_j^{(k)} \| / \| {}_j^{(k)} \|_1$
0	-	1.84 (4)	-
1	1.0	1.10 (3)	1.53 (-2)
2	1.0	1.91 (2)	7.90 (-4)
3	1.0	4.36 (1)	2.29 (-4)
4	1.0	3.61 (-2)	6.04 (-5)
5	1.0	7.61 (-5)	7.65 (-8)

It is clear that the initial guess for N = 299 is already in the region of superlinear convergence, and one damped Newton iteration would have been sufficient in terms of computing an approximation to the solution of (5.1)

REFERENCES

1. I. Babuška and A. K. Aziz, The Mathematical Foundations of the Finite Element Method with Applications to Partial Differential Equations, Academic Press, NY, 1972.
2. I. Babuška and W. C. Rheinboldt, "Error Estimates for Adaptive Finite Element Computations" SINUM, 15, (1978), 736-754.
3. I. Babuška and W. C. Rheinboldt, "Analysis of Optimal Finite Element Meshes in R^1" Math. of Comp., 33, (1979), 435-464.
4. R. E. Bank, "A Comparison of Two Multi-Level Iterative Methods for Non-symmetric and Indefinite Finite Element Equations", submitted SINUM.
5. R. E. Bank and T. Dupont, "An Optimal Order Process for Solving Finite Element Equations," to appear Math. of Comp.
6. R. E. Bank and A. H. Sherman, PLTMG User's Guide CNA 152, Center for Numerical Analysis, University of Texas at Austin, September 1979.
7. R. E. Bank and D. J. Rose, "Global Approximate Newton Methods" submitted, SINUM.
8. R. E. Bank and D. J. Rose, "Parameter Selection for Newton-like Methods Applicable to Nonlinear Partial Differential Equations," to appear SINUM.
9. A. Brandt, "Multi-level Adaptive Solutions to Boundary Value Problems" Math. of Comp., 31, (1977), 333-390.
10. J. W. Daniel, The Approximate Minimization of Functionals, Prentice-Hall, NJ, 1971.
11. R. S. Dembo, S. C. Eisenstat and T. Steihaug, "Inexact Newton Methods" submitted, SINUM.
12. J. E. Dennis and J. J. Moré, "Quasi-Newton Methods - Motivation and Theory", SIAM Review, 19, (1977), 46-89.
13. W. Hackbusch, On the Convergence of a Multi-grid Iteration Applied to Finite Element Equations, Report 77-8, Universität Köln, July 1977.
14. W. Hackbusch, "On the Fast Solutions of Nonlinear Elliptic Equations", Num. Math., 32, (1979), 83-95.
15. O. A. Ladyzenskaja, Linear and Quasi-Linear Elliptic Equations, Academic Press, NY, 1968.
16. R. A. Nicolaides, "On the ℓ^2 Convergence of an Algorithm for Solving Finite Element Equations", Math. of Comp., 140, (1977), 892-906.
17. J. T. Oden and J. N. Reddy, An Introduction to the Mathematical Theory of Finite Elements, Interscience, NY, 1976.
18. J. M. Ortega and W. C. Rheinboldt, Iterative Solution of Nonlinear Equations in Several Variables, Academic Press, NY, 1970.

19. A. H. Schatz, "An Observation Concerning Ritz-Galerkin Methods with Indefinite Bilinear Forms", Math. of Comp., 28, (1974), 959-962.
20. G. Strang and G. Fix, An Analysis of the Finite Element Method, Prentice-Hall, NJ, 1973.

SOLVING ELLIPTIC PROBLEMS: 1930-1980

Garrett Birkhoff

Department of Mathematics
Harvard University
Cambridge, Massachusetts

1. THE STATE OF KNOWLEDGE IN 1930

It is a pleasure to review the enormous practical progress that has been made in solving elliptic boundary value problems during the last 50 years, and to pay tribute to those most responsible for this progress.

By 1930, much was already known theoretically about solving elliptic problems, especially those involving the Laplace, Helmholtz, and biharmonic equations. A reasonably balanced picture of this knowledge may be obtained by reading Kellogg's Potential Theory (1929), the path-breaking paper by Courant, Friedrichs and Lewy (1928), and the incisive analysis of practical computational methods by von Mises and Geiringer (1929).

Indeed, already by 1910, building on earlier work of Fourier, Gauss, Green, Dirichlet, and Riemann, potential theory had been made rigorous by Poincaré, Fredholm, Hilbert and other mathematicians. In the colorful language of 1900, "Analysis had been arithmetized" in principle. Methods were known which, in a finite number of steps, could solve many elliptic problems (especially those involving linear partial DE's with constant coefficients) to any desired degree of accuracy. Hilbert's Problems 19 and 20, proposed in 1900, are concerned with extending these methods, and progress in solving them has recently been reviewed by Serrin and Bombieri.[1]

[1]See pp. 507-36 of Proc. XXVIII Symp. Pure Math., Felix Browder (ed.), Am. Math. Soc., 1976.

ISBN 0-12-632620-7

However, my main concern today is with practical progress, and I shall say no more about this important theoretical progress. Instead, I will emphasize how much was also known in 1930 about numerical algorithms for solving elliptic problems. Two classes of such algorithms were known: variational methods and finite difference methods, the former being precursors of the 'finite element' methods (FEM) that are so popular today.

Variational methods had been used to compute numerical results by Rayleigh (1873), Ritz (1908), and Galerkin (1915). Concerning them, Courant wrote in 1925 (alluding to the Dirichlet Principle):[2]

"Riemann attracted all eyes to the suggestive idea that a boundary value problem can be solved directly [by trying to minimize the Dirichlet integral] without digressing to first find the general solution of the corresponding DE." After recalling that Weierstrass later showed the incompleteness of Riemann's proofs, Courant continued: "Through Hilbert's fundamental work, the Dirichlet Principle was reinstated and brilliantly justified. [As a result], such direct methods seem called upon to break paths to new fruitful developments in the calculus of variations, perhaps even more closely tied to mathematical physics than the classical calculus of variations." His paper concluded with a discussion of important areas for future theoretical work, including those suggested by Hilbert's Problems 19 and 20, already mentioned above.

Likewise, difference methods had been studied by Runge (1908), L. F. Richardson (1910), R. G. D. Richardson (1918), and Phillips and Wiener (1922), and their theory notably extended by Courant, Friedrichs, and Lewy (1928). It was realized, of course, that to apply difference methods one would have so solve 'large' systems of linear algebraic equations. Much was known theoretically in 1930 about this problem also, important contributors being Gauss (1823), Gerling (1843), Jacobi (1845), Seidel (1874), and L. F. Richardson (1911). Specific references to their pioneer papers are given in the well-known books by Forsythe-Wasow, Householder, Varga, and Young; I will not repeat them here. Those interested in the pre-1930 history of the subject should read the 1929 paper by von Mises and Geiringer, as well as papers by Ostrowski which often include scholarly discussions of the earlier literature.

---

[2] Jahresb. Deutsche Math. Ver. 34 (1925), 90-117.

In spite of all this theoretical knowledge, only very simple problems were solved in practice in 1930. This was because, even using the electrically powered desk calculators then available, to solve the linear source problem

$$(1.1) \qquad -(pu_x)_x - (pu_y)_y + q(x,y)u = f(x,y) \quad ,$$

on a 10×8 mesh by band elimination (for example), required 50 to 100 man hours, and was therefore rarely worth the effort.

2. Three Significant Advances: 1930-1945

Although deeply depressing from a human standpoint, the years 1930-45 saw at least three major advances which paved the way for the "great leap forward" during 1945-60. Namely: (1) the first substantial programmable computers were designed and made operational, (2) the solution of linear elliptic difference equations with 100-500 unknowns by hand relaxation was developed into a fine art under the leadership of R. V. Southwell, and (3) the discretization errors associated with elliptic difference approximations became better understood. I will next briefly review these three advances.

Programmable computers. During the years 1930-45, the most widely used devices for engineering computations were still analog computers. Of these, the most popular was the "differential analyzer" of Vannevar Bush, an electrically powered mechanical device having planimeters for central processors. Thus, at the 1945 Canadian Mathematical Congress, a high point was Douglas Hartree's explanation of how the "method of lines" could be used to semi-discretize parabolic initial-boundary problems, and then a differential analyzer made to integrate the resulting system of ordinary DE's. More directly relevant to elliptic problems were electrolytic tanks, very different analog "machines" which solved the Laplace equation in plane and cylindrical geometries, with acceptable engineering accuracy.

Potentially far more important were the first programmable digital machines designed and built during the years 1930-45. Aiken's Mark I at Harvard, the ENIAC at the Moore School of Electrical Engineering, and the Zuse machines in Germany represented the dawn of a new era. (For the story of these machines, see N. Metropolis, J. Howlett, and G.-C. Rota (eds.), "History of

Computing in the Twentieth Century", Academic Press, 1980.)

Error bounds. Another major advance was Gershgorin's 1930 generalization of the theoretical error bounds of Courant, Friedrichs and Lewy (and their predecessors) from the Poisson equation to elliptic DE's of the form

$$Au_{xx} + Cu_{yy} + Du_x + Eu_y + qu = f \quad . \tag{2.1}$$

Gershgorin's technique consisted in majorizing the discrete Green's function of difference approximations to (2.1) on rectangular meshes.[3] For the Poisson DE on a square mesh, Gershgorin's error bounds simplify to

$$|e_{ij}| = hM_1/2 + h^2 M_4 \alpha^2 \beta^2/12(\alpha^2 + \beta^2) \tag{2.2}$$

in any plane domain contained in the ellipse $x^2/\alpha^2 + y^2/\beta^2 = 1$, M_k standing for the maximum absolute value of a k-th derivative of u.

An excellent exposition of this and related results can be found in Kantorovich and Krylov (1958), pp. 231-9. Many other advances in our theoretical understanding of approximate numerical solutions of elliptic problems can be found in the same book. These include Gershgorin's integral equation for computing the conformal maps of a given simply connected domain onto a circle.

Improved techniques for solving elliptic eigenproblems numerically (i.e., for computing eigenvalues and eigenfunctions) were also developed in 1930-45. An excellent survey of such techniques, as of 1945, is given in L. Collatz, "Eigenwertprobleme und ihre numerische Behandlung", Leipzig, 1945.

Relaxation methods. However, the number of unknowns in the problems routinely solved by the preceding authors rarely exceeded 20. Using highly intuitive and unorthodox "relaxation methods" based on mechanical analogy, Southwell and his school were able to solve elliptic difference equations having 100-500 unknowns. These iterative methods were variational in concept, systematically reducing the "constraints" or "residuals" needed to maintain mechanical equilibrium. The success of 'relaxation' was largely due to the intuitive skill and ingenuity of the experts using it. These experts developed flexible techniques of local mesh refinement, block relaxation, and over-

[3] ZaMM 10 (1930), 373-82. For his work on conformal mapping, see Mat. Sbornik 40 (1933), 48-58.

relaxation, techniques that were to become automated a decade later, as we shall see.

3. Iterative Methods: 1947-1962

The years 1947-62 were notable for the invention of four ingenious new iterative methods for solving linear systems approximating the linear source problem

$$L[u] = -\nabla \cdot [p(x,y)\nabla u] + q(x,y)u = f(x,y) \tag{3.1}$$

on a rectangular mesh, through the standard discretization

$$L_h[u] = \alpha_{ij}u_{i,j} + \beta_{ij}u_{i+1,j} + \gamma_{ij}u_{i,j+1} + \delta_{ij}u_{i-1,j} + \epsilon_{ij}u_{i,j-1} = \phi_{i,j} \quad . \tag{3.2}$$

As will be explained below, all four methods are more efficient by an order of magnitude than the earlier iterative methods of Jacobi, Gauss-Seidel, and Richardson. Moreover, their modest storage requirements made them at that time also preferable (for large problems) to any direct method.

Two of these methods (SOR and ADI) achieved widespread adoption for nuclear power reactor design and petroleum reservoir exploitation, respectively, saving hundred of millions of dollars. They have been brilliantly analyzed in the books by Forsythe-Wasow (1960) and Varga (1962), to which I refer you for details -- including the formulas expressing the coefficients of (3.2) in terms of the functions p, q, and f in (3.1). Using them, the approximate solution of linear source problems involving 10^4-10^5 unknowns became routine after the 'first generation' of large computers became available around 1955.

The 'model problem'. To gain insight, many numerical analysts around 1950 concentrated on the 'model problem' defined by the Poisson equation $-\nabla^2 u = f(x,y)$ in the unit square $S : [0,1] \times [0,1]$, with boundary conditions $u \equiv 0$ on ∂S, and its 5-point difference approximation on a square mesh with side $h = 1/n$. (For the Poisson equation, $\alpha_{ij} = 1$ and $\beta_{ij} = \gamma_{ij} = \delta_{ij} = \epsilon_{ij} = -1/4$.)

For this model problem, Frankel (1950) showed that the rate of convergence of the ("first-order") Richardson and Liebmann[4] methods could be increased from $O(1/n^2)$ to $O(1/n)$ by using instead Frankel's own "second-order Richardson" and "extrapolated Liebman" methods.

I will discuss Frankel's "second-order Richardson" method in §8 ; his "extrapolated Liebmann" method is the "successive overrelaxation" (SOR) method of Young (1950, 1954), who also derived its general theory. I will discuss the SOR method next.

SOR. I had proposed to Young in 1947 the problem of automating Southwell's "relaxation methods" as a Ph. D. Thesis topic. I suggested that he should concentrate on the Dirichlet problem in a general domain, having as his objective the solution of such problems on sequential digital computers like Aiken's Mark I. This had just become operative at Harvard.

Of course, I realized that this was a formidable problem. When I asked Southwell for advice about it in 1948, he said he thought its practical solution was impossible: the computer could not match the subtleties of human intuition. However, some of his junior staff told me that what are today called 'over-relaxation' and 'block relaxation' often accelerated convergence. Needless to say, I passed on this information to Young.

By 1949, after mastering the mathematical theory of the Jacobi and Gauss-Seidel methods, and relating it to such new concepts as 'Property A' and 'consistent ordering', Young had proved that using a suitable 'optimum over-relaxation factor ω_b' speeded up the asymptotic rate of convergence by an order of magnitude. See Young (1954) for his most significant results; his Thesis Young (1950) contains many more.

ADI. The 'elliptic ADI' methods of Peaceman and Rachford (1955) and Douglas and Rachford (1956) had a very different background. They were outgrowths of "parabolic ADI' methods for solving diffusion problems in plane regions which, in turn, were extensions of an earlier method of Crank and Nicolson (Proc. Camb. Phil. Soc. 43 (1947), 50-67) for solving heat conduction problems in one space dimension.

These extensions decomposed the 5-point difference operator L_h of (3. 2) into two 3-point horizontal and vertical difference operators, H and V . For

---

[4] The "Liebmann" method is the Gauss-Seidel method for the Laplace operator on a square mesh.

any sequence of scalars ρ_n, we have (for example)

$$(3.3)\qquad \begin{aligned} (H + \rho_n I)\underline{u}^{(n+\frac{1}{2})} &= \underline{k} - V\underline{u} \\ (V + \rho_n I)\underline{u}^{(n+1)} &= \underline{k} - H\underline{u}^{(n+\frac{1}{2})} \end{aligned} ,$$

provided that $L_h = H + V + \rho_n I$. Moreover each of the two Eqs. (3.3) is easily solved, since its coefficient-matrix is triangular.

4. Chebyshev Methods

Whereas the SOR and ADI methods were specifically designed to solve linear systems arising from second-order, self-adjoint linear source problems (3.1), the Chebyshev and Frankel schemes can be used to solve $A\underline{u} = \underline{b}$ for any positive definite symmetric matrix A. They are most closely related to the "first-order Richardson" method, which I will now recall.

Richardson method. Clearly, the (exact) solution of $A\underline{u} = \underline{b}$ minimizes the quadratic functional

$$(4.1)\qquad J[\underline{u}] = \tfrac{1}{2}\langle \underline{u}, A\underline{u}\rangle - \langle \underline{b}, \underline{u}\rangle ,$$

whose gradient is the affine vector field

$$(4.1')\qquad \operatorname{grad} J = \nabla J = A\underline{u} - \underline{b} .$$

Hence, at any point $\underline{u}$, the residual $\underline{r} = \underline{b} - A\underline{u}$ is the negative gradient of J and, unless this is $\underline{0}$, points in the "direction of steepest descent". The "path of steepest descent" issuing from any initial value $\underline{u}_0$ is therefore a solution of the vector DE

$$(4.2)\qquad \underline{u}'(t) = \underline{b} - A\underline{u} .$$

Richardson suggested solving (4.2) approximately by the Cauchy polygon method. This is defined recursively by

$$(4.3)\qquad \underline{u}_{m+1} = \underline{u}_m + \Delta_m \underline{r}_m , \quad \underline{r}_m = \underline{b} - A\underline{u}_m .$$

An easy calculation gives

(4.3') $$\underline{r}_{m+1} = \underline{b} - A\underline{u}_m = \underline{b} - A\underline{u}_m - \Delta_m A\underline{r}_m = (I - \Delta_m A)\underline{r}_m \quad .$$

Using induction on m, there follows

THEOREM 1. The final residual $\underline{r}_m = \underline{b} - A\underline{u}_m$ of any m-step Cauchy polygon approximation to the solution of (4.2) for the initial value $\underline{u}_0$ is given by

(4.4) $$\underline{r}_m = (I - \Delta_m A)(I - \Delta_{m-1} A) \cdots (I - \Delta_1 A)\underline{r}_0 \quad ,$$

where $\underline{r}_0 = \underline{b} - A\underline{u}_0$ is the initial residual.

Clearly, the set of all possible such residuals lies in the (in general,[5] m-dimensional) subspace K_m of vectors

(4.4') $$\underline{r} = p(A)\underline{r}_0 = \underline{r}_0 + q(A)A\underline{r}_0 \quad ,$$

where $p(\lambda)$ is a polynomial of degree m with $p(0) = 1$. We will call this the Krylov subspace of degree m generated by (4.2) and (4.3); Eq. (4.4) shows that the possible residuals $\underline{r}_m$ constitute the subset of this subspace, associated with the real polynomials p with $p(0) = 1$ whose irreducible factors are all linear.

Conjugate gradient method. Perhaps because of uncertainty about the choice of Δ_k and the build-up of roundoff errors, Richardson's method lay dormant until its potentialities were made clear by M. R. Hestenes and E. Stiefel (1953). They invented a "conjugate gradient" algorithm for computing recursively the $\underline{u}_m$ having the smallest residual in the Krylov subspace constructed above. Their algorithm is based on the following lemma.

LEMMA. If $\underline{u}_j$ minimizes $\underline{r}_j = \underline{b} - A\underline{u}_j$ in K_j, where $K_1 \subset K_2 \subset K_3 \subset \cdots$, and $\underline{u}_{m+1} = \underline{u}_m + \Delta_m \underline{s}_m$, where Δ_m is a scalar, then $\underline{s}_{m+1}$ is A-orthogonal to K_m.

In the absence of roundoff, the conjugate gradient algorithm would give the exact solution in N or fewer steps, where N is the number of unknowns.

---

[5] In the exceptional case that some first $\underline{r}_j = \underline{0}$, $j < m$ the subspace $K_m = K_j$ is j-dimensional.

Because of this, it was first proposed as a direct method, having certain advantages over Gauss elimination (it is related to the QR-algorithm).

Chebyshev semi-iteration. As Stiefel pointed out in 1954,[6] the algorithm can also be viewed as an iterative method; it provides a relaxation scheme of "best strategy" (i. e., optimal). More precisely, if the (positive) eigenvalues of A span the interval $[\alpha, \varkappa\alpha]$, then $\rho = (\varkappa - 1)/(\varkappa + 1) < 1$, then the optimal Richardson scheme is the semi-iterative 'Chebyshev' scheme defined by

$$\underline{u}_{m+1} = \omega_{m+1}\{A\underline{u}_m - \underline{u}_{m-1} + \underline{b}\} + \underline{u}_{m-1}, \tag{4.5}$$

where $\omega_{m+1} = 2T_m(1/\rho)/\rho T_{m+1}(1/\rho)$, the $T_m(x)$ being the Chebyshev polynomials.

More precisely (see Varga (1962), p. 138), it can be shown that this choice of the ω_m minimizes the spectral radius of the transformation $\underline{r}_0 \mapsto \underline{r}_m$.

In spite of this, as Varga showed in 1957,[7] optimum SOR is preferable in practice for solving (3.2) because it requires less storage and less computation per iteration cycle and has a smaller spectral radius.

CCSI. Four years later, Golub and Varga (1961) showed how to combine Chebyshev 'optimality' with Young's Property A so as to obtain the best of both. By using the 'red-black ordering', the Jacobi iteration matrix $B = D^{-1}A$ (D – diag A) assumes the weakly 2-cyclic form $B = \begin{bmatrix} 0 & F \\ G & 0 \end{bmatrix}$. Hence, for some $\underline{u} = \left[\frac{\underline{v}}{\underline{w}}\right]$, the Jacobi scheme $\underline{u}_{m+1} = B\underline{u}_m + k$ is equivalent to

$$\underline{w}_{m+1} = F\underline{v}_m + \underline{k}, \quad \underline{v}_{m+1} = G\underline{w}_{m+1} + \underline{g}, \tag{4.6}$$

requiring about half as many operations in each complete iteration.

Golub and Varga (1961) showed that the resulting 'cyclic Chebyshev semi-iterative' (CCSI) method "can be viewed as a variant of the SOR method, which gives improved average rates of convergence" (Varga (1962), p. 153). It quickly became the method of choice for solving large problems of the form (3. 2), at least in nuclear reactor design, and may be regarded as the 'last word' in a very exciting era of research.

---

[6] Proc. Int. Math. Congr. Amsterdam (1954), vol. i, pp. 384-91; Comm. Math. Helv. 29 (1955), 157-79; J. SIAM 4 (1956), 63-85.

[7] J. SIAM 5 (1957), 39-46.

5. Finite Element Methods

The most dramatic advance in elliptic problem solving during the 1960's was unquestionably the development of finite element methods. Conceptually, these methods exploit old ideas similar to those expressed by Courant in 1925 (see §1), and again in 1943 (Bull. Am. Math. Soc. 49 (1943), 1-23). Moreover other distinguished analysts such as Bickley and Temple, Pólya and Szegö, Prager and Synge, and Weinstein and Diaz had made further scholarly contributions, applying their new analytical formulas to particular cases. Pólya-Szegö's Isoperimetric Inequalities in Mathematical Physics (Princeton Univ. Press, 1951), and Synge's The Hypercircle in Mathematical Physics (Cambridge Univ. Press, 1957) were especially notable.

But generally speaking, these mathematicians applied the ideas of Rayleigh and Ritz only to spaces of piecewise linear and bilinear trial functions. These same spaces also formed the backbone of the important NASTRAN package developed in the early 1960's, under the leadership of R. H. MacNeal, to help in designing vehicles for manned Moon-landings. It was MacNeal whose stimulus was acknowledged by Frankel (1950), and who had published a suggestive paper on the network analogy in the Quar. Appl. Math. 11 (1953), 295-310.

But none of the above scientists studied or used higher-order piecewise polynomial approximations that characterized the finite element methods (FEM) of the 1960's. It was research-oriented engineers like Argyris, Clough, Irons, and Zienkiewicz who invented, analyzed, and used these to fit solutions of their problems. Felippa and Clough have given an admirable survey of FEM developments prior to 1968 in Birkhoff-Varga (1970), pp. 210-54; and the latest (1978) edition of Zienkiewicz's popular The Finite Element Method in Engineering Sciences up-dates this survey from an engineering standpoint.

In contrast to the difference methods reviewed in §§3-4, the FEM of the 1960's (and 1970's) tended to use triangular 'elements' instead of rectangular meshes, and the structural problems to which they were typically applied corresponded to fourth-order rather than second-order elliptic problems. The resulting replacement of 5-point stencils by more complicated equations, taken with the use of fewer unknowns to achieve the needed accuracy and the availability of second generation computers with more core storage, tended to make direct methods more efficient than iterative methods for solving linear FEM problems; see §7 for some consequences of this change.

Experience to date suggests that FEM are superior to difference methods for problems in solid mechanics, but no such superiority has been demonstrated in fluid mechanics.

Experience also suggests that quadratic and biquadratic elements are adequate for many problems, but of course higher-order elements are more interesting mathematically; see §6.

Multivariate splines. On rectangular meshes, multivariate Hermite and (linearized)[8] spline functions constitute very good finite elements. Univariate 'splines' have long been used by naval architects; they are just point-loaded thin beams. They may have been first studied by Navier and Clapeyron, but it was I. Schoenberg who first analyzed their mathematical properties in depth,[9] assuming a uniform mesh.

Multivariate splines were discovered in connection with a project at the General Motors Research Laboratories for computerizing the manufacture of dies for sheet metal car body surfaces. This eventuated in an INCA package somewhat analogous to NASTRAN, and developed at about the same time. A year after H. L. Garabedian and I had proposed a more complicated rectangular 'element', Carl de Boor proposed in 1961 using tensor products of univariate splines to approximate car body surfaces. For an analysis of this idea, see de Boor's and my report in pp. 164-90 of Garabedian (1965). Our report concludes with a brief discussion of bicubic splines as possible 'finite elements' (the name had not yet become current); see also our paper with Swartz and Wendroff in J. SIAM Numer. Anal. 3 (1966), 188-203.

Since then, many authors have studied multivariate splines as finite elements. Surveys of this subject, with applications to elliptic and parabolic problems and further references, are contained in Strang and Fix (1972) and Schultz (1972); see also §6.

---

[8]Nonlinear splines were the subject of Max Born's Ph. D. Thesis (1906); see Love, Mathematical Theory of Elasticity (4th ed.), pp. 411-12. For the Navier-Clapeyron theory, see pp. 22, 373.

[9]I. Schoenberg, Quar. Appl. Math. 4 (1946), 45-99.

Global 'elements'. Although piecewise smooth approximations by 'finite' elements provide the most versatile variational methods for solving elliptic problems, classical global approximations and related integral equation methods are effective for some problems. Already in Math. Annalen 100 (1928), 503-21, Runge's son-in-law E. Trefftz had anticipated modern "penalty methods" by proposing least squares fitting of boundary data (e. g., by harmonic polynomials for the Laplace equation, relying on classical theorems by Runge, Bergman and Walsh). From 1947 on, Bergman and various collaborators had adapted this idea to (then) current computing techniques.[10] A serious technical problem (as with splines; see §6) is the construction of reasonably well-conditioned bases; for its solution, see P. Davis and P. Rabinowitz, Adv. in Computers 2 (1961), 53-133. A superb demonstration of the power of this method (for special problems problems) was the accurate computation by Fox, Henrici and Moler of the eigenvalues of an L-shaped membrane in SIAM J. Numer. Anal. 4 (1967), 89-102; for a penetrating theoretical analysis, see Bramble and Schatz in pp. 107-30 of Hubbard (1971).

6. Some Intriguing Mathematical Questions

Lars Ahlfors once said, half-jokingly: "Applied mathematics is very useful; it provides pure mathematicians with many fascinating problems." Indeed, each new major "applied" advance seems to leave in its wake a host of new theoretical techniques and a handful of nagging, purely mathematical questions. Finite element methods are a case in point.

Before 1960, the theory of multivariate interpolation and approximation was largely confined to extensions of the Weierstrass Approximation Theorem, and a handful of formulas for numerical cubature. The dramatic successes described in §5 have stimulated an enormous theoretical literature, concerned primarily with questions of functional analysis.

The many distinguished contributors include Babuska, Bramble, Strang, Thomee, Widlund, and Zlamal in addition to those named in the preceding section. Above all the French school of Lions, Aubin, Cea, Ciarlet and Raviart has been

---

[10] Quar. Appl. Math. 5 (1947), 69-81; Proc. VI Symp. Appl. Math. (1953), 11-29. J. P. Aubin and H. Burchard attribute the 'method of the hypercircle' to Trefftz in pp. 1-67 of Hubbard, 1971.

productive; Ciarlet's book (1978) summarizes admirably their view of the present position of the subject. Based on "Lagrangian" (polynomial interpolants of degree n in uniformly subdivided m-simplices) and "Hermitian" (odd degree interpolants in m-dimensional boxes), a powerful and smooth theory of approximation by continuous functions having arbitrarily high order of accuracy in spaces of any dimension has been constructed. Moreover this theory has been extended by the French school to parabolic Stefan and optimal control problems.

I will not try to summarize this work. Instead, I will mention a few elementary unsolved problems in multivariate approximation theory, which are prototypical of large families of questions whose answers we do not know.

Conforming elements. It is classic that the solution of the Dirichlet problem minimizes the Dirichlet integral

$$J_1[v] = \iint \nabla v \cdot \nabla v \, dx\, dy \quad , \tag{6.1}$$

with respect to all continuous functions v having the same boundary values. Likewise, the solution of any clamped plate problem minimizes the integral

$$J_2[v] = \iint (\nabla^2 v)^2 \, dx\, dy \quad , \tag{6.2}$$

with respect to all continuously differentiable, piecewise smooth functions having the same boundary values and normal derivatives. Loosely speaking, bicubic Hermitian elements fit together to form globally defined functions in $C^1(\Omega)$, and thus 'conform' to both restrictions; but Lagrangian elements are merely continuous across interfaces.

This raises the question: what is the simplest family of 'conforming' Lagrangian elements, that can be smoothly fitted together so as to constitute a globally defined, continuously differentiable function? For triangles in the plane, Zenisek showed that a (known) 18-parameter family of quintic polynomials is smallest, but a 12-parameter family of rational triangular elements can also be constructed. For tetrahedra in space, it seems to be impossible to use less than a 220-parameter family of ninth degree polynomials, although a 76-parameter family of rational functions suffices.[11] But what is the least number of

[11] A. Zenisek, J. Approx. Theory 7 (1973), 334-51; Lois Mansfield, J. Math. Anal. Appl. 56 (1976), 137-64.

parameters that will allows one to fit together simplicial elements in $\mathbb{R}^m$ so as to produce a function $v \in C^k(\Omega)$, for general m and k?

Interpolating to edge data. A second class of intriguing mathematical problems concerns exact interpolation to Dirichlet boundary conditions. To solve this problem, and avoid the complication of having to specify 'penalty' criteria, Irons and Zienkiewicz invented a simple but effective class of isoparametric elements, designed to match arbitrary smooth values on one curved side, and conventional polynomial values on the other straight sides of a boundary region.

A similar scheme, which interpolates to specified (smooth) edge values on all four sides of any rectangle R, was studied by Mangeron in 1932. His is the unique interpolant satisfying the hyperbolic DE $u_{xxyy} = 0$; it approximates every other smooth interpolant with $0(h)^4$ accuracy, and minimizes $\iint_R u_{xy}^2 \, dx\, dy$. Analgous schemes were invented independently for use in computer-aided design by the late S. A. Coons, and analyzed by W. J. Gordon and the author in J. Approx. Theory 1 (1968), 199-208.

This approach was promptly generalized by Gordon to 'spline-blending' in a series of articles,[12] and a few years later to schemes for interpolating to edge values in triangles and other simplices by R. E. Barnhill, Gordon, and the author (J. Approx. Theory 8 (1973), 114-28). In barycentric coordinates, the interpolant is characterized by the third-order hyperbolic DE $u_{xyz} = 0$, and it approximates any other smooth interpolant with $0(h^3)$ accuracy. However, we do not know a variational principle which characterizes it, even in the simplest case.

In my view, the prceding papers constitute just the entering wedge into a vast new theoretical area: optimal interpolation to data on specified k-cells in a general simplicial or, more generally, polyhedral complex (in the sense of Poincaré), subject to specified smoothness constraints. Some advances in this area may prove very useful practically.

For instance, since the theory of univariate interpolation (e. g., by cubic

[12] W. J. Gordon, SIAM J. Numer. Anal. 8 (1971), 158-77; also in Schoenberg (1969), pp. 233-77.

splines) is quite well understood, it is easy to think of many good schemes for interpolating along mesh lines in a general rectangular complex, in any number of dimensions. Interpolating to the resulting edge values in mesh rectangles and 'bricks' provides a promising scheme for constructing 'finite element' approximations associated with local mesh refinements.[13]

Multivariate spline mysteries. Univariate spline interpolation and approximation are now well understood, and it seems clear that splines are more efficient than Hermite approximants for elliptic problems in boxes. Minimizing $u^{(m)2}dx$ gives piecewise polynomial interpolants of degree $2m - 1$ in $C^{2m-2}[a,b]$, which approximate with $O(h^{2m})$ accuracy. One can 'clamp' any desired set of nodes x_i by specifying any set of $u_i^{(k)}$, $k < m$, including endpoints. Before 1960, the functions x^k $(k < 2m-2)$ and $(x-x_i)_+^{2m-1}$ provided an ill-conditioned basis. But if a basis of B-splines is used, the dependence of the interpolant $u(x)$ on nodal values is given by a uniformly sparse, well-conditioned matrix (by quasi-interpolation).[14]

Using tensor products, one can extend many of the preceding results to box-shaped regions. Thus, the bicubic spline interpolant in a rectangle R minimizes

$$\iint_R u_{xxyy}^{\;2}\, dx\, dy + \iint_{\partial R} u_{ss}^{\;2}\, ds \quad . \tag{6.3}$$

However, multivariate splines are still mysterious; it is not even clear how they should be defined in general rectangular polytopes.[15]

L-shaped regions. It is not even clear what is the 'natural' well-posed interpolation problem for bicubic splines in L-shaped regions. Ralph Carlson and C. A. Hall have proposed one recipe in J. Approx. Theory 4 (1971), 37-53; Lois Mansfield has proposed a different recipe based on a variational principle which is, however, not the same as (6.3) (ibid., 12 (1974), 99-112). Perhaps one should just minimize $\iint u_{xxyy}^{\;2}dx\, dy$ in the space of Hermite interpolants,

---

[13] G. Birkhoff, J. C. Cavendish, and W. J. Gordon, Proc. Nat. Acad. Sci. USA 71 (1974), 3423-5.

[14] H. B. Curry and I. J. Schoenberg, J. d'Analyse Math. 17 (1966), 71-107; C. de Boor and G. Fix, J. Approx. Theory 8 (1973), 19-45.

[15] C. de Boor, A Practical Guide to Splines, Springer, 1978, p. 362. See also my review of the problem in Schoenberg (1969), pp. 211-16.

and in the affine subspace of functions satisfying the specified interpolation conditions on the $u_i^{(k)}$, $k < 2$. This condition generalizes easily to multivariate splines in any rectangular polytope.

7. Direct sparse matrix methods

From the start, proponents of iterative methods realized that their efficiency could be increased by combining them with direct methods. Thus line and 2-line SOR utilized factorizations of tridiagonal and quindiagonal matrices.

The 'regular splittings' introduced by Varga in 1960,[17] extending earlier ideas of Fiedler and Ptak (Math. Nachr. 5 (1956), 31-8), represented a more ambitious step in the same direction: accelerating iterative methods by interspersing them with direct methods. The 'strongly implicit' methods of H. L. Stone[18] and others constituted an important further advance in the same direction. Today, one speaks of preconditioning the matrix A of $A\underline{u} = \underline{b}$ by an easily 'inverted' matrix M such that $A\underline{u} = \underline{b}$ is equivalent to $M\underline{u} = B\underline{u} + \underline{k}$, where B is sparse and $\rho(B)$ is small.

Important though these developments were, I do not wish to focus attention on them. Neither will I comment on the important improvements in roundoff error control and understanding, in Wilkinson played such a large role.[19] Instead, I want to say something about the enthusiasm for new direct sparse matrix methods which burned so brightly in the early 1970's.

Fast Poisson solvers. In this connection, the Fast Fourier Transform (FFT) should be mentioned first. Rediscovered by Hockney in 1065, the method has been traced back to Gauss.[19] Its wide applicability as a "preprocessor" for solving partial DE's with variable coefficients in "irregular" regions was demonstrated by Buzbee, Dorr, George, and Golub in 1971 (see SIAM. J. Numer. Anal. 8, pp. 722-38).

Minimizing fillin. An equally fascinating and very different stimulus originated in efforts to solve network problems. These efforts can be traced

---

[17] pp. 121-42 of R. E. Langer (ed.), Boundary Problems in Differential Equations, Univ. of Wisconsin Press, 1960.

[18] H. L. Stone, SIAM J. Numer. Anal. 5 (1968), 530-58.

[19] H. H. Goldstine, A History of Numerical Analysis ..., Springer, 1977, p. 238; R. W. Hockney, J. ACM 12 (1965), 95-113.

back at least to Grabriel Kron (cf. B. K. Harrison, J. SIAM 11 (1963), 256-80), and have been developed into a very useful code package by Fred Gustavson. Since many difference approximations to second-order elliptic DE's can be interpreted (via Stieltjes matrices) as networks, such efforts are directly relevant to our central concern. The challenging question is order the nodes of a network (graph) so as to minimize fillin during the process of Gauss elimination (alias LU or Cholesky decomposition). This is a beautiful combinatorial problem, the 'computational complexity' of whose optimal solution is also fascinating.

Nested dissection. A third, again very different and very original direct method for solving linear source problems in plane domains was provided by Alan George's method of nested dissection. This has the same order of computational complexity as the iterative methods discussed in §3, and suggests the heady idea that iterative methods are superfluous for treating such problems. However, it is much harder to program (for general regions) and requires much more storage, and I think this conclusion is unrealistic.

Marching methods. Finally, I mention the extremely fast, but inherently unstable 'marching method' for solving the same class of problems. Mentioned years ago by von Mises, the method has been partially stabilized using Richardson extrapolation by Bank and Rose.[20] Very recently, E. C. Cartland has shown in unpublished work that 'marching' is a very good procedure for stabilizing the convection-diffusion equation

$$Uu_x = \epsilon(u_{xx} + u_{yy}) + f(x,y) \quad , \tag{7.1}$$

in a channel, provided that the mesh length exceeds the boundary layer thichness.

Fortunately, a substantial fraction of the most important work on the three preceding methods may be found in Rose and Willoughby (1972), Bunch and Rose (1976), and Traub, (1973) and (1976). Those who wish to learn more about the very active area of research which they typify, should study the contents of these books. They will find that much is left for them to do.

---

[20] See their article in pp. 201-50 of Traub (1976); for the background of the method (and an exposition of nested dissection), see Alan George's and my article in Traub (1973), pp. 221-70.

8. Current trends

I hope I have conveyed a sense of enormous progress in my hasty review of advances in the art of elliptic problem solving since 1930. However, I am well aware that I have barely scratched the surface. My discussion of current trends will necessarily be even more superficial, and is intended primarily to whet your curiosity about the talks to come. Nevertheless, I want to comment briefly on three topics which, in my opinion, merit the closest high-level attention.

Subroutine libraries. The first is the development of comprehensive software packages such as ELLPACK and ITPACK. Packages such as these should make it much easier for us and our students to stand on the shoulders of our predecessors, primarily by making it much easier to try out different methods and estimate the cost of solving our problems by various commonly used algorithms. As an extra dividend, they can also help to bring out the strong and weak points of such algorithms.

It would therefore have been most interesting to compare the subroutines in the above packages with those contributed to the Club Modulef organized in Western Europe by IRIA as a collaborative enterprise. Whereas ELLPACK and ITPACK tend to stress difference and collocation methods, as you know, the Club Modulef features finite element methods. I am very disappointed that we could not attract a leading representative of that group or of NAG. IMSL, of course, distributes ELLPACK and ITPACK and is represented here.

Although I believe that comprehensive software packages will play a vital role in accelerating future progress in elliptic problem solving, as in many other areas of scientific computing, I also question the finality of general conclusions about the comparative merits of algorithms based on any limited set of tests, however objectively and skillfully designed. In particular, there is almost inevitably a tradeoff between efficiency and generality: the more comprehensive a package or algorithm is, the higher the overhead! Yet we always want both efficiency and generality.

Three-dimensional domains. Until this year indeed, because of overhead, ELLPACK could not handle 'in core' on its CDC-6600 problems larger than those represented by a 22×22 grid. (I have been informed that much larger problems can now be treated on the VAX with the Unix system.) As David Young has remarked, he could treat problems of this size in 1950!

In the next decade, with faster computers having cheaper core storage, I cannot see why solving three-dimensional elliptic problems should not become routine. Of course, ELLPACK already has this capability for box-shaped domains. And many other regions can be transformed into box-shaped domains in suitable parameter spaces.

The domain problem. However, I don't think one can solve most elliptic problems efficiently in this way. I think it much more efficient to use a three-dimensional grid, often a uniform grid, with 'irregular stars' on the boundary in the spirit of Southwell and Fox (1945), not to mention Courant, Friedrichs and Lewy. Using the network analogy, one can approximate linear source problems on such a grid by a 7-point difference approximation. (One can, of course, use a 27-point or HODIE formula having a higher order of accuracy, but I am not convinced of the truth of John Rice's contention that "higher order methods are always at least as good" in his experience!)

It will presumably, be most efficient to solve the linear algebraic equations obtained from such a discretization by semi-iterative methods having a strong iterative component; because, generally speaking, the rate of convergence of iterative methods does not decrease with the dimension.

With the preceding scheme, it should not be too difficult to solve the indexing problem of storing coefficients and nodal values in geometric (i, j, k) form, correlated with the sequential $\kappa \longleftrightarrow (i, j, k)$ order commonly associated with matrix elimination. (Of course, the variety of topologically distinct boundary 'stars' increases rapidly with the dimension.)

The indexing problem and algebraic solution of finite element approximations will be harder to achieve with the same generality. Also, I know of no way to construct three-dimensional 'singular elements' analogous to those used so successfully by Bruce Kellogg and George Fix, to achieve higher-order accuracy in the presence of corners. However, FEM have compensating advantages.

Parallel processors. Some optimists claim that, by necessity, computing speeds will continue to rise and storage costs continue to fall by orders of magnitude. For solving many kinds of partial differential equations more quickly and cheaply by orders of magnitude, parallel processors of the kind envisioned in the ILLIAC IV have always seemed to me to offer the most promise, in principle, of making such further progress possible. Now that the CRAY-1 and STAR have demonstrated that vector machines are cost-effective for scientific computing, while even the ILLIAC IV has helped to show that R_{317} (the decimal number consisting of 317 1's) is a very large prime, perhaps a cautious return to the optimism implicit in Traub (1973) (but more muted in Traub (1976)) is in order.

Of course ELLPACK, ITPACK and YALEPACK were not designed for vector or parallel processors, and will become obsolete if sequential computers (and Fortran) do. This would be a minor tragedy, but let's look at the optimistic side, and think of what parallel computers with 1024 arithmetic units could accomplish, and how.

For a truly parallel computer, Frankel's algorithm appeals to me as a very attractive basic iterative method, because it has roughly the same rate of convergence as SOR, and is a method of 'simultaneous displacements'. Hence, whereas a vector machine might be 50 times as fast as a sequential machine with the same cycle time, such a parallel computer would be 20 times faster still. However, Frankel's algorithm has complex eigenvalues, and so is harder to 'accelerate' than SSOR. Hence the improvement over SSOR would be smaller.

This is just one sample of a rich variety of interesting new questions, whose answers will be so important if and when parallel computers become generally available.

REFERENCES

A. K. Aziz (ed.), (1972). The Mathematical Foundations of the Finite Element Method, Academic Press.

G. Birkhoff and R. S. Varga (1959), "Implicit alternating direction methods", Trans. Am. Math. Soc. 92, pp. 13-24.

G. Birkhoff and R. S. Varga (1970), (eds.), Numerical Solution of Field Problems in Continuum Physics, SIAM-AMS Proc., vol. II.

G. Birkhoff, R. S. Varga, and D. M. Young (1962), "Alternating direction implicit methods", Advances in Computers 3, pp. 190-273.

J. R. Bunch and D. J. Rose (eds.), (1976). Sparse Matrix Computations, Academic Press.

P. G. Ciarlet (1978), The Finite Element Method for Elliptic Problems, North-Holland.

L. Collatz (1945), Eigenwertprobleme und ihre numerische Behandlung, Leipzig, 1945.

R. Courant, K. Friedrichs and H. Lewy (1928), "Uber die partielle D'gleichungen der mathematischen Physik", Math. Annalen 100, pp. 32-74.

Jim Douglas and H.H. Rachford (1956), "On the numerical solution of heat conduction problems... ", Trans. Am. Math. Soc. 82, pp. 421-39.

G. E. Forsythe and W. R. Wasow (1960), Finite Difference Methods for Partial Differential Equations, Wiley.

S. P. Frankel (1950), "Convergence rates of iterative treatments of partial differential equations", Math. Tables Aids Comp. 4, pp. 65-75.

S. Gerschgorin (1930), "Fehlerabschätzung fur die Differenzenverfahren... ", Zeits. ang. Math. Mech. 10, pp. 373-82.

Gene H. Golub and R. S. Varga (1961), "Chebyshev semi-iterative methods, successive overrelaxation ... and second order Richardson iterative methods", Numerische Math. 3, pp. 147-68.

M. H. Hestenes and E. Stiefel (1952), "Method of conjugate gradients for solving linear systems", J. Res. Nat. Bur. Standards 49, pp. 409-36.

Bert Hubbard (ed.) (1971). SYNSPADE 1970, Academic Press.

L. V. Kantorovich and V. I. Krylov (1958), Approximate Methods of Higher Analysis, Interscience. (First Russian edition 1936.)

R. von Mises and H. Pollaczek-Geiringer (1929), "Praktische Verfahren der Gleichungsauffösung", Zeits. ang. Math. Mech. 9, pp. 58-77 and 152-64.

D. W. Peaceman and H. H. Rachford (1955), "The numerical solution of parabolic and elliptic differential equations", J. SIAM 3, pp. 28-41.

D. J. Rose and R. J. Willoughby (eds.), (1972). Sparse Matrices and their applications, Plenum Press.

I. J. Schoenberg (1969), (ed.), Approximation with Special Emphasis on Spline Functions, Academic Press, 1969.

Martin H. Schultz (1973), Spline Analysis, Prentice-Hall.

R. V. Southwell (1940, 1946). Relaxation Methods (2 vols.), Clarendon Press, Oxford.

E. Stiefel (1955). "Relaxationsmethoden bester Strategie... ", Comm. Math. Helv. 29, pp. 157-79.

Gilbert Strang and George J. Fix (1973), An Analysis of the Finite Element Method, Prentice-Hall.

J. E. Traub (ed.) (1973). Complexity of Sequential and Parallel Numerical Algorithms, Academic Press.

R. S. Varga (1962). Matrix Iterative Analysis, Prentice-Hall.

David M. Young (1950), "Iterative methods for solving partial difference equations of elliptic type", Harvard Ph. D. Thesis.

——— (1954). "Iterative methods for solving partial difference equations of elliptic type", Trans. Am. Math. Soc. 76, pp. 92-111.

——— (1971), Iterative Solution of Large Linear Systems, Academic Press.

MULTIGRID SOLVERS ON PARALLEL COMPUTERS

Achi Brandt

Weizmann Institute of Science
Rehovot, Israel

CONTENT

1. INTRODUCTION

To discuss multigrid solvers on parallel computers (as suggested by the conference organizers) seems first to be untimely. Massively parallel computers as assumed here are not yet available (but see Hunt et al, 1980). It is not even clear what will be their main features. They are likely to be very cost-effective in the sense that it is much less expensive to build an array of n simple processors than to build one processor n times faster. But it is not clear how expensive the interconnections between processors will be, what limitations they will pose, how the overall control mechanism will work, etc. On the other hand a large-scale parallel computer cannot usefully be designed before we study the hypothetical algorithms that will use it, their effectiveness and the interconnection requirements they pose--especially in fields where massive calculations are required, such as the numerical solution of partial differential equations.

Indeed, parallel PDE solvers have been studied quite extensively. See for example the surveys of Ortega and Voigt (1977) and Heller (1978). Most of these studies are based, however, on solution techniques much slower (on sequential machines) than multigrid methods. The latter, moreover, are highly parallelizable. Each of their processes can simultaneously be made at all grid points. The multigrid gain on large-scale parallel machines is sometimes not as great as on sequential or vector machines (since coarse grids may not use all processors), but the potential is still very high.

For example, the standard 5-point Poisson equation on an $(2\ell + 1) \times (2^\ell + 1)$ grid is solved by a multigrid algorithm in $9\ell^2$ parallel *additions*, using one processor per two grid points (Section 3.3). FFT solves these problems in only $O(\ell)$ operations. (On sequential and vector machines multigrid is faster than FFT even for these problems: See at the end of Brandt (1980a). But parallel multigrid solutions achieve basically the same efficiency for *general* classes of linear and nonlinear, elliptic as well as non-elliptic, partial differential equations on general domains.

The purpose of the paper is thus to preliminarily explore the potential of multigrid, or, more generally, Multi-Level Adaptive Techniques (MLAT) on computers with many processors, and the interconnection requirements they pose. Section 2 is a survey of the basic MLAT processes. The description is in terms of finite-difference formulations, but analogous methods exist also in finite-element formulations (see Bank, 1980, Brandt, 1977a and 1980a, Hackbusch, 1980, and Nicolaides, 1977). We emphasize some of the less known features of FAS multigrid processing, such as the high efficiency in solving nonlinear and eigenvalue problems as well as chains of many similar problems, the flexibility to inexpensively produce optimal local refinements combined with high-order approximations, the ability to use a small computer storage, etc. These features should be taken into account both in designing the parallel algorithms and in comparing the multigrid performance to other methods.

Section 3 then describes how the basic processes are to be implemented on parallel machines, with the main conclusions surveyed in Sec. 3.1. Assuming the number of processors to be on the order n (the number of grid points), the difference equations are solved in $O(|\log n|^2)$ parallel operations. The same efficiency, with the same interconnection schemes, is obtained even when flexible local refinements are incorporated. The same operation count is also obtained in solving initial-value problems (where n now is the total number of grid points at all time steps combined). The constants in these operation counts are likely to be dominated by the amount of processing required at grid points on or near the boundaries.

A previous study of multigrid solutions on parallel computers has been conducted by Grosch (1978, 1979. See Sec. 3.7).

Much of the present study is relevant also to multigrid processes on *vector computers*, especially the discussion of relaxation techniques.

2. BASIC PROCESSES

To understand the basic multigrid processes, consider first the usual situation where a partial differential equation

$$LU(x) = F(x), \quad \text{for} \quad x = (x_1,\dots,x_d) \in \Omega , \tag{2.1}$$

with suitable boundary conditions, is discretized in a preassigned manner on a given uniform grid G^h, with mesh-size h, yielding the finite-difference equations

$$L^hU^h(x^h) = F^h(x^h), \qquad (x^h \in G^h) . \tag{2.2}$$

Here $U = (U_1,U_2,\dots,U_q)$ and its discrete approximation U^h are q-dimensional vectors of unknown functions, L is a linear or nonlinear differential operator and $L^hU^h(x^h)$ is, correspondingly, a linear or nonlinear expression involving values of U^h at x^h and at neighboring grid points. At various instances of the solution process, we have on G^h an approximation to U^h, which we will generally denote by u^h.

2.1 Coarse-Grid Approximation

To obtain a fast solution to equation (2.2) via the multi-grid method, we add to G^h a sequence of courser uniform grids. Let G^{2h} be such a coarser grid; e.g., let the grid-lines of G^{2h} be every other grid-line of G^h, so that its mesh size is $2h$. (More general situations will be discussed below.)

One way of inexpensively obtaining an approximate solution u^h to (2.2) is first to obtain an (approximate) solution u^{2h} to the corresponding coarser problem

$$L^{2h}U^{2h}(x^{2h}) = F^{2h}(x^{2h}), \qquad (x^{2h} \in G^{2h}) , \tag{2.3}$$

(which is much less expensive to solve since it contains far fewer unknowns) and then to interpolate u^{2h} to the fine grid:

$$u^h = I_{2h}^h u^{2h} \quad . \tag{2.4}$$

The symbol I_{2h}^h stands for the operation of interpolating from G^{2h}

to G^h. For example, linear interpolation in the one-dimensional (d=1) case is defined by

$$(I_{2h}^h u^{2h})(x^h) = \begin{cases} u^{2h}(x^h), & \text{if } x^h \in G^{2h}, \\ \frac{1}{2}\left[u^{2h}(x^h-h) + u^{2h}(x^h+h)\right], & \text{otherwise.} \end{cases} \tag{2.5}$$

Interpolations in higher dimensions can be performed as a sequence of one-dimensional interpolations. Higher-order interpolations should sometimes be used in (2.4) to produce a better approximation u^h. (The optimal order depends in a simple manner on the order of the differential equations, the order of the discretization error, and the order of the derivatives which we seek to approximate.)

How good the approximation (2.4) is depends on the smoothness of the solution U^h. In some cases U^h is so smooth that, if the interpolation I_h^{2h} and the coarse-grid operator L^{2h} are of order high enough to exploit that smoothness, then u^h obtained by (2.4) satisfies

$$\| u^h - U\| \lesssim \| U^h - U\| , \tag{2.6}$$

in some suitable norm. This means that u^h solves (2.2) "to the level of the truncation error", which is all we can meaningfully require in solving (2.2). In such cases, however, the fine grid G^h is not really needed: the coarser grid G^{2h} already yields a solution with the required accuracy. If G^h is at all needed, our first approximation (2.4) will require a considerable improvement.

Can we compute a correction to u^h again by some interpolation from the coarse grid G^{2h} ? Namely, can we somehow approximate the error $V^h = U^h - u^h$ by some V^{2h} computed on G^{2h} ? Normally, the answer is no. If u^{2h} in (2.4) is a good enough approximation to U^{2h} , then V^h will be a rapidly oscillating function that cannot meaningfully be described on the coarse grid G^{2h} . Therefore, before we can reuse coarse grids, the error V^h should be smoothed out.

2.2 Relaxation

An efficient error smoothing is obtained by *relaxation sweeps*. A standard example is the Gauss-Seidel relaxation sweep. This is a process in which all points x^h of G^h are scanned one by one in some prescribed order. At each point the old value $u^h(x^h)$ is replaced by a new value, which is computed so that (2.2) is satisfied at that particular point x^h (or nearly satisfied, in case (2.2) is nonlinear at that point; e.g., one Newton step of solving (2.2) for the unknown $u^h(x)$ is enough). Having completed such a sweep, the system (2.2) is not yet solved, because its equations are coupled to each other; but the new approximation u^h is hopefully "better" than the old one.

In fact, a well known, and extensively used, method for solving (2.2) is by a long sequence of relaxation sweeps. The trouble is the slow convergence of such a sequence. The larger the grid, the slower the convergence. The reason is the *local* nature of the relaxation process. It solves the equations locally, with only slight global effect. In other words, relaxation can have only a small impact (per sweep) on smooth errors, since such errors exhibit only small local errors (small residuals) compared with their own magnitude. On the other hand, relaxation sweeps can be very efficient in reducing errors which oscillate on the scale of the grid, because such errors are local; i.e., they can be approximated locally from the residuals.

Thus, relaxation is generally not an efficient solver, but it is a very efficient error smoother. Quantitatively this can best be seen by considering difference equations with constant coefficients in the infinite space, in which case the error in the discrete solution $V^h = U^h - u^h$ can be expanded in Fourier integrals, and the amplification factor (by one relaxation sweep) of each Fourier component can easily be calculated. The relaxation *smoothing factor* $\bar{\mu}$ is defined as the worst (i.e., the maximum) magnitude of an amplification factor $\mu(\underline{\theta})$ of a high-frequency component $\underline{\theta}$, where we define as *high-frequency components* all those which are visible on G^h but not on G^{2h} (i.e., on G^{2h} they coincide with lower frequencies). With a proper choice of the relaxation schemes it can be shown that the smoothing rate is uniformly bounded away from 1 if and only if the difference equations are stable. More precisely, $0 \le \bar{\mu} \le \mu' < 1$ where μ' depends on a certain measure of stability, not on h (Brandt, 1980). Moreover, in all practical problems considered, ranging from the simple Poisson equation to Navier-Stokes, singular-perturbation and non-elliptic boundary-value problems, we found relaxation schemes which yield $\bar{\mu} \le .5$ (Brandt, 1977a and 1979; Brandt and Dinar, 1979; see also Sec. 3.3 through 3.6). This means that rapidly oscillating error components are practically liquidated in just a few sweeps.

2.3 Coarse-Grid Corrections

As soon as the convergence of relaxation slows down (e.g., the convergence factor becomes larger than $\bar{\mu}$), we know that the error $V^h = U^h - u^h$ in our current solution u^h must be smooth. (More precisely, we know that $\| L^h(u^h+V^h) - L^h u^h \| \ll \| L^h \| \, \| V^h \|$ in suitable norms). Hence a good approximation V^{2h} to the error can inexpensively be computed on the coarser grid G^{2h}. For this purpose, the fine-grid equation

$$L^h(u^h + V^h) - L^h u^h = r^h , \tag{2.7}$$

where

$$r^h \equiv F^h - L^h u^h , \tag{2.8}$$

is approximated by the coarse-grid equation

$$L^{2h}\left(\bar{I}_h^{2h} u^h + V^{2h}\right) - L^{2h}\left(\bar{I}_h^{2h} u^h\right) = I_h^{2h} r^h . \tag{2.9}$$

Here I_h^{2h} and $\bar{I}_h^{2h}$ are interpolation operators (not necessarily the same) from G^h to G^{2h}; i.e., each of them represents either injection or some local averaging.

Observe that at this stage we could *not* approximate the equation $L^h(u^h + V^h) = F^h$ by the simpler approximation

$$L^{2h}\left(\bar{I}_h^{2h} u^h + V^{2h}\right) = I_h^{2h} F^h ,$$

since the error of this approximation depends on the high-frequency components of u^h. Such components may be large compared with V^h. On the other hand the error in the approximation (2.9) depends only on the high-frequency components of V^h.

In case L is linear, we could write (2.9) in the form

$$L^{2h}V^{2h} = I_h^{2h} r^h . \tag{2.10}$$

Generally, however, to avoid complicated linearizations in solving (2.9), a new unknown function

$$\bar{U}^{2h} = I_h^{2h} u^h + V^{2h} \tag{2.11}$$

is introduced (instead of the unknown V^{2h}), in terms of which (2.9) becomes

$$L^{2h} \bar{U}^{2h} (x^{2h}) = \bar{F}^{2h}(x^{2h}) , \tag{2.12}$$

where

$$\bar{F}^{2h} = L^{2h}(\bar{I}_h^{2h} u^h) + I_h^{2h} r^h . \tag{2.13}$$

The advantage of (2.12) is that it has the same form as (2.2) and (2.3). Hence the same routines (e.g., the same relaxation routine), can be used in treating all of them, even if L is nonlinear.

The new unknown $\bar{U}^{2h}$ represents, on the coarse grid, the sum of the basic approximation u^h and its correction V^h. This scheme of forming the coarse-grid problem is therefore called the *Full Approximation Scheme* (*FAS*) - to be distinguished from the *Correction Scheme* (*CS*) which directly uses V^{2h}. The Correction Scheme is less attractive in nonlinear problems and cannot be used on composite grids (Sec. 2.8).

Once (2.11) is approximately solved, its approximate solution u^{2h} is used to correct the basic approximation u^h. In doing so we should keep in mind that $V^{2h} = \bar{U}^{2h} - I_h^{2h} u^h$, and not $\bar{U}^{2h}$ itself, is the function which approximates a smooth fine-grid function (V^h). Hence, the coarse-grid correction to the fine-grid solution u^h is given by

$$u^h_{NEW} = u^h_{OLD} + I^h_{2h} \left(u^{2h} - \bar{I}^{2h}_h u^h_{OLD} \right) . \tag{2.14}$$

Since $I^h_{2h} I^{2h}_h u^h \neq u^h$, (2.14) is not equivalent to

$$u^h_{NEW} = I^h_{2h} u^{2h} . \tag{2.15}$$

The important difference is that (2.14) preserves the high-frequency content of u^h_{OLD}, while (2.15) does not use u^h_{OLD}, and thus looses that content, i.e., looses the most expensive-to-compute part of the solution.

The order of the I^h_{2h} interpolation in (2.14) can sometimes be lower than in (2.4). But $\bar{I}^{2h}_h$ in (2.14) must be identical with $\bar{I}^{2h}_h$ in (2.13).

2.4 Full Multi-Grid Algorithms

We still have to specify how the coarse-grid equations, (2.3) as well as (2.12), are to be solved. (2.3) is, of course, solved in the same way as (2.2); namely, by obtaining an initial approximation from a still-coarser grid G^{4h}, then relaxing on G^{2h}, and then obtaining a coarse-grid correction using G^{4h} again. In solving (2.12), on the other hand, we start with the approximation

$$u^{2h} = \bar{I}^{2h}_h u^h , \tag{2.16}$$

the same $\bar{I}^{2h}_h$ as in (2.13) and (2.14). Hence we do not usually need a first approximation from a coarser grid. A satisfactory approximation to V^{2h} is obtained if, starting from the initial solution (2.16), we relax (2.12) and then obtain a correction from the next coarser grid G^{4h}.

The full algorithm has several variations. One is flowcharted here as Fig. 1. Others are given in Brandt (1977a), where a Fortran program of 146 statements is appended, together with its output. Still another is shown in Fig. 1 of Brandt (1979).

Level (grid)	Mesh size
1	h_1
2	$h_2 = \frac{h_1}{2}$
3	$h_3 = \frac{h_1}{4}$
⋮	⋮
M-1	$h_{M-1} = 2h_M$
M	h_M

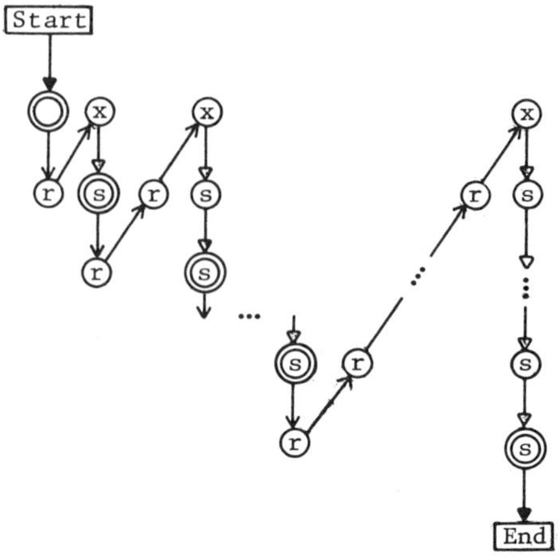

FIGURE 1. Full Multigrid Algorithm

Shown in each circle is the number of relaxation sweeps performed at each stage. The grid being swept and its meshsize are indicated at left on the same row. In most applications $(r,s) = (1,1), (2,0), (2,1)$ *or* $(3,0)$, *while* x *depends on the problem and on* h_1. *For some problems the equations on grid 1 should be solved by a direct method. A double circle indicates the stage where the error obtained satisfies (2.6), i.e., it is below the truncation error for that grid. The transfer "↓" from grid* $2h$ *to* h *is the first-approximation interpolation (2.4). "⇣" is the correction interpolation (2.14). The transfer "↗" from meshsize* h *to* $2h$ *involves the calculation of the new right-hand side (2.13) and the first approximation (2.16).*

Taking the work in one relaxation sweep on grid M as our work unit, the relaxation sweep on level $M-k$ costs about 2^{-kd} work units (on a in mind that $V^{2h} = \bar{U}^{2h} - I_h^{2h} u^h$, and not $\bar{U}^{2h}$ itself, is the function which approximates a smooth fine-grid function (V^h). Hence, the coarse-grid correction to the fine-grid solution u^h is given by

$$u_{NEW}^h = u_{OLD}^h + I_{2h}^h \left(u^{2h} - \bar{I}_h^{2h} u_{OLD}^h \right) . \tag{2.14}$$

Since $I_{2h}^h I_h^{2h} u^h \neq u^h$, (2.14) is not equivalent to

$$u_{NEW}^h = I_{2h}^h u^{2h} . \tag{2.15}$$

The important difference is that (2.14) preserves the high-frequency content of u_{OLD}^h, while (2.15) does not use u_{OLD}^h, and thus looses that content, i.e., looses the most expensive-to-compute part of the solution.

The order of the I_{2h}^h interpolation in (2.14) can sometimes be lower than in (2.4). But $\bar{I}_h^{2h}$ in (2.14) must be identical with $\bar{I}_h^{2h}$ in (2.13).

2.4 Full Multi-Grid Algorithms

We still have to specify how the coarse-grid equations, (2.3) as well as (2.12), are to be solved. (2.3) is, of course, solved in the same way as (2.2); namely, by obtaining an initial approximation from a still-coarser grid G^{4h}, then relaxing on G^{2h}, and then obtaining a coarse-grid correction using G^{4h} again. In solving (2.12), on the other hand, we start with the approximation

$$u^{2h} = \bar{I}_h^{2h} u^h , \tag{2.16}$$

the same $\bar{I}_h^{2h}$ as in (2.13) and (2.14). Hence we do not usually need a first approximation from a coarser grid. A satisfactory approximation to V^{2h} is obtained if, starting from the initial solution (2.16), we relax (2.12) and then obtain a correction from the next coarser grid G^{4h}.

The full algorithm has several variations. One is flowcharted here as Fig. 1. Others are given in Brandt (1977a), where a Fortran program of 146 statements is appended, together with its output. Still another is shown in Fig. 1 of Brandt (1979).

Level (grid)	Mesh size
1	h_1
2	$h_2 = \frac{h_1}{2}$
3	$h_3 = \frac{h_1}{4}$
⋮	⋮
M-1	$h_{M-1} = 2h_M$
M	h_M

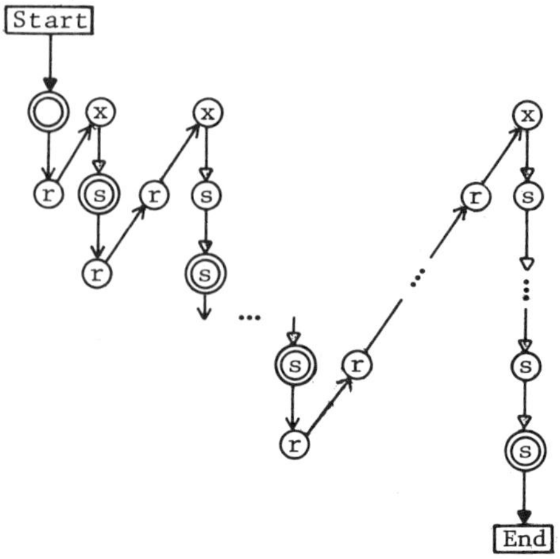

FIGURE 1. Full Multigrid Algorithm

Shown in each circle is the number of relaxation sweeps performed at each stage. The grid being swept and its meshsize are indicated at left on the same row. In most applications $(r,s) = (1,1)$, $(2,0)$, $(2,1)$ *or* $(3,0)$, *while* x *depends on the problem and on* h_1. *For some problems the equations on grid 1 should be solved by a direct method. A double circle indicates the stage where the error obtained satisfies (2.6), i.e., it is below the truncation error for that grid. The transfer "↓" from grid* $2h$ *to* h *is the first-approximation interpolation (2.4). "⇩" is the correction interpolation (2.14). The transfer "↗" from meshsize* h *to* $2h$ *involves the calculation of the new right-hand side (2.13) and the first approximation (2.16).*

Taking the work in one relaxation sweep on grid M as our work unit, the relaxation sweep on level $M-k$ costs about 2^{-kd} work units (on a sequential machine); hence the solution algorithm of Fig. 1 costs about $(r+s)/(1-2^{-d})^2$ work units, in addition to a smaller amount of work involved in the inter-grid transfers. On parallel computers the relations are different (see Secs. 3.1, 3.3). Usually $r+s = 2$ or 3 is what one needs to get (2.6), independently of the shape of the domain, the type of the boundary conditions, the smoothness of the solution, or the mesh size. Similar efficiency is obtained even for problems with strongly discontinuous coefficients, provided more sophisticated inter-grid transfers are employed (Alcouffe et al, 1980).

The storage requirement of multigrid solvers is minimal. Certain versions can operate with a storage much smaller than the size of the finest grid, without using external memory. Indeed, only a small portion of G^h is needed at a time in order to locally smooth the error in u^h so that $\bar{F}^{2h}$ can be calculated to the desired accuracy. Three kinds of small-storage algorithms are described in Brandt (1977a, Sec. 7.5; 1979a, end of Sec. 2.2) and Hackbusch (1980b).

2.5 Nonlinear Problems and Eigenvalue Problems

Nonlinearily hardly matters in FAS multigrid solutions. The same process described above works, with the same efficiency, whether the problem is linear or not. No global linearization is made, and no linearization error is involved in the coarse-grid corrections. Hence Newton steps should not be used: Solving the linear equations in each Newton step would cost as much as the FAS solution of the nonlinear problem itself (and usually even more, since the linearized equations are normally more complicated; their relaxation requires more calculations and/or more storage than the relaxation of the original nonlinear equations).

The only place in the algorithm involving a sort of linearization is at each relaxation step. This, however, is not a global linearization. In fact, the equation at each relaxation step may be linear(in the relevant unknown), even if the differential equations are nonlinear. For example, in Navier Stokes equations, if the derivative u_x in the nonlinear term uu_x is approximated by a central difference quotient, then the equation at any point $\underline{x}^h$ becomes linear in the corresponding unknown $u^h(\underline{x}^h)$, so that even pointwise linearization is not needed. Moreover, even when pointwise linearization is made in relaxation, it is not really different than what we do in linear problems: In each relaxation step, even for linear problems, we freeze some of the (pointwise) unknowns, while changing others. When the pointwise equation is nonlinear, we may simply regard linearization as adding some occurences of the corresponding unknown to the set of frozen unknowns, while changing its other occurences.

For multigrid continuation techniques for nonlinear problems, see Sec. 2.7.

Eigenvalue problems can be viewed as nonlinear problems, since the differential equations involve a product of one unknown (the eigenvalue λ) with another (the eigenfunction), as well as a nonlinear (L_2) normalization. Using FAS, relaxation of the problem on each level proceeds without changing λ, since λ is a constant and hence its error is always smooth. λ is changed only in solving the coarsest-grid problem. Numerical experiments indeed show that such FAS solutions of the eigenvalue problem require exactly the same work as the solution of the corresponding boundary-value problem. (Brandt McCormick and Ruge, 1980).

Equally successful has been the application of FAS to various nonlinear flow problems, such as *transonic flows* (South and Brandt, 1976; Brandt, 1977a; Jameson, 1979), even in a three-dimensional production code (McCarthy and Reyhner, 1980), and *incompressible Navier-Stokes equations* (Brandt and Dinar, 1979; Dinar, 1979). It has also been applied to the *constrained minimization problem*

$$\min_{u\geq 0} \int \left(U_x{}^2 + U_y{}^2 + 2FU\right) dx\, dy$$

arising from free boundary problems, yielding essentially the same efficiency as in solving the Poisson equation $-\Delta U = F$, to which the <u>un</u>constrained problem (dropping the $U \geq 0$ contraint) is equivalent (Brandt and Cryer, 1980).

2.6 The Fine-to-Coarse Correction τ

Using (2.8) and (2.13), we can rewrite the coarse-grid equation (2.12) in the form

$$L^{2h} \bar{u}^{2h} = F^{2h} + \tau_h^{2h} , \tag{2.17}$$

where

$$\tau_h^{2h} = L^{2h} \bar{I}_h^{2h} u^h - I_h^{2h} L^h u^h . \tag{2.18}$$

Comparing (2.17) with (2.3) we can now reverse our viewpoint. Instead of regarding the coarse grid as a device for calculating the correction (2.16) to the fine grid *solution* we can view the fine grid as a device for calculating the correction τ_h^{2h} to the coarse-grid *equation*. This correction makes the solution of the coarse-grid equation identical, at convergence, with the fine-grid solution: When $V^h = 0$ then $\bar{u}^{2h} = I_h^{2h} u^h$. We thus call τ_h^{2h} the *fine-to-coarse correction function*. It is a kind of defect correction. The main point of the multigrid algorithm is that this particular defect correction depends essentially on high-frequency components, and should therefore be obtained by relaxation sweeps on the finer grid.

The fine-to-coarse correction function τ_h^{2h} also represents the local truncation error on meshsize $2h$ relative to meshsize h; i.e., the discrepancy exhibited when U^h is substituted in the coarse-grid equations (2.3). It approximates the true local truncation error

$$\tau^{2h} = L^{2h} U - LU , \tag{2.19}$$

which is the discrepancy exhibited when the true differential solution U is substituted into those equations. More precisely it is easy to see that $\tau_h^{2h} \approx \tau^{2h} - \tau^h$. We can, therefore, use τ_h^{2h} to estimate τ^h, and hence to derive natural grid-refinement criteria. Or, alternatively, we could use τ_h^{2h} to improve our accuracy: If the order of accuracy of L^h is p, we can increase it simply by replacing τ_h^{2h} in (2.17) by $2^p(2^p-1)^{-1}\,\tau_h^{2h}$. This is called τ-*extrapolation*.

Other applications of τ_h^{2h} are described in the following sections.

2.7 Chains of Problems and Time-dependent Problems

We often need to solve not just one isolated problem but a sequence of (many) similar problems, depending on some parameter. For example, we may be studying the effect of changing some physical parameters on the "Performance" of a system, which we may moreover seek to optimize. Chains of similar problems are also encountered in continuation (or "embedding") processes for solving nonlinear problems, especially in following the graphs of bifurcating solutions. Even a linear problem may require a continuation process if its proper discretization (e.g., its local resolution at some specific regions) is not known in advance (Brandt, 1977a). Or, most frequently, the sequence of algebraic systems may represent the implicit discretization of an evolution problem, one algebraic system per time-step.

Such chains of problems can be solved very efficiently, using the *frozen-*τ *technique*. For each new problem in the chain we not only use the solution of the previous problem as our first approximation, but also employ the values of the previous coarse-to-fine correction functions, in the following manner. Before the algorithm has even reached G^h, we can use (2.17) instead of (2.3) as our equation on G^{2h}, with τ_h^{2h} taken from the previous problem. The error in u^{2h} so produced will not depend on the high-frequency components (as in solving (2.3)), but on the *changes* in those components from their values at the previous problem. Since these changes are normally small compared with the components themselves, we can obtain a solution with errors comparable to τ^h without even working at meshsize h. We may have to work also on G^h only once in many steps in order to update the values of τ_h^{2h}.

Moreover, the changes in quantities such as τ_{2h}^{4h} can serve as natural criteria for deciding at what stage we need to work on G^h (see Sec. 3.9 in Brandt, 1980a).

The above is especially true in solving parabolic time-dependent problems, where high-frequency components settle quickly into quasi steady-state. It is then possible to make most time-steps on coarser grids only, seldom visiting the finest grid at all.

2.8 Local Refinements and Transformations. High-order Approximations

An important feature of FAS multi-grid is the possibility of creating *local* refinements; that is, any grid, G^h say, may cover only part of the domain covered by the next coarser grid G^{2h}. The FAS equations (2.17) can still be used, the only change then being that $\tau_h^{2h}(x^{2h}) = 0$ for every point x^{2h} which is an interior point of G^{2h} but not an interior point of G^h; i.e., the fine-to-coarse correction is simply dropped wherever the fine grid does not exist. The FAS solution process can therefore proceed essentially as before (some of the returns to coarser grids may be avoided), with basically the same efficiency. In this manner increasingly finer levels may be added, confined to increasingly smaller subdomains, thus providing higher resolution only where desired. Natural optimal criteria for deciding where a further refinement, $G^{h/2}$ say, is needed, can be based on the relative local truncation errors τ_h^{2h} (Brandt, 1977a).

Moreover, each of these localized finer grids may be a uniform grid in its own *local system of coordinates*, chosen so as to fit curved boundaries or to approximate directions of interior interfaces and thin layers. Unlike global coordinate transformation, these local coordinates do not complicate the difference equations throughout the domain, nor do they introduce singularities (at corners, etc.). Also, they are not used for producing stretching or any desired pattern of meshsizes - this is done more efficiently by the local refinement technique mentioned above. Since their only use is for producing desired grid *directions* and because of their local nature, these transformations are particularly simple. Each of them can be stored in terms of a small set of lower-dimensional arrays (Brandt, 1977b).

Near boundary layers, shocks and other thin layers or discontinuities, very fine meshsizes are sometimes needed in one direction (roughly perpendicular to the layer, to resolve its thin width), but not in others. Even when the required meshsize is very small, not many grid prints will be needed (since the layer is correspondingly very thin), provided the fine meshsize is introduced only in the desired direction. Such *unisotropic local refinements* can be created by combining the local coordinate transformations with the above technique of local refinements, applied in the desired direction only. That is, the local grids become increasingly finer only in that direction, employing uniform rectangular instead of square meshes (Brandt, 1979a. See related remarks in Sec. 3.5 below).

All these patches of local grids are not tailored to each other along their boundaries. Each patch serves only to transfer τ corrections to a coarser level (as in (2.17) - (2.18)), from which it will in turn get a correction of the type (2.14). The patches thus interact through the FAS multigrid process, which provides fast solution to their difference equations (an important advantage over other methods of patching grids or using transformations).

This structure, in which non-uniform discretization is produced through a sequence of uniform grids is highly flexible, since changing local meshsizes is affected by the relatively simple operations of extending or contracting uniform grids. Also, since in this structure only equidistant differencing is used, it is easy to employ *high-order difference approximations*, despite the non-uniformity.

A convenient way of introducing high-order approximations into the multigrid processing is by defect corrections, which can usually be made on coarser grids only; that is, increasing the approximation order of L^h in (2.18) (or, more generally, using a higher-order approximation in calculating the residuals (2.8) which are used in (2.12) - (2.13)), while keeping the low-order L^h in the relaxation sweeps. This technique exploits the fact that in many cases the high-order correction is desired only for lower-frequency components of the solution. Its advantages are that the higher-order operator L^h need not be stable, and that the extra computer work needed for obtaining the higher-order approximation is small (Brandt, 1980b).

3. PARALLEL MULTIGRID PROCESSING

3.1 General Statements

An almost immediate observation concerning each of the basic multigrid processes mentioned above is that in principle it can be made simultaneously at all grid points, and hence it can effectively benefit from as many parallel processors as the number of points on any one grid. Sections 3.2 through 3.8 will discuss this kind of parallelism and the inter-processor connections it requires. We will see that with proper connections, one can execute a relaxation sweep (or a residual transfer) in a small, bounded number, K say, of machine cycles. (K is ideally the number of operations, plus the number of data transfers, per grid point. In practice it is likely to be the maximal number of operations at any one grid point, which may be quite large due to complexity near boundaries. See Sec. 3.8). A complete multi-grid solution will then typically require $2K(\log_2 N)^2$ cycles of the parallel machine, where $N = O(1/h)$ is the number of fine-grid points in any one direction, because the number of grids is about $\log_2 N$ (and see Fig. 1). For comparison, solving the same difference equations just by relaxation would typically cost $KO(N^2)$ cycles of the same parallel computer although this can be reduced to $KO(N)$ for particularly simple operators and geometries by employing optimal relaxation parameters. In addition, multigrid can in principle handle local refinements with essentially the same efficiency (Section 3.7)

A small number of parallel processors can also be fully exploited. One way of doing this is to partition the domain into subdomains, each of which is assigned to one processor. This is indeed a special case of the methods described in Sections 3.2 - 3.8 below. Alternatively, each processor can perform a different task: While the first one performs the first relaxation sweep the next processor can already work on the second sweep, properly lagging behind, and the third processor, with an additional delay, can work on transferring residuals to the coarser grid. Or several processors can work in parallel to perform the relaxation step at each grid point. This is particularly effective and easy to design in the case of a system of several field equations, resulting in several equations per mesh cell. Each processor can then be assigned to the relaxation of one of these equations. Proper care should, of course, be taken to coordinate their data flow. (A simple, but not the only, way is to delay substituting results).

Another interesting situation is where the number of processors is much larger than the number of grid points. In multigrid solutions this it likely to occur, at least in the stage of working on coarser grids. One can then use several processors at each grid point, performing in parallel the relaxation step or the residual calculation. Again, this can be particularly efficient in the case of a system of several equations with several unknown functions. It cannot remain efficient, however, when the number of processors per grid point becomes too large. We may therefore have periods of severe unemployment among processors, especially when working on coarser grids. It still pays, however, to keep those coarser grids in the algorithm since for some smooth error components, the coarser grids are more efficient not only per (serial machine) operation, but also per sweep. This, in fact, is the very reason for the superior performance of the multigrid *parallel* algorithm over relaxation *parallel* algorithms (despite the full employment offered by the latter).

Processors unemployed in the coarse-grid operations can actually be used in a number of ways. One way is to perform additional relaxation sweeps on fine grids in parallel to the coarse-grid operations. This is discussed in Section 3.9. Another way is to solve simultaneously several problems, so that each of them uses the fine grid in its turn, but on coarse grids they work in parallel (see Sec. 3.7).

3.2 Relaxation and Interconnections

Assume first we have as many processing elements as grid points, or, more precisely, as many processors as difference equations. We can then relax all these equations in parallel, provided the corresponding data flow is efficiently implementable. In particular it is necessary that the results of each processor will at the next step be available to all processors corresponding to neighboring equations, where a difference equation i is called a "neighbor" of equation j if the function values changed when relaxing j participate in equation i. (In various situations "distributive" relaxation schemes are used in which several unknowns are changed per each difference equation. See Brandt and Dinar 1979). This may imply that the inter-processor connections depend on the specific difference equation and relaxation scheme being used. For most purposes it is enough, however, to arrange the processors in a uniform d-dimensional square lattice and connect each processor to its 2d immediate neighbors. The 2-dimensional case is shown in Fig. 2. A "Connection" between two processors means that data can in one machine "step" be transferred from the memory of one to the other. If the relaxation scheme requires data transfer between farther neighbors, this can then be done in several such steps; rarely more than three are required. This will hopefully require not much time compared with the time of arithmetic operations per relaxation step.

(a)
(b)
(c)

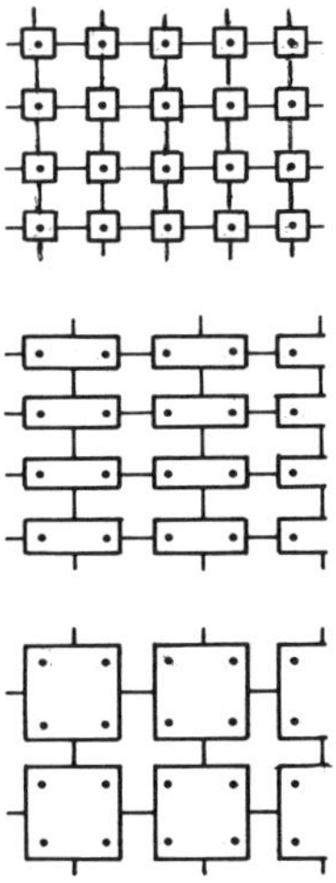

FIGURE 2.

Two-dimensional lattice of processors and their nearest-neighbor connections. (a) One grid point per processor. (b) Two grid points per processor. (c) Four grid points per processor.

Arrays of processors, although limited in size, have been in operation for several years now, starting with Illiac IV (Barnes et. al., 1968; see Bouknight et. at., 1972), and then the Distributed Array Processor (DAP; Parkinson, 1976; Gostick, 1979). Others are in various stages of design or construction, such as the Wisconsin Parallel Array Computer, (WISPAC; Cyre et. al, 1977) and the Massively Parallel Processor (MPP; Goodyear, 1979). All these are single-Instruction-Multiple-Data (SIMD) machines, i.e., all processors execute the same instruction at the same time, except for WISPAC, which is designed to be a Multiple-Instruction-Multiple-Data (MIMD) machine. The execution of relaxation on such machines has been described by various authors (see Ortega and Voigt, 1977, and references therein).

It is not, however, necessary that the processors should be arranged in such lattices. In the most important problems for which parallel processing is sought, such as weather prediction and aerodynamic calculations, the amount of operations per grid point is so large that less efficient communication between processors may be acceptable. This is even more so in case more grid-points are assigned to each processor (Fig. 2. (b), (c)), where the amount of arithmetic operations per inter-processor transfer is even larger. In SMS 201, designed for weather prediction, 16 grid points were assigned to each of the 128 parallel microprocessors, hence a data exchange mechanism using a single data bus was found to be adequate (Nagel, 1979). Many other interconnection methods have been proposed, some of them designed so that any data permutation between the processors can be made in just $O(dog_2 n)$ steps, where n is the total number of processors, and each processor is connected to just a few others (Parker, 1980, Sullivan and Bashkow, 1977, and references therein). Among these methods are the shuffle-nearest-neighbor schemes, like the one discussed in Sec. 3.7 below. These more universal networks are useful for many other algorithms, not necessarily related to differential equations. Moreover, they are geometrically more flexible, since data transfers required by grids of different dimensionality and varying sizes can be programmed, and more computing power can be assigned to special gridpoints. They may also offer greater reliability: In a rigid lattice one faulty processor can degrade the entire system much more than in a lattice programmed on more universal connections. A further study should be made, however, to determine, for each of the universal networks, how efficiently it can emulate d-dimensional lattices of flexible sizes.

Avoiding further discussion here of the interconnection issue, let us simply assume that we can relax simultaneously at all grid points. What simultaneous relaxation schemes are available, and which offer the most efficient error smoothing required by the multigrid algorithm?

3.3 Relaxation Schemes. Poisson Equation

The classical simultaneous relaxation is the Jacobi relaxation. For the standard example of the 5-point Poisson equation $-\Delta_h u = F$ it has the form

$$\bar{u}_{ij} = (1-\omega)\left(u_{ij} + \frac{\omega}{4} \quad u_{i-1,j} + u_{i,j-1} + u_{i,j+1} + u_{i+1,j}\right) + h^2 F_{ij} \qquad (3.1)$$

where u_{ij} is the approximate solution at grid point (i,j) before relaxation, $\bar{u}_{ij}$ its new value, h is the meshsize and ω is the "relaxation parameter." The best parameter is $\omega = .8$, for which the smoothing factor (see Sec. 2.2) is $\bar{\mu} = .6$. This is worse than the smoothing factor $\bar{\mu} = .5$ obtained for the Gauss-Seidel relaxation on sequential machines, where the points (i,j) are taken in lexicographic order and $\omega = 1$. Gauss-Seidel costs 5A (five Additions per grid point; a multiplication by a power of 2 is counted as an addition), while Jacobi costs 5A + 2M (five additions and two multiplications per grid point); but the latter can be done simultaneously at all grid points. (The number of additions in Jacobi relaxation can be reduced by one if the sum $u_{i-1,j} + u_{i,j-1}$

can serve also as the sum $u_{(i-1),(j-1)+1} + u_{(i-1)+1,(j-1)}$ in calculating $\bar{u}_{(i-1),(j-1)}$. But this trick may require an additional interprocessor transfer.

A generalization of Jacobi relaxation is the Weighted Simultaneous Displacement (WSD) relaxation (Brandt, 1977a), given generally be

$$\bar{u}_\nu = u_\nu + \sum_{\gamma \varepsilon \Gamma} \omega_\gamma \, \delta u_{\nu+\gamma} , \tag{3.2}$$

where $\nu = (\nu_1,\cdots,\nu_d)$ are grid points (ν_j integers), ω_γ are the weights (or the relaxation parameters), $\gamma = (\gamma_1,\cdots,\gamma_d)$, γ_j integers, and Γ is a small set near $(0,\cdots,0)$. u_ν is the approximate solution at ν before the relaxation step, and $\bar{u}_\nu$ is its value after the step. δu_ν is the correction which, replacing u_ν by $u_\nu + \delta u_\nu$ and keeping u_α unchanged at all points $\alpha \neq \nu$, would satisfy the difference equation at ν. For the 5-point Poisson equation and the set $\Gamma = \{(\gamma_1,\gamma_2): |\gamma_1| + |\gamma_2| \leq 1\}$, the optimal weights are $\omega_{00} = 48/41$, $\omega_{0,\pm 1} = \omega_{\pm 1,0} = 8/41$, and the smoothing factor is $\bar{\mu} = 9/41 = .22$. This costs $10A + 2M$ (reducible to $8A + 2M$ with the above trick). Hence, per operation, it is more efficient than Jacobi.

Another well-known type of simultaneous relaxation is the Red-Black (RB, or checker-board, or chess-board, or odd-even) relaxation, which is like Jacobi (with ω=1) scheme, except that (3.1) is performed in two stages. In the first stage (3.1) is calculated at all "red" (or "even") points (i.e., points (i,j) where i+j is even), and the new values $\bar{u}_{ij}$ replaces the old ones. In the second stage (3.1) is performed at "black" (or "odd") points using on the right-hand side of (3.1) the results from the first stage. Note that, since any red-point value u_{ij} participates in only one red-point equation, after the first stage all the red-point equations are simultaneously satisfied; similarly for the black points after the second stage. The method is equivalent to Gauss-Seidel relaxation with red points relaxed before black points, and is sometimes called "Gauss-Seidel relaxation with red-black ordering."

The RB relaxation introduces a fixed coupling between any frequency $\underline{\theta} = (\theta_1, \theta_2)$ and the frequency $\underline{\theta}' = (\theta_1+\pi, \theta_2+\pi)$; i.e., a fixed ratio between the amplitude of the Fourier component $\exp[i(\theta_1 x_1+\theta_2 x_2)/h]$ corresponding to $\underline{\theta}$, and the one corresponding to $\underline{\theta}'$. It was therefore believed that full weighting should be used in transferring residuals to a coarser grid (Brandt, 1977b). In fact, however, one can use weighted injection, injecting just half the residuals ($I_h^{2h} \equiv 1/2$ in (2.13)), in case coarse-grid points are red points (Stueben, Trottenberg and Foerster, 1980). The smoothing factor $\bar{\mu} = \max \mu(\underline{\theta})$ for this relaxation should be calculated with max ranging over the set

$$\left\{ \underline{\theta} = (\theta_1,\theta_2) \;\middle|\; \min(|\theta_1|,|\theta_2|) \le \frac{\pi}{2} \le \max(|\theta_1|,|\theta_2|) \le \pi \right\}, \tag{3.3}$$

which is the set of high-frequency components not determined by low-frequency components. Since $\mu(\underline{\theta}) = (\cos\theta_1 + \cos\theta_2)^2/4$, and since $\cos\theta_1$ and $\cos\theta_2$ have opposite signs on the set (3.2), it follows that $\bar{\mu} = .25$.

The number of operations for RB relaxation is just 5A (4A with the above trick) at each stage, totalling 10A (8A) per sweep. Hence with less operations than WSD it yields the same smoothing. More precise comparison between them should be made by a two-level analysis (Brandt and Dinar, 1979) and numerical experiments. The numbers in Appendix C of Brandt (1977a), compared with those in Stueben, Trottenberg and Foerster (1980), indicate that a cycle with 2 sweeps of WSD (with suitably optimized weights) on each grid is about equivalent to a similar cycle of RB. But RB has an important advantage over WSD: Its efficiency does not result from a careful choice of parameters (it uses ω=1), hence it is more robust, and will work equally well for many other problems. Also note that for the above performance RB relaxation requires only half the number of processors; the configuration of Fig. 2(b) can be used by it.

(Concerning the advantage of RB over Gauss-Seidel relaxation for sequential multigrid algorithms - See Stueben et. al., 1980.)

Thus *RB relaxation is recommended in solving Poisson equations.* (It was so recommended also by Grosch, 1978, although without realizing its combination with weighted injection.) Two-level mode analyses, as well as numerical experiments, show that using this relaxation in the multigrid algorithm of Fig. 1, with r=s=1, yields a solution with algebraic error less than half the truncation error. With the Correction Scheme (Sec. 2.3), *this entire algorithm for an* $(2^{\ell}+1) \times (2^{\ell}+1)$ *grid will cost about* $9\ell^2$ *parallel additions.* Such a count is, of course, only partially meaningful, since no account is taken of interprocessor transfers. Depending on machine architecture, these transfers may cost more than the arithmetic. Also, this count neglects the cost of distinguishing at each processor between red and black. (Note in Fig. 2(b) that the location of the red point in half of the processors is on the left, and in the other half it is on the right. A similar need for distinguishing between several kinds of points occur in the linear interpolation). Furthermore, on more general domains more complex difference equations will be formed near boundaries, at least on coarser grids, and the operation count will consequently increase (cf. Sec. 3.8).

A generalization of RB relaxation can be used in case the number of processors is less than half the number of grid points. We can use configurations like Fig. 2(c), with k grid points (more precisely: k grid equations) per processor, and relax in k stages, where the points (or equations) relaxed in the j-th stage constitute a uniform subgrid. In analogy to the red-black denomination, we can imagine each such subgrid to have its distinct "color", and we call any such relaxation scheme pattern relaxation. Its smoothing rate, in fact all the numbers it produces, are identical to that of RB relaxation (in the case of Poisson equation), if the k colors are ordered so that those corresponding to "red" are relaxed before all those corresponding to "black". Hence, if enough processors are available, RB is preferred to any other pattern relaxation.

Analogous simultaneous relaxation schemes exist in higher dimensions (d>2), and again RB relaxation (relaxing first points $\nu = (\nu_1,\cdots,\nu_d)$ where $\nu_1+\cdots+\nu_d$ is even, then those where it is odd) is perhaps the best smoother for the Poisson and many other equations. The set (3.3) generalizes to

$$S = \left\{ \underline{\theta} = (\theta_1,\cdots,\theta_d) \,\middle|\, \min_{1\le j\le d} |\theta_j| \le \frac{\pi}{2} \le \max_{1\le j\le d} |\theta_j| \le \pi \right\} . \tag{3.4}$$

For the Poisson equation $\mu(\theta) = (\Sigma_1^d \cos\theta_j/d)^2$, hence the smoothing factor is $\bar{\mu} = \max_S \mu(\theta) = [(d-1)/d]^2$. Note that the higher the dimension the worse is $\bar{\mu}$. This is true also for other relaxation schemes. For Jacobi relaxation, the optimal relaxation paramer $\omega = 2d/(2d+1)$ yields $\bar{\mu} = (2d-1)/(2d+1)$. For the three-dimensional (d=3) Poisson equation, the RB smoothing factor is $\bar{\mu} = 4/9$, implying that in applying the multigrid algorithm of Fig. 1 one should use $r+s = 3$, so that to solve an $(2^\ell+1) \times (2^\ell+1) \times (2^\ell+1)$ grid problem would cost about $12\ell^2$ parallel additions and $2\ell^2 + \ell$ parallel multiplications. (Here we have multiplications by 1/6, which cannot be counted as additions.)

In higher dimensional (e.g., d=3) problems, the situation will frequently be that the machine processors correspond to only lower-dimensional (e.g., planar) sections of the grid, either because of their limited number or because of their interconnections. Each processor corresponds then to, e.g., a line of grid points. We can still perform the RB relaxation. Moreover, in this case we can use as many processors as there are lines, for while half of them relax the red points of one (planar) section, the other half can relax the red points of the next section. We can also use zebra (or line RB) relaxation, in which we change all the unknowns corresponding to one processor so as to simultaneously satisfy all the corresponding equations. For this purpose the processor has to solve a tridiagonal system of equations, which is easily done since the system is diagonally dominant. This is done simultaneously by all processors corresponding to "red" lines and then, in a second stage, by all those corresponding to "black" lines. In case of the Poisson equation, however, the pointwise RB relaxation has a better smoothing rate per (parallel) operation.

3.4 Pattern Relaxation of Other Equations

It requires further research to select the best schemes for parallel relaxation of other classes of difference equations. A quick look at some representative examples shows, however, that RB relaxation, and more general pattern relaxation, line-pattern relaxation and weighted-pattern relaxation, will be among the best schemes for most purposes. (RB-type schemes, including zebra, were studied and recommended by South Keller and Hafez, 1980).

The smoothing analysis of such schemes is generally somewhat more involved than in the above cases. Only if RB relaxation is used with 2d-point-star equations is it true that some high frequencies are uniquely determined by the low frequencies, and $\bar{\mu}$ can therefore be determined over a set like (3.3) or (3.4). This is no longer true when more general stencils are used. In such cases, after the first stage of RB relaxation (relaxing at red points), the red equations are not satisfied, because they are coupled: Each red-point equation involves more than one red-point unknown. Consequently, the simultaneous *displacement* of new values does not imply simultaneous *solution* of the equations.

The terminology may create some confusion. RB relaxation means simultaneous *displacement* at all red points, not simultaneous *solution*. We will thus say that the red-point equations are solved *in parallel*, not *simultaneously*. The zebra (or line-RB) relaxation described above does *not* mean simple parallel solution (or simultaneous displacement) at all red-line points. Rather, it means that the simultaneous *solution* of the equations on each individual line is done in parallel (i.e., simultaneously *displaced*) on all red lines. For certain three-dimensional equations, like the Poisson, but *not* generally, this would imply simultaneous *solution* of all red-line equations.

Any pattern relaxation with the property that the equations solved in parallel are thereby solved simultaneously (i.e., after each stage all the equations corresponding to that stage are simultaneously satisfied) will be called *decoupling pattern relaxation*. RB and zebra for the Poisson equation are examples of decoupling schemes. Note that this property depends not only on the difference equations and the order in which they are taken (or grouped into stages), but also on the scheme of distributing and/or weighting of corrections. (In distributive relaxation each equation is satisfied by changing several unknowns; Brandt and Dinar, 1979. Weighted relaxation is explained above). Thus, a distributive RB relaxation for 5-point-star equations (e.g., as recommended below for singular perturbation equations) is generally *not* a decoupling scheme.

For any stencil of difference equations (and any distribution scheme) it is, of course, easy to find a decoupling pattern by taking it large enough, i.e., with enough stages. Such a decoupling pattern relaxation is not necessarily the best parallel scheme for the given equation. For example, if the difference equation differs from a 5-point-star equation, but not by much, then RB relaxation is not decoupling, but it may still by much better than any decoupling scheme (provided enough processors are available). Moreover, the decoupling pattern relaxation is not easier to analyze.

Generally, the smoothing analysis of a k-stage (or k-"color") pattern relaxation should involve k coupled Fourier frequencies $(\underline{\theta}^1, \underline{\theta}^2, \cdots, \underline{\theta}^k)$ at a time. The k-vector of their error amplitudes $\underline{A} = (A_1(\underline{\theta}^1), A_2(\underline{\theta}^2), \cdots, A_k(\underline{\theta}^k))^T$ is transformed by the k-stage sweep into a new vector of error amplitudes $B\underline{A}$, where $B=B(\underline{\theta}^1)$ is a $k \times k$ matrix. Let $C\underline{A}$ be the vector obtained from $\underline{A}$ when $A_\alpha(\underline{\theta}^\alpha)$ is replaced by zero for any low-frequency $\underline{\theta}^\alpha$; that is, for any frequency $\underline{\theta}^\alpha = (\theta_1^\alpha, \cdots, \theta_d^\alpha)$ such that $|\underline{\theta}^\alpha| \underset{D}{=} \max |\theta_j^\alpha| \le \frac{\pi}{2}$. (Such frequencies do not alias with lower frequencies on the $2h$ grid -- hence the coarse-grid correction can approximately liquidate their error). We can then define the smoothing rate as

$$\bar{\mu} = \max \sigma(CB) , \tag{3.5}$$

where $\sigma(CB)$ is the spectral radius of CB and the max is taken over all coupled k-tuples $(\underline{\theta}^1, \cdots, \underline{\theta}^k)$, $|\underline{\theta}^\alpha| \le \pi$. A rough idea about the efficiency of various relaxation schemes can quickly be obtained by calculating $\sigma(CB)$ for some extremal $(\underline{\theta}^1, \cdots, \underline{\theta}^k)$. The convergence factor per multigrid cycle is approximately $\bar{\mu}^{r+s}$, where $r+s$ is the number of sweeps on each level (cf. Fig. 1). An improved prediction is given by the modified definition

$$\bar{\mu}^{r+s} = \sigma(CB^{r+s}) . \tag{3.6}$$

For more accurate prediction one should, of course, use the full two-level analysis (Brandt and Dinar, 1979). RB and zebra are so analyzed in Stueben et al (1980). But (3.6), as smoothing analysis in general, gives us an additional hint: It tells us what convergence factor per cycle we should try to get by a proper choice of the inter-grid transfers.

Another general remark about pattern relaxation schemes concerns the residual transfers that should follow them. Since high frequencies are coupled with low ones, full residual weighting is generally needed; i.e., I_h^{2h} in (2.13) should satisfy $\Sigma\left(I_h^{2h} r^h\right)(x^{2h}) = \Sigma\, r^h(x^h)$ for any function r^h. (A modification of this requirement near boundaries will be given elsewhere). But, as observed in Stueben et al (1980) for RB relaxation, we can sometimes use a simplified weighting. If the equation we solve has smooth coefficients, then the k-stage pattern relaxation smoothes the residual function on each one of the k "colors". Hence, if full residual weighting is given by

$$\left(I_h^{2h}\, r^h\right)(x) = \Sigma\, W_{\underline{\alpha}}\, r^h(x+\underline{\alpha}h), \qquad (\underline{\alpha} = (\alpha_1,\cdots,\alpha_d)) ,$$

we can instead sometimes use a simplified weighting such as

$$\left(I_h^{2h}\, r^h\right)(x) = \sum_{j=1}^{k} W_j\, r^h(x + \underline{\alpha}_j\, h) , \tag{3.7}$$

where $x + \underline{\alpha}_j h$ is in the j-th color and W_j is the sum of $W_{\underline{\alpha}}$ for all $\underline{\alpha}$ such that $x + \underline{\alpha}h$ is in the j-th color. Furthermore, if the pattern relaxation is decoupling, the last term (j=k) in (3.7) can be omitted, since $r^h(x + \alpha_k h) = 0$ after the k-th stage. Two-level analysis can tell us when it pays to use simplified weightings.

3.5 Degenerate, Singular Perturbation and Non-Elliptic Equations

Degenerate elliptic equations are equations in which a small parameter $\varepsilon>0$ multiplies *some* (not all) of the higher-order terms, so that the equations are almost parabolic. The typical example is the equation

$$\varepsilon u_{xx} + u_{yy} = F . \tag{3.8}$$

such equations are usually obtained from regular, isotropic equations (like the Poisson equation, in case of (3.8)) by coordinate stretching. Most such stretchings create degenerat equations in certain parts of the domain. The main trouble with such equations is that point-wise relaxation, whether parallel or sequential, does not efficiently smooth the error; e.g., Gauss-Seidel, RB, and any other point relaxation of (3.8) efficiently smooth error oscillations in the y direction only, and not in the x direction.

This trouble does not arise if multi-grid methods of local refinement

(Section 2.8) are used instead of coordinate stretching. To cover large (or unbounded) domains, the multigrid approach is to introduce increasingly coarser grids covering increasingly larger domains, thus completely avoiding unisotropic grids. When unisotropic grids are desired, in boundary layers for example, the sequence of grids being used has increased refinements (mesh halving) only in one desired direction. The difference equations thus obtained are similar to the (isotropic-grid) differencing of (3.8), so that point relaxation smoothes the error only in one direction. But this exactly corresponds to the only direction of refinement; smoothing in the other direction is not needed since the next coarser grid is not coarser in that direction.

If, however, we still want to fully smooth degenerate difference equations, we have to use line (or plane) relaxation; e.g., for a square-grid approximation to (3.8) we should use y-line relaxation schemes (solving simultaneously all equations on the same grid line parallel to the y axis; or at least solve simultaneously segments of such lines, where the length of each segment is $O(h\varepsilon^{-1/2})$). Some line-Jacobi methods were analyzed in Brandt (1976, 1977a). But the best scheme seems to be line-RB (zebra) relaxation. It can be performed as described above (Sec. 3.3). If, however, the number of processors is large enough, we may want to use parallel processing in simultaneously solving each line's equations. For small ε the appropriate methods are similar to those mentioned in Sec. 3.10 below. (For $\varepsilon = O(1)$, the line equations are also very efficiently solved by relaxation. But exactly for this reason line relaxation is not at all needed for that case).

Singular-perturbation difference equations arise when the small parameter multiplies the entire high-order part of the differential equation. The typical example is the discretization of

$$\varepsilon\Delta U + aU_x + bU_y = F \qquad (3.9)$$

on a square grid with meshsize h, when

$$\varepsilon \leq O(h) \max(|a|, |b|) . \tag{3.10}$$

Even if ε is not very small, in multigrid programs (3.10) can occur on some of the coarser grids. For stable discretization of such equations (as defined in Brandt, 1980), the smoothing rate can be much improved by sequential relaxation marching in the convection direction; e.g., if a and b in (3.9) are non-negative, Gauss-Seidel forward relaxation (taking grid points in their lexicographic order) yields $\bar{\mu} < .5$ (Brandt, 1976). If a and b change signs over the domain, several marching directions should be used sequentially. These methods cannot be used on parallel machines (nor can they always be used on a sequential machine if one wants to avoid the programming of several marching directions). Instead one can use increased artificial viscosity (Brandt, 1980) and/or distributive relaxation schemes (Brandt, 1979a; Dinar, 1979), which yield good smoothing even in marching <u>against</u> the convection direction, let alone in parallel relaxation.

A combination of these methods with pattern relaxation seems very promising. For example, taking (3.9) with $a = b = 1$ and $\varepsilon << h$, stable differencing is given by

$$\beta h \, \Delta_h \, u^h + \partial_x^h \, U^h + \partial_y^h = F , \tag{3.11}$$

where Δ_h is the 5-point Laplacian and ∂_x^h and ∂_y^h are the usual <u>central</u> differencing of the corresponding derivatives. The first term in (3.11) represents artificial viscosity, and the case $\beta = 1/2$ is equivalent to the usual upstream differencing. RB relaxation of (3.11) for $\beta = 1/2$ gives the smoothing factor $\bar{\mu} = \mu(\pi/2, \pi/2) = 1$, which is, of course, bad. Increasing the artificial viscosity to $\beta = 1$, however, yields $\bar{\mu} = \mu(\arcsin 1/3, \frac{\pi}{2}) = 1/3$, an excellent factor. If one insists on solving (3.11) with $\beta = 1/2$ (although there is nothing sacred about this value), or even with $0<\beta<1/2$, Distributive Red-Black (DRB) relaxation can be used. This means that the difference equations (3.11) are swept in two stages, the red ones first, the black, second. The equation at the (i,j) grid point is satisfied in its turn by changing not only u^h_{ij} , but also some neighboring values, in a fixed proportion. For example, one can change u^h_{ij} to $u^h_{ij} + 2\delta$, $u^h_{i,j+1}$ to $u^h_{i,j+1} + \delta$, and $u^h_{i+1,j}$ to $u^h_{i+1,j} + \delta$, δ being chosen so that (3.11) at (i,j) is satisfied by the new values. This yields a good smoothing factor even for $\beta = 1/2$.

Non-elliptic boundary-value problems can be treated as singular perturbation equations (Brandt, 1980). For example, the treatment of the equation $aU_x + bV_y = F$ is identically the same as that of (3.9) for $\varepsilon << h \max(|a|, |b|)$.

3.6 Relaxation of Cauchy-Riemann and Navier-Stokes Equations

The Distributive Gauss-Seidel (DGS) relaxation schemes introduced for various elliptic systems (Brandt and Dinar, 1979; Brandt, 1980c) can easily be modified to Distributive Red-Black (DRB) schemes, which can be executed in parallel and which yield excellent smoothing rates. In fact, even in sequential execution DRB is often preferrable to DGS; e.g., in relaxing each of the Cauchy-Riemann equations, or in relaxing the continuity equation of the Navier-Stokes system, and also the momentum equations in case the cell Reynolds number Rh is small.

The change from DGS to DRB in each case is just a change in the order the grid points are taken: In the first stage all the red ones are relaxed, then in the second, the black. In staggered grids, the division into red and black points is separately done for each field equation; e.g., the coloring of the difference equations approximating the first Cauchy-Riemann equation $U_x + V_y = F_1$ is independent of the coloring of the difference approximation to the second equation $U_y - V_x = F_2$. The improvement in the smoothing factor obtained by this change in all the above-mentioned cases is the same as in the Poisson case, since the smoothing factors are the same. Also, in all these cases one can compute in parallel the corrections due to different field equations; e.g. the correction due to red continuity equations can be calculated in parallel to the corrections due to red momentum equations. Only in adding these corrections to the old velocities, the continuity-equation corrections should be added separately from (sequentially to) any momentum-equation correction, since they correct the same velocity field.

For higher Reynolds numbers $(R \geq O(h^{-1}))$, each of the d momentum Navier-Stokes equations can be discretized and relaxed by the procedures described above for equation (3.9). We can still relax all d of them in parallel, and essentially also in parallel (except for the substitution of new velocities) to the DRB relaxation of the continuity equation. The smoothing factor is $\bar{\mu} = \max(\bar{\mu}_1, \bar{\mu}_2)$, where $\bar{\mu}_1$ is the smoothing factor of RB relaxation for Poisson equation, and $\bar{\mu}_2$ is the smoothing factor in relaxing (3.9). This can be shown in the same way as in Brandt and Dinar (1979).

3.7 Inter-Grid Transfers

The coarse-to-fine interpolation I_{2h}^h in (2.4) and (2.14), as well as the fine-to-coarse transfers (possibly with averaging) I_h^{2h} and $\bar{I}_h^{2h}$ in (2.10), (2.13) and (2.14), are static operations. Their results at different grid points are independent, and can therefore be calculated fully in parallel. (For the implications on the residual transfers from parallel relaxation - see end of Sec. 3.4). The remaining issue is the inter-processor connections required for efficient parallel implementation of these transfers.

A natural idea is the "pyramid" structure proposed by several people

(Reutersberg, Trottenberg and Witch, 1980). The one-dimensional case is shown in Fig. 3(a); the extension to higher dimensions is obvious. Each grid has its own array of processors, one processor per grid point, say, with nearest-neighbor connections (cf. Fig (2)). This array is used to perform relaxation on the grid and compute the residuals. A subset of this array (those processors connected to the coarser-grid array) is employed to average the residuals and transfer them to the coarser grid. The transfer $\bar{I}_h^{2h}$ and the calculation of $u^{2h} - \bar{I}_h^{2h} u^h$ are also done by this subset, and the interpolation (2.14) is then completed by the entire array. For problems whose finest grid has more points than in the finest array of processors (or even if this is the case in just one of the grid directions), the finest array should represent several grids. For example, in two dimensions, each finest-array processor may represent four finest-grid points (Fig. 2(c)), as well as one second-finest-grid point. The transfers between the finest and the second finest grid are then done within this array.

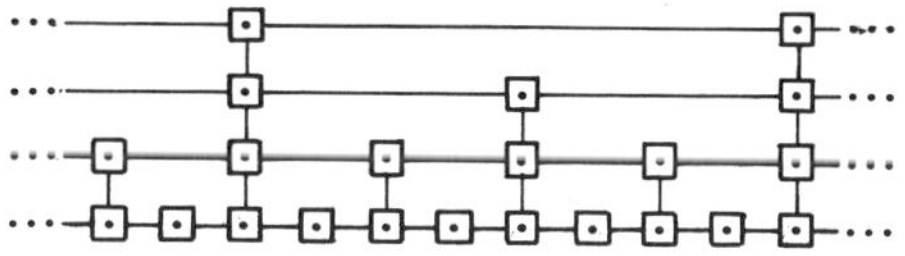
(a)

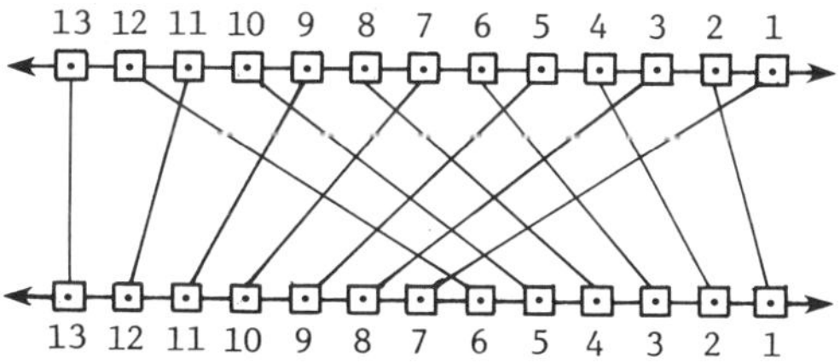

(b)

FIGURE 3. Inter-grid connections in one dimension:

(a) The pyramid structure in one dimension. (Each row shows part of a grid, lower rows corresponding to finer grids.) (b) Perfectly-shuffled-Nearest-Neighbor (PSNN) array. (Each processor is shown twice to make the drawing of interconnections easier.)

It would be a waste of most of its processing power if the pyramidal topology is used in the solution of one problem. Note in Fig. 1 that the algorithm spends most of its steps on coarser grids. The processing power would fully be used only if several problems are solved simultaneously, and if there are several arrays of processors per each coarser grid (about k arrays for the k-th-finest grid). Coordinating the flow of all these problems may prove very difficult, though. Also, the structure lacks flexibility in its usage. Most importantly, parallel local refinements are not implementable.

Another promising, more flexible topology is the Perfectly-Shuffled-Nearest-Neighbor (PSNN) array (Grosch, 1979). The one dimensional case is depicted in Fig. 2 (b): There is an odd number n of processors. The shuffle connections connects processor i to processor $S(i) = 2i (\mathrm{mod}\ n)$, thus interlacing the elements of the first half of the array with elements of the second half as in the perfect shuffle of a deck of cards (Pease, 1968; Stone 1971). In addition each processor i is connected to its nearest neighbors $i \pm 1$, and the last processor to the first. Let any consecutive (mod n) set of processors corresponds to a certain grid. The grid points representing the next coarser grid correspond then to processors $k, k+2, \cdots, (\mathrm{mod}\ n)$, say. Then the next coarser grid can be assigned to processors $\ell, \ell+1, \cdots,$ where $\ell = S^{-1}(k)$, i.e. $2\ell = k\ (\mathrm{mod}\ n)$, hence $\ell+j = S^{-1}(k+2j)$. Thus the shuffle connections will serve as efficient coarse-to-fine and fine-to-coarse connections.

Observe that instead of assigning processors $(\ell, \ell+1, \cdots)$ to the coarser grid, we could use $(\ell', \ell'+1, \cdots)$, where $\ell' = S^{-1}(k+1) = \ell + (n+1)/2$. We can do it either by arranging $(k+1, k+3, \cdots)$ to be the coarse-grid positions on the fine-grid, or, if the positions $(k, k+2, \cdots)$ are insisted upon, by using two inter-processor transfers for each inter-grid transfer. The sequences $(\ell, \ell+1, \cdots)$ and $(\ell', \ell'+1, \cdots)$ are disjoint (except possibly when the fine grid extends over the entire array of processors), hence we can use them in parallel. Thus we can solve two problems simultaneously, where fine-grid processes are made sequentially (first for problem 1, then for problem 2), but coarse-grid processing is in parallel. On still coarser grids we can process still more problems in parallel. This is important since coarser grids are used more than fine grids (cf. Fig. (1)).

Moreover, the shuffle connections allow the efficient introduction of ever finer grids, including local refinements (see Sec. 2.8). Let $(k, k+1,\cdots,m)$ be the sequence of processors assigned to a certain grid, or a section of a grid, which we want to refine. In case the number of points in the finer grid is greater than n, one should assign two processors to each fine grid point, hence the same sequence of processors $(k, k+1,\cdots,m)$ can also serve the finer grid, and the efficient inter-grid connections are trivial. If the number of fine-grid points is not greater than n, we like to assign one processor to each point of the refinement, with efficient inverse-shuffle connection to the coarser-grid processors $(k, k+1,\cdots,m)$. So we simply assign processors $(S(k), S(k) + 1, S(k) + 2 = S(k+1), S(k) + 3, S(k) + 4 = S(k+2),\cdots, S(m))$ to the fine grid. Notice that for this to be possible, for arbitrary subgrid $(k, k+1,\cdots,m)$, we do need the circular character of the array (the connection between processors 1 and n).

Another advantage of the PSNN connections is their universality. They permit transfer of information from any processor to any other in at most $\lceil \log_2 n \rceil$ steps, where $\lceil \alpha \rceil$ denotes the smallest integer not smaller than α. Moreover, in at most $3\lceil \log_2 n \rceil$ steps one can make any permutation of data between the n processors (Parker, 1980). Indeed this can be done using only part of the PSNN connections. Pease (1968) and Stone (1971) showed that the perfect-shuffle connections allow the performance in parallel of various important algorithms, such as Fast Fourier Transform, polynomial evaluation, sorting and matrix transposition.

The generalization of the PSNN array to higher dimensions is immediate. In an $n_1 \times n_2$ lattice of processors, for example, we simply require PSNN connections in each row and in each column. We call this a $PSNN^2$ array. Again it alows us efficient inter-grid transfers (each requiring 2 inter-processor transfers, though, one in rows and one in columns), as well as using coarser grids in parallel for several problems, and arbitrary local refinements. Moreover, we can make unisotropic local refinements (e.g., refining in only one direction; cf. Sec. 2.8).

Another way of generalizing PSNN to an $n_1 \times n_2$ array is to require nearest-neighbor-circular connections in each row and column, together with truely 2-dimensional shuffle connection; i.e., connecting (i_1, i_2) to $S(i_1, i_2) = (2i_1 (\mathrm{mod}\ n_1)\ ,\ \ 2i_2 (\mathrm{mod}\ n_2))$. We denote this array by $PSNN_2$. It requires less physical inter-connections than $PSNN^2$. Also, the inter-grid transfers of usual two-dimensional multigrid processes are slightly more efficient, requiring one machine step instead of two. On the other hand unisotropic refinements are not efficiently implementable.

Another approach is to use any universal interconnection system, which can emulate the above connections, substituting $O(\log_2 n)$ transfers for each of the above. This can be rather efficient, especially if the inter-connections can be made very fast (e.g., by transferring in parallel all the bits of a word). Such universal interconnections are not only useful for more problems, but they are also more flexible and more reliable (see Sec. 3.2).

3.8 Troubles: Set Up and Boundaries

Once efficient and reliable grid and inter-grid connections are implemented, what are the main factors that may cause degradation of the system? One bottleneck may be input and output, especially the set-up of geometrical and numerical data into the array of processors. The main stumbling block is likely to be related to boundaries.

The amount of processing required at grid points near boundaries is usually much larger than elsewhere. To set up the special geometrical information (distance to boundary, normal direction, available interior neighbor grid points) and boundary conditions may require long calculations near boundaries, with no counterpart at interior points. The solution procedures themselves (relaxation, interpolations) are also usually much more involved near boundaries than elsewhere, and so are the post-solution processing (e.g., calculations of some integrals of the solution along the boundary).

The work near the boundary increases even much further in case of *general-domain programs*; that is, programs whose users can specify the domain at will. Even on sequential machines, multigrid general-domain programs run typically twice as long as special-domain programs solving the same simple problem (e.g., Poisson equation with Dirichlet boundary conditions). Especially expensive, unless programmed with special care, turn out to be the general-domain higher-order (e.g., cubic) interpolation routines. If the fine-grid and the coarse-grid domains are both general (to provide general local refinements, as in Sec. 2.8), the interpolation between them requires many checks near boundaries (to determine what coarse-grid data can best be used to interpolate at each fine-grid point. The number of possible cases, even for cubic interpolation in two dimensions, is large).

Moreover, in many problems there are *special points*, whether near the boundary or not, where the amount of calculations is especially large; for example, singular points of a structure, re-entrant corners, leading and trailing edges of airfoils, small regions with special subgrid (e.g., chemical or turbulent) interactions, etc.

On parallel computers the near-boundary and special points calculations will easily become the overriding factor. If the same processing power is assigned to every grid point, most processors will be assigned to interior points but will remain idle most of the time because of the much longer calculations performed by the processors assigned to near-boundary and special points. This indicates the need for a very flexible MIMD architecture, allowing more processors per point where needed, as well as the possibility of assigning special, much faster processors to special grid points. The problem of efficient set-up and control will thus be aggravated.

Some of these problems can sometimes be avoided by coordinate transformation, transforming the domain of the problem into a rectangle, and use only rectangular local refinements (as in Sec. 2.8). The availability of such local refinements simplify the coordinate transformation, by freeing it from stretching requirements. The *local* coordinate transformations (Sec. 2.8) are not suitable to parallel execution, except perhaps when universal interconnection scheme is in use.

3.9 Relaxation in Parallel to Coarse-grid Calculations

Since the parallel multigrid algorithm spends most of its time on coarser grids, where many processors may be unemployed, it is natural to employ them by executing, in parallel to the coarse-grids calculations, additional relaxation sweeps on the fine grid. Too many such sweeps would not be useful, since they will cause smoother components to change and hence to get wrong corrections from the coarser grid. The relaxation done in parallel should therefore be as highly *separative* as possible, i.e., its amplification factors for high frequencies should be as large as possible *compared with* those for low frequencies. This is usually obtained by slower relaxation schemes, such as under-relaxed fully-distributive (like Kaczmarz, but with simultaneous displacement) relaxation schemes: They are slow for high frequencies, but still much slower for low frequencies.

In collaboration with Klaus Stueben we have preliminarily studied a multigrid cycle for the five-point Poisson equation which includes r relaxation sweeps before transferring the residuals to the coarse grid, additional r_0 sweeps after the transfer but before adding the coarse-grid corrections (hence these r_0 sweeps can be parallel to the coarse-grid calculations), and finally s more sweeps after the coarse-grid corrections. The coarse-to-fine interpolation I_{2h}^{h} is linear. Some results of the two-level analysis are summarized in Table 1. It shows a certain (not impressive) improvement of the convergence factor per cycle if, in addition to the regular RB relaxation, one or two Jacobi sweeps are made in parallel to coarse-grid calculations.

TABLE 1

Type of Relaxation: Before & After	Type of Relaxation: Parallel	Type of I_h^{2h}	r+s	Convergence Factors Per Cycle: $r_0 = 0$	$r_0 = 1$	$r_0 = 2$	$r_0 =$
Jacobi	Jacobi	Full	0	1.00	.60	.64	.78
$\omega = .8$	$\omega = .8$	Full	1	.60	.36	.38	.47
		Full	2	.36	.22	.23	.28
		Inj.	2	.36	.54	.59	.64
RB	RB	Full	0	1.00	.50	.90	.98
		Full	1	.25	.31	.38	.44
		Full	2	.074	.11	.21	.29
		Wt. Inj.	2	.125	.21	.30	.37
RB	Jacobi	Full	1	.250	.250	.203	.219
	$\omega = .5$	Full	2	.074	.074	.051	.055

3.10 Initial-Value Problems and Ordinary Differential Equations

Multi-level adaptive methods for time-dependent problems will be reported elsewhere. (See also Sec. 2.7 above). They are fully parallelizable. Here we only emphasize the following unique property of parallel multi-grid solutions: Although time-dependent algorithms are naturally sequential in time, advancing one time step after the other, the multigrid technique makes it possible to solve *simultaneously in time*. The algorithm of Figure 1, for example, can be used, where each grid is a full space-time grid; the time step of a coarser grid is proportionately larger, too. The relaxation methods for singular perturbation problems (Sec. 3.5) can be used over these grids. As a result the number of computer steps will depend only logarithmically on the number of time steps.

There is little motivation today to consider such techniques, since the array processors are not that large; one is satisfied with parallel calculation at each individual time step. Algorithms of the type described above have therefore not been tried so far. But there is no reason why a problem like $u_t + u_x = F$, for example, cannot be solved by methods that solve $-\varepsilon\Delta u + u_y + u_x = F$ very efficiently. There are, on the other hand, good reasons to use such methods even on sequential machines. For example, the multi-level method of higher-order coarse-grid corrections (Brandt, 1980b) can be very effective, whether on parallel or sequential machines. For only a fraction of extra computational work one can get higher-order schemes, without the need to devise stable high-order difference equations.

In case of ordinary differential equations, whether initial-value or boundary-value problems, there is, of course, much more current motivation for fully parallel solutions. The system of difference equations, if linear, form in this case a banded $n \times n$ matrix with small bandwidth b. The parallel solution of such systems has been studied by many authors (see Heller, 1978, and references therein). In the tridiagonal (b=1) case, it is interesting to observe that some of the best methods, such as odd-even elimination or odd-even reduction, can be regarded as one-cycle multigrid algorithms (Cycle C, with special relaxation and inter-grid transfers). Other kinds of multigrid algorithms are less efficient in this (b=1) case. The odd-even algorithms are also extended to the general case, by viewing the banded matrix as a block tridiagonal matrix, with bxb (dense) blocks. Using O(bn) processors one can solve such a system to the level of round-off errors in $O(b) + O((\log n) \log b)$ steps. A usual multigrid algorithm (as in Sec. 2.4) employing the same number of processors, would solve it to the level of *truncation* errors in $O((\log n)^2 \log b)$ steps. Thus, for larger b, the usual multigrid algorithm becomes faster. Moreover, with the same efficiency this algorithm would solve nonlinear and eigenvalue problems (without Newton iterations; see Sec. 2.5).

ACKNOWLEDGMENTS

The author is indebted to R. G. Voigt for valuable discussions and suggestions.

This report was prepared as a result of work performed under NASA Contract No. NAS1-15810 and NAS1-14472 at ICASE, NASA Langley Research Center, Hampton, VA 23665.

REFERENCES

Alcouffe, R. E., Brandt, A., Dendy, J. E., Jr., and Painter, J. W. (1980). *The multi-grid methods for the diffusion equation with strongly discontinuous coefficients*, LA-UR-80-1463, Los Alamos Scientific Laboratory, Los Alamos, New Mexico.

Bank, R. E. (1980). *Multi-level iterative methods - some recent results.* This Proceedings.

Bouknight, W. J., Denenberg, S. A., McIntyre, D. E., Randall, J. M., Sameh, A. H., and Slotnick, D. L. (1972). *The Illiac IV System*, Proc. IEEE 60, p. 369-388.

Brandt, A. (1976). *Multi-level adaptive technique (MLAT). I. The multigrid method.* IBM Research Report RC-6026, IBM T. J. Watson Research

Center, Yorktown Heights, New York.
Brandt, A. (1977a). *Multi-level adaptive solutions to boundary-value problems.* Math. Comp. 31, p. 333-390.
Brandt, A. (1977b). *Multi-level adaptive techniques (MLAT) for partial differential equations: Ideas and software.* Mathematical Software III (John R. Rice, Ed.), Academic Press, New York, p. 273-314.
Brandt, A. (1979). *Multi-level adaptive solutions to singular-perturbation problems.* Numerical Analysis of Singular Perturbation Problems, (P. W. Hemker and J. J. H. Miller, Eds.), Academic Press, p. 53-142.
Brandt, A. (1980a). *Multi-level adaptive finite-elements methods: I. Variational problems.* Special Topics of Applied Mathematics, (J. Frehse, D. Pallaschke and V. Trottenberg, Eds.), North Holland, Amsterdam. Also ICASE Report 79-8.
Brandt, A. (1980b). *Numerical stability and fast solutions of boundary-value problems.* Boundary and Interior Layers - Computational and Asymptotic Methods, (J. J. H. Miller, Ed.) Boole Press, Dublin.
Brandt, A. (1980c). *Multi-level adaptive computations in fluid dynamics.* Proc. AIAA 4th Computational Fluid Dynamics Conf. (Williamsburg, VA, July 1979). ICASE Report No. 79-19. AIAA J. (in press).
Brandt, A., and Cryer, C. W. (1980). *Multi-grid algorithms for the solution of linear complementarity problems arising from free boundary problems.* MRC Report, University of Wisconsin.
Brandt, A., and Dinar, N. (1979). *Multi-grid solutions to elliptic flow problems.* Numerical Methods for Partial Differential Equations, (S. Parter, Ed.) Academic Press, p. 53-147.
Brandt, A., McCormick, S., and Ruge, J. (1980). *Multigrid Solution of eigenvalue problems.* To appear.
Cyre, W. R., et. al., *WISPAC: A parallel array computer for large-scale system simulation,* Simulation, November 1977, p. 165-172.
Dinar, N. (1979). *Fast methods for the numerical solution of boundary-value problems,* Ph.D. Thesis, The Weizmann Institute of Science, Rehovot, Israel.
Goodyear Aerospace Corp. (1979). *Massively Parallel Processor (MPP).* Akron, Ohio.
Gostick, P. W. (1979). *Software and algorithms for the Distributed-Array Processors,* ICL Technical Journal, London.
Grosch, C. E. (1978). *Poisson solvers on large array computer,* Proc. 1978 LASL Workshop on Vector and parallel processors (B. L. Buzbee and J. F. Morisson, Eds.)
Grosch, C. E. (1979). *Performance analysis of Poisson solvers on array computers,* Report TR 79-3, Old Dominion Univ., Norfolk, VA.
Hackbusch, W. (1980a). *On the convergence of multi-grid iterations,* Beitr. Numer. Math. 9 (to appear).
Hackbusch, W. (1980b). *The fast numerical solution of very large elliptic difference schemes,* J. Inst. Math. Appl., to appear.
Heller, D. (1978). *A survey of parallel algorithms in numerical linear algebra,* SIAM Review 20, 740-777.
Hunt, D., Webb, S., and Wilson, A. (1980). *Applications of a parallel processor to the solution of finite difference problems.* This proceedings.
Jameson, A. (1979). *Acceleration of transonic potential flow calculations on arbitrary meshes by the multiple-grid method,* AIAA-79-1458. In Proc. AIAA Computational Fluid Dynamics Conf., July 1979, Williamsburg, VA., p. 122-146.
McCarthy, D. R., and Reyhner, T. A. (1980). *A multi-grid code for three-dimensional transonic potential flow about axisymmetric inlets at angle of attack,* AIAA-80-1365. In Proc. AIAA 13th Fluid and Plasma Dynamics Conf., July 1980, Snowmass, Colorado.

Nagel, K. (1979). *Weather simulation with the multi-microprocessor system* SMS 201. Proc. Military Electronics Defense Expo '78, Interavia S. A., Geneva, p. 60-67.

Nicolaides, R. A. (1977). *On the ℓ^2 convergence of an algorithm for solving finite element equations*, Math. Comp. 31, p. 892-906.

Ortega, J. M., and Voigt, R. G. (1977). *Solution of partial differential equations on vector computers*, Proc. 1977 Army Numerical Analysis and Computers Conf., ICASE Report 77-7.

Parker, D. S., Jr. (1980). *Notes on shuffle/exchange-type switching networks*, IEEE Trans. on Computers C-29, 2B-222.

Parkinson, D., *Technical Description of the Distributed Array Processor*, International Computers, Ltd. Document No. AP-2, 1976.

Pease, M. C. (1968). *An adaptation of the fast Fourier transform for parallel processing*, J. Assoc. Comput. Mach. 15, p. 252-264.

Reutersberg, H., Trottenberg, U., and Witsch, K. (1980). *Oncept of a multigrid computer with pyramidal topology*. To appear.

South, J. C., Jr., Brandt, A. (1976). *Application of a multi-level grid method to transonic flow calculations*, ICASE Report No. 76-8, NASA Langley Research Center, Hampton, VA

South, J. C., Jr., Keller, J. D., and Hafez, M. M. (1980), *Vector processor algorithms for transonic flow calculations*, AIAA J. 18, p. 786-792.

Stone, H. S. (1971). *Parallel processing with the perfect shuffle*. IEEE Trans. Comput. C-20, p. 153-161.

Stueben, K., Trottenberg, U., and Foerster, H. (1980). *Non standard multigrid techniques using checkered relaxation and intermediate grids*. This proceedings.

Sullivan, H. and Bashkow, T. R. (1977). *A large scale homogeneous fully distributed parallel machine*, 4th Annual Symposium on Computer Architecture Proceedings, 1977, p. 105-117.

IMPLEMENTING TECHNIQUES FOR ELLIPTIC PROBLEMS ON VECTOR PROCESSORS

B. L. Buzbee

Computer Science and Services Division
Los Alamos Scientific Laboratory
Los Alamos, New Mexico

I. INTRODUCTION

To provide the arithmetic power required by large-scale numerical simulations, the fastest computers today (for example, the Cray-1 and the Cyber 205) incorporate vector processing. In such computers, a vector is an n-tuple of numbers systematically stored in memory. Because these computers attain their highest level of performance when processing vectors, considerable effort has been spent in the past decade developing algorithms that can be expressed as operations on vectors.

In Section II of this paper we define two types of vector architecture. We then discuss the variation in performance that can occur on a vector processor as a function of algorithm and implementation, the consequences of this variation, and the performance of some basic operators on the two classes of vector architecture. We also discuss the performance of some higher level operators that should be used with caution. In Section III we review the implementation of techniques for elliptic problems using the operators discussed in Section II.

ISBN 0-12-632620-7

Included are Fast Poisson solvers, dissection, and point, line, block, and conjugant gradient schemes. In Section IV we note some areas of research.

II. VECTOR PROCESSORS

Vector Algorithms

A question that arises when considering work on vector algorithms is "Will recent developments in Josephson Junction (JJ) technology preempt further work on vector algorithms?" Figure 1 displays the execution bandwidth in scalar mode of supercomputers for the past 30 years. (We define a supercomputer to be the fastest computer available at any point in time.) This data shows the diminishing growth rate in scalar performance of supercomputers. It further suggests that there is an upper bound on the scalar performance of the computer.

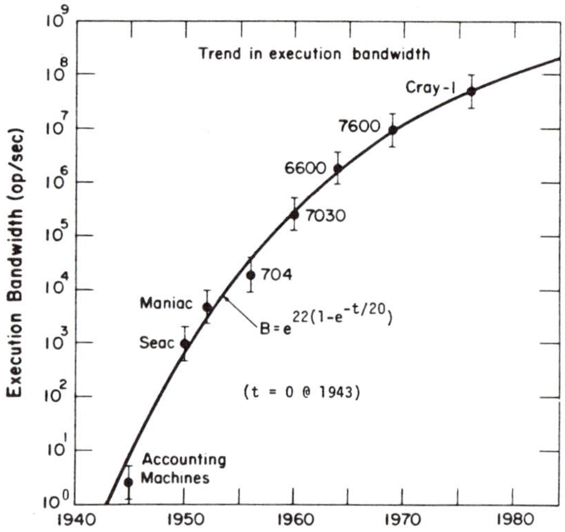

FIGURE 1. Execution bandwidth of supercomputers.

If such a bound exists, its precise value is yet to be determined. Nevertheless, this data is consistent with projected performance of computers based on JJ technology. For example, Robinson projects that JJ technology will produce scalar processors with a cycle time of 2-5 nanoseconds (1); Matisoo projects 3 nanoseconds (2). Because the cycle time of the Cray-1 is about 12-1/2 nanoseconds, JJ technology does not provide a quantum jump in performance. Further, because some three-dimensional calculations require a two order of magnitude increase in performance over the Cray-1 (3), JJ technology will not provide scalar processors with processing power sufficient for future needs. Provision of that power will probably be based on current trends to parallelism, and vector processing is one form of parallelism. Thus we anticipate a continuing need for, and interest in, algorithms that can exploit vector and multiprocessing capability.

Vector Architectures

Although a satisfactory taxonomy for vector architectures is yet to be devised, we define two classes of vector architecture: memory-to-memory (MM) and register-to-register (RR).

MM ARCHITECTURE. Typically, for an MM architecture to achieve its highest level of performance, it must process algorithms that satisfy the following boundary conditions.

- Operand and result vectors must be stored contiguously in memory; that is, successive elements of the vector must be stored in adjacent memory locations.
- Vectors must be long.

Examples of MM architecture are the CDC 7600 when processing vectors from large-core memory, the Texas Instrument ASC, and the CDC Cyber 200 series.

RR ARCHITECTURE. RR architecture typically involves some sort of cache between memory and the processing units, with arithmetic operations being performed on contents of the cache. These processors achieve their highest level of performance when processing algorithms that satisfy the following conditions:

- Data memory between memory and the cache is minimized.
- Parallel execution of the functional units is maximized.

Examples of this architecture are the 120B array processor of Floating Point Systems, the Cray-1 of Cray Research, Inc., and the BSP of Burroughs Corporation.

Performance of Vector Processors

Performance on a vector processor can vary widely as a function of algorithm and implementation. For example, when solving dense linear systems of equations, Cray-1 performance varies as follows.

Mode of Implementation	Performance Level
Scalar Fortran	2-6 megaflops[a]
Vector Fortran	2-30 megaflops
Vector Assembly Language	2-120 megaflops

[a]Megaflop = one million floating point operations per second.

The consequences of this variation are substantial, because on any computer

Execution Time = N*T ,

where N is the number of operations and T is the average time per operation.

On scalar computers there is very little variation in T. Thus, throughout the age of electronic computation, we have focused on minimizing N, i.e., optimal complexity. However, as we have just noted, there is a large variation in T on vector processors. In devising vector algorithms, therefore, we must minimize the product N*T. Consequently, on vector processors we may find it desirable to use algorithms that are nonoptimal in complexity, provided we can do the additional work at a sufficiently high rate (4,5). For example, a numerical simulation at the Los Alamos Scientific Laboratory requires the complimentary error to be evaluated millions of times. This function can be nicely approximated by the inverse of a sixth degree polynomial raised to the fourth power. Direct evaluation of this approximation on a scalar computer encompasses significant amounts of arithmetic operations. Thus, on scalar computers, it is tabulated and evaluation of the function is through interpolation on that table. This results in a small number of arithmetic operations but requires table lookup. On a vector processor, table lookup is relatively slow, and the faster thing to do is to evaluate the polynomial approximation. Although N is large, N*T is small.

The class of algorithms that minimize N*T is probably larger than the class that minimize N. Further, the variation in performance of vector processors as a function of vector length often leads to polyalgorithmic software. Although we have a large "forest" in which to seek algorithms, care must be used. A consistent algorithm is defined to be one that is optimal in arithmetic complexity. A little thought shows that for sufficiently large problems a consistent algorithm will always outperform an inconsistent algorithm, irrespective of the mode of implementation and computer architecture. Conversely, given two consistent algorithms on a vector processor, one cannot settle their performance by studying the coefficient of the low-order term in their complexity. In this case, the question is which one produces the smaller execution time.

Basic Operators

The following basic operators are available in either hardware or software for MM and RR architectures.

1. Vector Diadics
2. Inner Product
3. Matrix Multiply
4. Polynomial Evaluation

A discussion of matrix multiply and polynomial evaluation for RR architecture can be found in Ref. 6. Matrix multiplication is a high-performance operation on vector processors even for banded matrices; for example,

$$\begin{pmatrix} \ddots & \ddots & \emptyset \\ & d_1 \ \ddots \ d_2 & \\ \emptyset & \ddots & \ddots \end{pmatrix} x = d_1 \otimes x + d_2 \otimes \tilde{x} \ , \tag{1}$$

where $\tilde{x}$ is the vector x with the first element deleted (7). The following high-level operators also perform well on both architectures.

- Single FFT (Refs. 5, 8)
- Families of linear system solvers, for example, FFTs (Refs. 5, 8); tridiagonals (Ref. 6).
- Banded system solvers (Refs. 6, 9).

We have distinguished families of FFTs from a single FFT because the performance for a family is significantly higher than for a single FFT.

"Judicious Use" Operators.

The following operators should be used with caution on either architecture:

Recursions

Table lookup

Conditionals

One-to-many mappings

Many-to-one mappings

Caution is required with these operators because their overall performance on a vector processor will be poor if they are not combined with a sufficient amount of arithmetic. Included in recursions are solutions of a single tridiagonal system, which generally does not perform well on a vector processor (6, 10). Table lookup has implications for adaptive procedures. Poor performance of conditionals implies that we must forego our frequently used pointwise convergence test. See Ref. 11 for a discussion of programming considerations in the use of a vector processor.

III. IMPLEMENTING TECHNIQUES FOR ELLIPTIC PROBLEMS

In this section we discuss the implementation of classes of techniques for elliptic problems using the aforementioned operators. A similar discussion was given by Ortega and Voight (12).

FAST POISSON SOLVERS. Fast Poisson solvers are implementable on both MM and RR architectures by using operators to perform families of FFTs and to solve families of tridiagonals. On MM architecture, some care is required to ensure that the vectors manipulated are stored contiguously in storage. Workers in this field include Buzbee (13) and Kascic (14).

DISSECTION SCHEMES. One-way dissection schemes are implementable in any environment where there is good capability for solving small, dense, linear systems. Incomplete nested dissection is implementable on both types of architecture and is attractive if only a few dissections are done. Workers in this area include George, et al. (15), Voight (9), and Calahan (16).

POINT SCHEMES. Point Jacobi schemes are attractive on either class of architecture because they are implementable by vector diadics and banded-matrix multiply operators involving long vectors (17). These advantages have prompted renewed interest in these techniques (18) with an objective of producing Jacobi techniques with attractive convergence rates.

Point successive overrelaxation schemes can be implemented on either architecture with wave fronting on a natural ordering or by checkerboard ordering. Wave fronting involves processing the mesh by diagonals. Because vectors must be stored contiguously on MM architecures, this is not an attractive alternative on them. It is feasible on RR architecture. Using the checkerboard ordering, implementation on either architecture is feasible if the red and black points are stored as separate arrays (19, 20). It follows that multigrid schemes are implementable on either class of architecture. Of course, these techniques involve one-to-many and many-to-one mappings. These mappings occur in families and can be vectorized.

LINE SCHEMES. Line schemes are implementable by using operators for solving families of tridiagonal systems. This is true for both line relaxation schemes and alternating direction schemes. In the latter case some care must be used on MM architecture to produce vectors contiguously stored in memory. For example, Gaussian elimination can be used in one coordinate direction and odd-even reduction in the other to produce contiguously stored vectors (21). On RR architecture, line schemes are easily implemented in Fortran and achieve high rates of vectorization (22).

BLOCK SCHEMES. Block schemes are easily implemented on both architectures using operators to solve banded linear systems or block tridiagonal systems. Unfortunately they suffer from excessive arithmetic complexity and storage requirements. Progress is being made toward reducing their complexity (23).

CONJUGANT GRADIENT SCHEMES. Conjugant gradient techniques can be implemented on vector processors using any of the previously discussed schemes as the approximate factorization. Also, incomplete Cholesky vectorization can be implemented on either class of architecture. On MM architecture odd-even reduction has been used (17), and on RR architecture a variant of block elimination is used (24).

IV. RESEARCH ISSUES

We noted that matrix multiplication is an efficient operator on vector processors even in the banded case because the operator can be expressed in terms of vector diadics on the diagonals of the matrix. This raises the question of whether one can solve linear systems of equations by manipulating diagonals. Such an algorithm exists for tridiagonal systems (25), but not for the general banded case. If available, such techniques would be attractive, especially on MM architecture.

Block schemes are attractive on vector processors because of their ease of implementation and their amenability to vectorization. Faber (23) has shown that certain strategies of block relaxation produce high convergence rates. The problem remaining is to produce efficient techniques for dealing with the associated blocks.

Three-dimensional problems are occurring with increasing frequency. Typically, the associated mesh is held in secondary storage of some sort, with the result that overall execution time is a function of how often the mesh must be transmitted to and from the central processor. This suggests development of techniques that minimize the number of iterations, perhaps at the expense of additional arithmetic operations.

REFERENCES

1. A. L. Robinson, "Superconducting Electronics; Toward an Ultrafast Computer," Science 201 (August 1978).

2. J. Matisoo, "The Superconducting Computer," Scientific American (May 1980).

3. B. L. Buzbee, et al., "DOE Research in Utilization of High-Performance Computers," to appear as a Los Alamos Scientific Laboratory Report.

4. N. K. Madsen and G. H. Rodrigue, "Two Notes on Algorithm Design for the CDC-STAR-100," Numerical Mathematics Group Informal Technical Memorandum 75-1, Lawrence Livermore Laboratory (July 1975).

5. O. G. Korn and T. T. Lambiotte, Jr., "Computing the Fast Fourier Transform on a Vector Computer," Reprint No. 78-5, ICASE, NASA Langley Research Center (February 1978).

6. K. Fong and T. L. Jordan, "Some Linear Algebraic Algorithms and Their Performance on Cray-1," Los Alamos Scientific Laboratory Report LA-6774 (June 1977).

7. T. I. Karush, N. K. Madsen, and G. H. Rodrigue, "Matrix Multiplication By Diagonals on Vector/Parallel Processors," Lawrence Livermore Laboratory report UCID-16899 (August 1975).

8. C. Temperton, "Fast Fourier Transforms on Cray-1," European Centre for Median Range Weather Forecasts Report 21, (January 1979).

9. R. G. Voight, "The Influence of Vector Computer Architecture on Numerical Algorithms," ICASE, NASA Langley Research Center, Report 77-8 (March 1977).

10. N. K. Madsen and G. H Rodrigue, "A Comparison of Direct Methods for Tridiagonal Systems on the CDC-STAR-100," Lawrence Livermore Laboratory report NCRL-76993 (July 1975).

11. B. Brode, "How to Get More Out of Your Vector Processor," Proceedings of the 1978 LASL Workshop on Vector and Parallel Processors, Los Alamos Scientific Laboratory report LA-7491-C (September 1978).

12. J. M. Ortega and R. G. Voight, "Solution of Partial Differential Equations on Vector Computers," ICASE, NASA Langley Research Center, report 77-7 (March 1977).

13. B. L. Buzbee, "A Fast Poisson Solver Amenable to Parallel Computation," IEEE Transactions on Computers C-22 (August 1973).

14. M. T. Kascic, Jr., "A Direct Poisson Solver on STAR," Proceedings of the 1978 LASL Workshop on Vector and Parallel Processors, Los Alamos Scientific Laboratory report LA-7491-C (September 1978).

15. A. George, W. G. Poole, Jr., and R. G. Voight, "Analysis of Dissection Algorithms for Vector Computers," ICASE, NASA Langley Research Center, Report 76-17 (June 1976).

16. D. Calahan, "An Equation Solver Vectorized for Both Simultaneous Sparse and Single Dense Systems," these proceedings.

17. C. Hendrickson, M. Pratt, and G. Rodrigue, "The Numerical Solution of the LaGrangian Diffusion Equation on a Vector Processor," Lawrence Livermore Laboratory report UCRL-82930 (June 1979).

18. O. Johnson and G. Paul, "Vector Algorithms for Elliptic P.D.E.s Based On the Jacobi Method," these proceesings.

19. B. Buzbee, G. Golub, and J. A. Howell, "Vectorization for the Cray-1 of Some Methods for Solving Elliptic Difference Equations," Proceedings of the Symposium on High Speed Computer and Algorithm Organization, Urbana, Illinois (April 1977).

20. W. Gautzsch, C. Weiland, and D. Muller-Wichards, "Possibilities and Problems with the Application of Vector Computers," German Research and Testing Establishment for Aerospace (April 1980).

21. P. Dubois and G. Rodrigue, "Operator Splitting on the STAR Without Transposing," Lawrence Livermore Report UCID-17515 (June 1977).

22. D. Boley, B. L. Buzbee, and S. V. Parter, "Applications of Block Relaxation," Proceedings of AIME Symposium on Reservoir Simulation (February 1979).

23. V. Faber, "Block Relaxation Strategies," these proceedings.

24. T. L. Jordan and M. Mahaffy, private communication, 1980.

25. T. L. Jordan, "A New Parallel Algorithm for Diagonally Dominant Tridiagonal Matrices," Los Alamos Scientific Laboratory report LA-UR-74-1903 (1974).

ON SOME TRENDS IN ELLIPTIC PROBLEM SOLVERS

S. C. EISENSTAT AND M. H. SCHULTZ

1. Introduction

Elliptic boundary value problems are at the core of many systems of partial differential equations occurring in mechanics. Examples of applications include fluid dynamics, semiconductor device modelling, and structural analysis. In addition, they often appear as a result of applying an implicit difference approximation to the time derivative in a parabolic problem. Thus, it is important to have efficient and robust elliptic problem solvers. Moreover, we expect that many of the issues faced in the development of such solvers are typical of those to be faced in the development of large-scale scientific problem solvers.

The development of high-technology elliptic problem solvers must include serious consideration of algorithms, architecture, and applied mathematics, each traditionally the subject of an entire discipline. Recent and expected advances in algorithm and hardware technologies strongly suggest the possibility of both dramatically reducing the cost and dramatically increasing the scope of

---

This work was supported in part by the U.S. Office of Naval Research under Grant N00014-76-C-0277.

ISBN 0-12-632620-7

scientific computational modelling. But an integrated highly interdisciplinary approach will be necessary to realize this possibility.

Success in the development of high-technology elliptic problem solvers will significantly increase the effectiveness and productivity of engineers and scientists by decreasing the cost of their computing, by increasing the efficiency with which they can use the computer, and by allowing them to solve increasingly more complex problems, which are computationally intractable on even the fastest machines available today.

In this paper we discuss some of the issues involved in the design of a high-technology elliptic problem solver. In particular, we will concentrate our attention on the design of a modular, heterogeneous multi-processor elliptic problem solver consisting of a host computer and one or more peripheral processors.

We wish to acknowledge the help of our current and former students and colleagues in the Department of Computer Science of Yale University in developing the ideas presented here. In particular, our colleague Josh Fisher played a key role in formulating the ideas in the Architecture section of this paper.

2. Architecture

We propose a high-technology elliptic problem solver consisting of a heterogeneous multi-processor computer system. This system would have a host machine (such as a minicomputer with a

32-bit word length and an operating system supporting virtual memory) and one or more highly optimized peripheral processors. An important practical advantage to this type of architecture is the fact that one does not have to design a complete hardware-software system to gain great computational power.

Commercial high-performance, programmable peripheral processors, which we will refer to hereafter as attached processors, have become quite popular for signal processing. These processors feature tailored scientific instruction sets, direct user microprogramming, and very wide instruction words. This provides potentially very large speed-ups, but makes the processors extremely difficult to code for efficient performance. Clearly, programming tools such as optimizing FORTRAN compilers are a critical need if the use of these devices is to become commonplace. It seems quite likely that for the next five years such attached processors will provide a very attractive architecture for rapid and cost-effective scientific computation.

The value of dual-processor systems involving a general-purpose host and an attached processor for solving elliptic problems has yet to be convincingly demonstrated. Little is known about many fundamental issues. Perhaps the most critical is the identification of which key subalgorithms attached processors can do sufficiently well to insure that the cost of moving the data describing the problem does not overwhelm the gain obtained. Indeed, short of moving the entire problem to the attached processor, which may not be feasible because of programming and memory considerations, it is not a priori clear that such subalgorithms exist. If this multi-processor approach is to be successful, the host machine must be capable of supporting peripheral processors via a very fast specialized bus, rather than a slow general-purpose communications bus.

In thinking about the architecture of an elliptic problem solver, we should look beyond current commercial attached processors. VLSI-based peripheral processors, which provide extremely fast cycle time because of high density, the potential for specification of highly concurrent calculations, very low cost, and easy design with automated design tools, may turn out to be revolutionary.

In combination, these attractions will allow us to fabricate special-purpose processors with the potential for massively reducing computation time. Furthermore, they introduce the possibility of moving many of the complexities of large scientific software systems into hardware and in this way drastically reducing the cost of scientific programming. The potential of this technology is so great that it promises to revolutionize our concept of what is a computationally tractable problem.

As Kung [6,7] has emphasized, the key to cost-effective design and performance analysis of VLSI is the underlying high-level algorithm. In order to maximize chip density at any level of fabrication technology, it is necessary to have an architecture that involves simple and regular interconnections. Hence, underlying the chip should be an efficient parallel algorithm that has simple and regular movements of data. Moreover, to maximize the throughput of the peripheral processor, the algorithm should use pipe-lining techniques to overlap I/O with computation.

Despite its extraordinary potential, VLSI design and fabrication are still somewhat experimental. The current design process is lengthy and chip densities of the type we need are still well in the future. We believe, however, that is not too early to begin thinking about the basic issues.

3. Algorithms

As the peripheral processors and their algorithms become computationally more effective, we run an increasing risk that the whole system will founder on communications. The turn-around or wall-clock time for a given simulation will be bounded below by communication time, the time needed to move the data between the host and the peripheral processors.

We will focus our attention in this section of the paper on some of the implications algorithms have for communications. We believe that the interplay between architecture and algorithms for such multi-processor systems, and in particular their impact on the turn-around time for large problems, is ill understood. Much theoretical modelling and experimentation remains to be done. Our intent is to illustrate some of the issues.

Generally speaking, algorithms for elliptic problems at some stage reduce to solving very large sparse linear systems. These linear algebra subproblems have two distinct computational phases, the assembly (i.e., the computation) of the linear systems to be solved and the solution of the linear systems.

This leads to many questions. What is the class of algorithms for assembling linear systems? What are the implications of each assembly algorithm for computation and communication costs? What is the class of algorithms for solving the resulting highly structured sparse linear systems? What are the implications of each solution algorithm for computation and communication costs? What are the bandwidths for data flow between different levels of memory and processors? How much fast memory is needed to guarantee that algorithms will be compute-bound? Which subalgorithms can and should be implemented on attached processors? For which algorithms should special-purpose VLSI based processors be built?

How can we reduce turn-around time? Despite what the textbooks tell us, neither algorithms that are more efficient in arithmetic operations nor faster peripheral processors will in themselves automatically reduce turn-around time. The key is to do something about I/O. In the remainder of this section, we will illustrate four general techniques for addressing this issue:

1. Hide the I/O, i.e., overlap the I/O as much as possible with useful computation (under the assumption, of course, that the computing environment supports user-controllable parallel execution of I/O and computation).

2. Trade I/O off for incremental arithmetic computation until the two are completely overlapped.

3. Use linear algebra techniques to reduce the amount of required storage and hence the data flow.

4. Use analytic techniques to reduce the amount of required storage and hence the data flow by, for example, using adaptive high-order discretizations.

In order to illustrate these techniques, we consider the model problem of a second-order, linear, self-adjoint elliptic boundary value problem in the unit square. We discuss how each technique can be used to improve turn-around time.

For the first three techniques we consider solving the standard five-point central-difference approximation to the model problem on a uniform nxn grid [11]. To illustrate how we might overlap I/O with computation in a multi-processor system, we consider the case of a dual-processor system, a central processing unit and an I/O processor, with dual-ported fast primary memory and slow secondary memory, such as a disk. We assume that the I/O processor is user-programmable and that it can operate in parallel with the CPU.

Suppose we wish to compute the Cholesky factorization [5] of the model matrix, but that the mesh is arbitrarily large, i.e., the number of mesh points, n, in each direction is arbitrarily large, so that the band of the matrix cannot be stored in primary memory.

What can we say about communication costs and turn-around time? Along with Jack Perry, we formulated and proved a rather startling result [8] about "out-of-core" Cholesky factorization algorithms, which we state informally.

> Given the ratio of the speeds of the two processors, there exists a constant M such that for all systems with at least M words of primary memory there exists a block Cholesky factorization algorithm and a two-processor schedule so that the factorization is compute-bound independent of n.

It is important to observe that, in discussing software for multi-processor systems, one must give not only a computational algorithm in the classical sense for serial machines but also a specification for the flow of data between memories and processors and a schedule for the processors. The specific case under discussion is particularly startling in that using larger amounts of primary memory beyond the minimum M given by the preceding result actually increases the turn-around time. If we assume that the band of the matrix can fit into primary memory but is initially stored in secondary memory, we can prove the following result [8]:

> The turn-around time of the above "out-of-core" algorithm is shorter than the turn-around time required for (1) reading the band into primary memory and (2) doing an "in-core" Cholesky factorization.

Many currently available computers do not allow us to control their I/O processors explicitly, but do have a virtual memory system using a heuristic page replacement algorithm such as LRU (Least Recently Used). We have been able to prove the following result about this situation; see Perry [8] for a formal statement and proof.

> Given a fixed amount of primary memory and page size, there exists a block Cholesky factorization algorithm, independent of n, that minimizes the number of page faults.

Despite what the operating-systems experts of several commercial vendors claim, and despite the fact that automatic paging may minimize the number of times the data must be moved, given the choice between using automatic paging and scheduling the I/O, scheduling the I/O is bound to be more efficient in turn-around time. The point is that automatic paging schemes do not necessarily move the data at the right time to overlap it with useful computation. Such considerations certainly should be of major importance in designing systems for scientific computation.

The last three general techniques for minimizing turn-around time are based on two observations: (1) Reducing storage is likely to reduce I/O. (2) There often exists a trade-off between solve-time and storage (or I/O time).

To illustrate a trade-off between solve-time and storage, we again consider the model linear system. The classical band Cholesky factorization algorithm requires $1/2\ n^4 + O(n^3)$ multiplications and n^3 storage. However, along with Andrew Sherman we developed a divide-and-conquer band elimination or minimal storage band elimination algorithm [3] that requires $5/6\ n^4 + O(n^3)$ multiplications and $(n+1)^2$ storage. Thus we can gain an order-of-magnitude improvement in the required storage at the expense of a constant-factor increase in the required work. The reduction in primary memory occupancy is particularly dramatic.

The third general technique for reducing turn-around is to use sophisticated linear algebra methods. This is a place where more classical kinds of numerical analysis play a very important role. Four of our favorite linear algebra methods are (1) sparse elimination [10]; (2) minimal storage sparse elimination [4]; (3) preconditioned conjugate gradient methods [2]; and (4) multi-grid [1]. These four methods have the following asymptotic costs for the model problem:

	Multiplications	Storage
(1) sparse elimination	$O(n^3)$	$O(n^2 \log n)$
(2) minimal storage sparse elimination	$O(n^3)$	$O(n^2)$
(3) preconditioned conjugate gradient method	$O(n^{2.5} \log n)$	$O(n^2)$
(4) multi-grid	$O(n^2)$	$O(n^2)$

These methods have the following virtues:

(1)-(2): (a) Good, general codes available.
(b) Efficient for two-dimensional problems.

(3): (a) Generally applicable with promise of good codes.
(b) Efficient for two- and three-dimensional problems.
(c) Low storage requirement.

(4): (a) Asymptotically optimally efficient.

On the other hand, these methods have the following possible shortcomings:

(1)-(2): (a) Inefficient in three dimensions.
(b) Large storage requirements, especially in three dimensions.

(3): (a) In some cases, parameters need to be estimated.

(4): (a) Difficult to implement efficiently.
(b) Difficult to implement for general problems.

What can we conclude from these asymptotic results? Methods (2), (3), and (4) have storage proportional to the number of unknowns. In theory, multi-grid looks like the clear winner. In practice, however, it is so difficult to implement and has such complex logic and overhead that it may not be efficient on any existing peripheral processor -- maybe not even on one specially designed.

The last but perhaps most powerful way to reduce time and storage is analytic: the use of adaptive high-order finite difference or finite element discretization techniques. Because the techniques are high-order, for a given accuracy the discretization involves few parameters and little storage and so can be solved quickly. Because the techniques are adaptive, we can attain high convergence rates for smooth and nonsmooth problems alike.

As an example, consider the use of piecewise bicubic polynomials with the Rayleigh-Ritz-Galerkin method [9]. Such a method requires $c\, n^2 + k\, n^q$, $2 \leq q \leq 4$, $k \ll c$ multiplications, where the first term corresponds to assembling the matrix and the second term corresponds to solving it. The value of q depends on the method used to solve the linear system. In general, for a given fixed accuracy, the linear systems for high-order discretizations are much smaller, and hence have more "information content" per nonzero entry, than the linear systems for low-order discretizations. But the computation per nonzero to assemble the matrix is much higher. Thus, we may draw several conclusions:

1. Solving the linear system for a high-order approximation is relatively easy.
2. I/O is reduced and is often unnecessary for high-order methods.
3. The assembly times are relatively high.

Clearly we should focus on optimizing the assembly phase. The first thing to investigate in the use of a peripheral processor is how well the assembly phase can be implemented. What can be done algorithmically to improve the assembly phase? Our studies with Alan Weiser [12] indicate that it is advantageous to use smooth tensor-product, piecewise-polynomial basis functions and to take advantage of all possible symmetries. For example, we get the following assembly costs per interior element:

Smoothness	Multiplications
C^1 (using only symmetry of matrix)	8102
C^1	882
C^2	558.

Moreover, smoothness has a beneficial effect on the solve phase in terms of work and storage. The following table gives the relative operation counts and storage for the Cholesky factorization of the corresponding band matrix:

Smoothness	Relative Multiplication Counts	Relative Storage
C^0	364.5	81
C^1	32.0	16
C^2	4.5	3

This shows that symmetries can play an important role in the assembly phase. If possible, it is highly advantageous to map the problem domain onto a union of rectangles and to use smooth tensor-products of piecewise polynomials.

Unfortunately, actual computational experience for realistic problem sizes giving engineering accuracy is not quite as clear-cut as we would like. The adaptive high-order methods have a very large overhead because of to their logical complexity. In practice we are trading arithmetic operations and storage for logical complexity. Moreover, it is not at clear whether we can effectively implement these methods on a multi-processor system.

When it comes to turn-around time, we do not know whether it is better to use a simple low-order method that involves a lot of data but can be implemented very effectively on a multi-processor or a sophisticated adaptive high-order method that involves little data but might not be effectively implementable. We cannot yet answer this fundamental question.

REFERENCES

[1] A. Brandt.
Multi-level adaptive solution to boundary value problems.
Mathematics of Computation 31:333-390, 1977.

[2] R. Chandra.
Conjugate Gradient Methods for Partial Differential Equations.
PhD thesis, Yale University, 1978.

[3] S. C. Eisenstat, M. H. Schultz, and A. H. Sherman.
Minimal storage band elimination.
In D. J. Kuck, D. H. Lawrie, and A. H. Sameh, editors, Proceedings of the Symposium on High Speed Computer and Algorithm Organization, pages 273-286. Academic Press, New York, 1977.

[4] S. C. Eisenstat, M. H. Schultz, and A. H. Sherman.
Software for sparse Gaussian elimination with limited core storage.
In Proceedings of the Sparse Matrix Meeting, pages 135-153. SIAM, Philadelphia, 1978.

[5] G. Forsythe and C. Moler.
Computer Solution of Linear Algebraic Systems.
Prentice-Hall, 1967.

[6] M. J. Foster and H. T. Kung.
The design of special-purpose VLSI chips.
Computer 13:26-40, 1980.

[7] C. Mead and L. Conway.
Introduction to VLSI Systems.
Addison-Wesley, 1980.

[8] J. R. Perry.
Secondary Storage Methods for Solving Banded Linear Systems.
PhD thesis, Yale University, 1981.

[9] M. H. Schultz.
Spline Analysis.
Prentice-Hall, 1973.

[10] A. H. Sherman.
On the efficient solution of sparse systems of linear and nonlinear equations.
PhD thesis, Yale University, 1975.

[11] R. S. Varga.
Matrix Iterative Analysis.
Prentice-Hall, 1962.

[12] A. Weiser, S. C. Eisenstat, and M. H. Schultz.
Solving finite element equations to moderate accuracy.
SINUM 17:908-929, 1980.

CO-ENERGY METHODS FOR ELLIPTIC FLOW AND RELATED PROBLEMS

R. A. Nicolaides

Department of Mathematics
Carnegie-Mellon University
Pittsburgh, Pennsylvania

A complementary or dual variational approach may be used to discretize a class of elliptic problems including linear elasticity and the stationary Navier-Stokes equations as special cases. The advantages and disadvantages of these discretizations are discussed, along with some algorithms for solving the specially structured discrete systems which arise.

1. INTRODUCTION

This paper aims to discuss, sometimes in a purely formal way, the use of complementary formulations of elliptic problems as a basis for discretization. Topics of interest are the methods of discretization, the stability and accuracy of the resulting approximations and the solution algorithms for the discrete systems. For the most part, these topics are considered in the relatively simple setting of particular extremum problems, rather than in the general variational case, since the important issues are most clear in the former. In §2 an idea of the meaning of the phrase "complementary formulations" is given, along with rigorous statements of a complementary extremum problem, the general variational complementary problem, and representative physical examples. In §3, discretization of the extremum principles is introduced. In addition, questions of error analysis are considered. These

ISBN 0-12-632620-7

questions are harder and more subtle than those arising from the direct extremum formulation as will be shown. The last section, §4 contains a discussion of algorithmic issues relating to the solution of the algebraic equations resulting from the discretizations, and proposes a general class of algorithms for solving these equations.

2. DIRECT AND COMPLEMENTARY FORMULATIONS

2.1. Let Ω denote a region of R^d, with boundary $\Gamma \equiv \Gamma_1 \cup \Gamma_2$ and let $f : R^d \to R^1$ and $\underline{g} : R^d \to R^d$ be given, appropriately smooth functions. A simple direct energy formulation of a variational problem would be the following:

$$\min_{\underline{u},\underline{v},\varphi} \int_\Omega \frac{1}{2}\, \underline{u}\cdot\underline{v} - f\varphi \tag{2.1}$$

subject to

$$\underline{u} = \nabla\varphi + \underline{g} \qquad \varphi|_{\Gamma_1} = 0 \qquad v = A^{-1}\underline{u}. \tag{2.2}$$

The stationarity conditions here are

$$-\text{div}\, \underline{v} = f \qquad \underline{v}\cdot\underline{n}|_{\Gamma_2} = 0. \tag{2.3}$$

In (2.2), A is a d×d positive definite matrix whose elements are functions of $\underline{x} \in \Omega$.

The complementary formulation of this same problem is

$$\max_{\underline{u},\underline{v}} \int_\Omega - \frac{1}{2}\, \underline{u}\cdot\underline{v} + \underline{g}\cdot\underline{v} \tag{2.4}$$

subject to

$$-\text{div}\, \underline{v} = f \qquad \underline{v}\cdot\underline{n}|_{\Gamma_2} = 0 \qquad \underline{u} = A\underline{v}. \tag{2.5}$$

The stationarity conditions are

$$\underline{u} = \nabla\varphi + \underline{g} \qquad \varphi|_{\Gamma_1} = 0. \tag{2.6}$$

From (2.2)-(2.3) and (2.5)-(2.6), it follows that the same

function $\underline{u}$ solves both problems. Also, it is straightforward to show that (2.1) and (2.4) have the same extreme value. Thus, (2.1)-(2.3) and (2.4)-(2.6) may be regarded as distinct formulations of the same problem. The second of eqns. (2.3) is the usual natural boundary condition for the direct formulation. This suggests calling (2.3) the "natural" conditions for (2.1)-(2.2) and calling (2.2) the "essential" conditions for (2.1). Correspondingly, in (2.4)-(2.6), (2.5) are the essential conditions for (2.4), and (2.6) are the natural conditions for (2.4)-(2.5). The duality, or complementarity between the two formulations can then be expressed by saying that the essential conditions of one formulation are the natural conditions of the other.

With only slight changes, the two main variational principles of linear elasticity theory may be viewed in this way. The direct formulation corresponds to the minimum potential energy technique, and the alternative to the complementary energy approach. The necessary changes are essentially formal. For example $\underline{u}$ should be interpreted as the strain tensor, $\underline{v}$ as the stress tensor and ∇ as the symmetric gradient of the displacement vector φ. In the elasticity context, it is worth noting that the essential condition in the direct principle is kinematic and in the complementary principle is dynamic, and dually.

The basic variable in the direct formulation is $\underline{u}$, or, since $\underline{u} = \nabla\varphi + \underline{g}$, the scalar φ. Correspondingly, the basic variable in the complementary formulation is $\underline{v}$. This brings out the point that if the variable of interest is $\underline{v}$ (e.g., stress rather than strain), then rather than use the direct principle to find the potential φ, differentiating it and then finding $\underline{v}$ from the "constitutive relation" $\underline{v} = A^{-1}u$, it may pay to compute $\underline{v}$ directly by means of the complementary principle. This is one motivation for studying the complementary principles. Another will be given below.

2.2. The problems (2.1)-(2.3) and (2.4)-(2.6) are, as stated, purely formal, since, e.g., it is not said in what sense equality holds in (2.2). A rigorous statement of a co-energy extremal problem is as follows:

$$\max_{\underline{u}\in\mathcal{V}^0} - \|\underline{u}\|^2_{\mathcal{V}} \tag{2.7}$$

subject to

$$c(\underline{u},\psi) = \langle f,\psi\rangle \qquad \forall\ \psi \in \mathcal{S}^0 \subset \mathcal{S} \tag{2.8}$$

where $\mathcal{V}$ (the "vector fields") is a Hilbert space, containing the dense subspace $\mathcal{V}^0$, and $\mathcal{S}$ (the "scalar fields") is a Hilbert space containing the dense subspace $\mathcal{S}^0$. f denotes a bounded linear functional on $\mathcal{S}$ and $c(\cdot,\cdot)$ a bilinear function of its arguments $\underline{u} \in \mathcal{V}^0$ and $\psi \in \mathcal{S}^0$.

As an example, consider the formal problem (leading to Poisson's equation)

$$\max - \int_\Omega \underline{u}\cdot\underline{u} \tag{2.9}$$

subject to

$$-\text{div}\ \underline{u} = f, \qquad \underline{u}\cdot\underline{n}|_{\Gamma_2} = 0. \tag{2.10}$$

Letting $\mathcal{V} = \vec{L}^2(\Omega)$, and $\mathcal{S} = L^2(\Omega)$, $\mathcal{V}^0$ may be taken as

$$\{\underline{u} \in L^2(\Omega)\ |\underline{u}\cdot\underline{n}|_{\Gamma_2} = 0,\quad \text{div}\ \underline{u} \in L^2(\Omega)\}$$

and $\mathcal{S}^0$ as $\mathcal{S}$ itself. The form $c(\underline{u},\psi)$ may be defined as

$$c(\underline{u},\psi) = \int_\Omega \psi\ \text{div}\ \underline{u} \qquad \forall\ \psi \in L^2(\Omega),\quad \underline{u} \in \mathcal{V}^0. \tag{2.11}$$

(2.9)-(2.10) may also be solved by handling the first constraint by a Lagrange multiplier φ [e.g. [6]], which leads to the formal system of equations for $(\underline{u},\varphi)$

$$\begin{aligned} (\underline{u},\underline{v}) + c(\underline{v},\varphi) &= 0 \;\forall\; \underline{v} \\ c(\underline{u},\psi) &= \langle f,\psi\rangle \;\forall\; \psi \end{aligned} \tag{2.12}$$

where $c(\cdot,\cdot)$ is given in (2.11) and $(\underline{u},\underline{v})$ is the $\vec{L}^2(\Omega)$ inner product.

Another often seen formal expression of (2.12) is the "operator" form

$$\begin{aligned} A\underline{u} + C\varphi &= 0 \\ C^*\underline{u} &= f \end{aligned} \tag{2.13}$$

in which, for the present case, $A \equiv I$, $C \equiv -\text{grad}$, $C^* \equiv \text{div}$.

Many problems of interest, linear and otherwise, arise in the form (2.12) or some generalization of it, rather than from an extremal calculation. This provides a second motivation for studying complementary formulations. A good example is furnished by the Navier-Stokes equations

$$\begin{aligned} -\Delta\underline{u} + \text{Re}(\text{grad}\;\underline{u})\underline{u} &= -\nabla p + \underline{f} \\ \text{div}\;\underline{u} &= 0 \\ u|_\Gamma &= 0 . \end{aligned} \tag{2.14}$$

A rigorous formulation of (2.14), obtained by multiplying by test functions and integrating by parts in the usual way [e.g., [8,14]] is as follows: find $\underline{u} \in \vec{H}^1_0(\Omega)$, $p \in \bar{L}^2(\Omega)$ such that

$$a(\underline{u},\underline{v}|\underline{u}) + b(\underline{v},p) = \langle f,\underline{v}\rangle \;\forall\; \underline{v} \in \vec{H}^1_0(\Omega)$$

$$b(\underline{u},q) = 0 \;\forall\; q \in \bar{L}^2(\Omega)$$

where

$$a(\underline{u},\underline{v}|\underline{w}) \equiv \int_\Omega \text{grad}\;\underline{u}:\text{grad}\;\underline{v} + \text{Re}(\text{grad}\;\underline{u})\underline{w}\cdot\underline{v}$$

$$b(\underline{w},s) \equiv -\int_\Omega \text{div}\;\underline{w}s$$

and where $\overline{L}^2(\Omega)$ consists of those scalar valued functions in $L^2(\Omega)$ with zero mean over Ω. In (2.14), grad $\underline{u}$ denotes the second order tensor which is the gradient field of the velocity vector $\underline{u}$.

2.3. The previous considerations motivate the following definition of a co-energy problem:

Definition

Let $\mathcal{V}_1, \mathcal{V}_2$ be a pair of Hilbert spaces, and $\mathcal{S}_1$ and $\mathcal{S}_2$ another pair. Let $a(\cdot,\cdot)$, $b(\cdot,\cdot)$ and $c(\cdot,\cdot)$ be bilinear forms defined as indicated below, and g and f linear functionals on $\mathcal{V}_2$ and $\mathcal{S}_2$ respectively. The general co-energy problem is defined as that of finding $u \in \mathcal{V}_1$ and $\varphi \in \mathcal{S}_1$ such that

$$
\begin{aligned}
a(u,v) + b(v,\varphi) &= \langle g,v\rangle \ \forall\, v \in \mathcal{V}_2 \\
c(u,\psi) &= \langle f,\psi\rangle \ \forall\, \psi \in \mathcal{S}_2 .
\end{aligned}
\tag{2.15}
$$

(2.15) generalizes the problems already considered by 1) allowing distinct forms $b(\cdot,\cdot)$ and $c(\cdot,\cdot)$ and 2) allowing different test and trial spaces for both the $\mathcal{V}$ and $\mathcal{S}$ fields. Existence and approximations for (2.15) in the special case $b(\cdot,\cdot) \equiv c(\cdot,\cdot)$ and $\mathcal{V}_1 \equiv \mathcal{V}_2$, $\mathcal{S}_1 \equiv \mathcal{S}_2$ were considered by Brezzi [3], in terms of the graph norm

$$|||(u,\varphi)||| \equiv \|u\|_{\mathcal{V}} + \|\varphi\|_{\mathcal{S}} .$$

The general case was analyzed in [13], using separately the norms $\|\ \|_{\mathcal{V}}$ and $\|\ \|_{\mathcal{S}}$. The use of these norms in this way is usually more natural physically, and besides, the results for the graph norm may be trivially obtained by addition, whereas the converse of this statement is false in general. The results referred to are obtained in both [3] and [13] under certain fairly standard "continuity" and "stability" assumptions analogous to the "continuity" and "ellipticity" assumptions of the direct energy theory.

The desire for $\mathcal{V}_1 \neq \mathcal{V}_2$ in (2.15) is motivated partly by fluid mechanics problems with significant transport terms, where it is known that using different test and trial velocity fields can contribute in an important way to the calculation. An example of a problem with $b(\cdot,\cdot) \neq c(\cdot,\cdot)$ is given in [12]. This example is too lengthy to reproduce here, but an admittedly artificial example may be easily given as follows: to solve

$$\operatorname{div}(a\nabla\varphi) = f \qquad \varphi|_\Gamma = 0$$

write it as the system

$$\operatorname{div} \underline{u} = f \qquad \underline{u} = a\nabla\varphi \qquad \varphi|_\Gamma = 0$$

which in turn may be written weakly as

$$(\underline{u},\underline{v}) + \int_\Omega \varphi \operatorname{div}(a\underline{v}) = 0 \qquad \forall\ \underline{v} \in \vec{L}^2(\Omega)$$

$$\int_\Omega \psi \operatorname{div} u = \int_\Omega f\psi \qquad \forall\ \psi \in L^2(\Omega)$$

Here, it is clear that $b(\cdot,\cdot) \neq c(\cdot,\cdot)$.

3. APPROXIMATION AND ERROR ANALYSIS

3.1. Discretization of the extremum problem (2.7)-(2.8) may be carried out in the natural way, namely, by choosing subspaces $\mathcal{V}^h \subset \mathcal{V}^0$ and $\mathcal{S}^h \subset \mathcal{S}^0$ and replacing (2.7)-(2.8) by the approximate problem

$$\max_{\underline{u}^h \in \mathcal{V}^h} - \|\underline{u}^h\|^2_{\mathcal{V}} \tag{3.1}$$

subject to

$$c(\underline{u}^h,\psi^h) = \langle f,\psi^h\rangle \qquad \forall\ \psi^h \in \mathcal{S}^h \tag{3.2}$$

For the purposes of analysis, it is necessary to distinguish two broad subclasses of problems.

1) $c(\cdot,\cdot)$ is a continuous function of its arguments in the sense

$$|c(u,\psi)| \le K\|\underline{u}\|_{\mathcal{V}}\|\psi\|_{\mathcal{S}} \tag{3.3}$$

2) All other cases.

Examples of the two cases are as follows:

1) $c(\underline{u},q) = \int_\Omega \operatorname{div} \underline{u}q \qquad \mathcal{V} = \vec{H}_0^1(\Omega), \quad \mathcal{S} = \overline{L}^2(\Omega).$

The natural condition for this particularization of (2.7)-(2.8) (generalized to allow nonzero body force $\underline{g}$) is the Stokes equation

$$-\Delta\underline{u} = -\nabla p + \underline{g} \tag{3.4}$$

Clearly,

$$|c(\underline{u},q)| \le \|\operatorname{div} \underline{u}\|_{L^2(\Omega)}\|q\|_{L^2(\Omega)}$$

$$\le \|\underline{u}\|_{\vec{H}_0^1(\Omega)}\|q\|_{L^2(\Omega)}$$

so that (3.3) holds in this case.

2) $c(\underline{u},\psi) = \int_\Omega \operatorname{div} \underline{u}\psi \qquad \mathcal{V} = \vec{L}^2(\Omega), \quad \mathcal{S} = L^2(\Omega)$

Clearly, it is false in general for this case that $\exists\, K \ne K(u)$ such that

$$|c(\underline{u},\psi)| \le K\|\underline{u}\|_{\vec{L}_2(\Omega)}\|\psi\|_{L^2(\Omega)}.$$

Of the two cases, the first is the easier to deal with, and in fact is treated from both the existence and approximation viewpoints in [3] and [13]. It is sometimes possible to choose a norm on $\mathcal{V}$ in such a way that an ostensible second case becomes of the first type. But this approach may disguise certain forms of instability. This was pointed out in [6]. Error analysis for both cases for a simple context is treated below.

3.2. An analysis of the error for the first, continuous case may be given as follows. Under the further hypothesis of stability of the $c(\cdot,\cdot)$ form, the space $\mathcal{V}$ may be orthogonally decomposed as

$$\mathcal{V} = Z \oplus \mathcal{W} \tag{3.5}$$

where $Z = \mathfrak{n}(D)$ and $\mathcal{W} = \mathfrak{R}(D^*)$, and where D is the Reisz operator $D : \mathcal{V} \to \mathcal{S}'$, so that

$$c(\underline{u},\psi) = \langle D\underline{u},\psi\rangle \qquad \underline{u} \in \mathcal{V}, \quad \psi \in \mathcal{S}.$$

The existence of decomposition (3.5) is heavily dependent on the continuity of $c(\cdot,\cdot)$. The particular characterization of $\mathcal{W}$ as $\mathfrak{R}(D^*)$ depends on the stability hypothesis. Details of the construction may be found in [13]. Usually, (3.5) corresponds to some well known expression of an element of $\mathcal{V}$ as a sum of other elements, for example, the Helmholtz type decomposition of a vector field into a solenoidal and a gradient part. Roughly speaking, in such a case, D would be the divergence operator, and Z its null space of solenoidal fields.

The approximating space $\mathcal{V}^h$ has a corresponding orthogonal decomposition, obtained similarly. Thus, let $\tilde{D}_h$ denote the Reisz operator associated with $c(\underline{u},\psi^h)$, so that

$$c(\underline{u},\psi^h) \equiv \langle\tilde{D}_h\underline{u},\psi^h\rangle \qquad \underline{u} \in \mathcal{V}, \quad \psi^h \in \mathcal{S}^h.$$

Note that $\tilde{D}_h : \mathcal{V} \to (\mathcal{S}^h)'$. Let D_h denote the restriction of $\tilde{D}_h$ to $\mathcal{V}^h$. Then in terms of the null space of D_h and range of D_h^*

$$\mathcal{V}^h \equiv Z^h \oplus \mathcal{W}^h \tag{3.6}$$

where $\mathcal{Z}^h = \mathcal{N}(D_h)$ and $\mathcal{W}^h = \mathcal{R}(D_h^*)$. In general,

$$\mathcal{Z}^h \not\subset \mathcal{Z} \qquad \mathcal{W}^h \not\subset \mathcal{W}.$$

3.3. To make these ideas more concrete consider the case with Ω a square. To define $\mathcal{S}^h$, subdivide Ω into equal subsquares by repeated halvings by lines parallel to the co-ordinate axes and then define $\mathcal{S}^h$ to be the class of all piecewise constant functions on the subsquares, with average value zero over Ω. $\mathcal{V}^h$ is defined on the triangulation of Ω obtained by drawing the rightward sloping diagonal in each subsquare, as the class of piecewise linear continuous fields both of whose components vanish in $\partial\Omega$. These spaces are appropriate for solving the Stokes problem (3.4). Here, $D \equiv \text{div}$; however

$$\begin{aligned} c(\underline{u},\psi^h) &= \int_\Omega \psi^h \text{div}\, \underline{u} \\ &= \int_\Omega \psi^h \Pi_{\mathcal{S}^h}(\text{div}\, \underline{u}) \end{aligned}$$

where $\Pi_{\mathcal{S}^h}$ is the orthogonal projector of $\bar{L}^2(\Omega)$ onto $\mathcal{S}^h$. Thus,

$$\tilde{D}_h \underline{u} = \Pi_{\mathcal{S}^h}(\text{div}\, \underline{u}).$$

Incidentally, the above choice of $\mathcal{V}^h$ and $\mathcal{S}^h$ differs qualitatively from the choices of these spaces made in the literature [2,6,9,11,14,15] in that a different triangulation is used for $\mathcal{V}^h$ and $\mathcal{S}^h$. There is no a priori reason for using the same triangulation.

3.4. Let $w \in \mathcal{W}$ denote the exact, and $w^h \in \mathcal{W}^h$ denote the approximate solutions. Then from the definitions

$$\langle D\underline{w},\psi\rangle = \langle f,\psi\rangle \qquad \forall\ \psi \in \mathcal{S}$$
$$\langle D\underline{w}^h,\psi^h\rangle = \langle f,\psi^h\rangle \qquad \forall\ \psi^h \in \mathcal{S}^h,$$

it follows that $\forall\ \hat{w}^h \in \mathfrak{w}^h$

$$\langle D(\underline{w}^h - \hat{w}^h),\psi^h\rangle = \langle D(w - \hat{w}^h),\psi^h\rangle \qquad \forall\ \hat{w}^h \in \mathfrak{w}^h$$

i.e.,

$$\langle D_h(\underline{w}^h - \underline{\hat{w}}^h),\psi^h\rangle = \langle D(w - \hat{w}^h\ ,\psi^h\rangle$$

Since $\mathfrak{w}^h = \mathcal{R}(D_h^*)$, choose

$$D_h^*\psi^h = w^h - \hat{w}^h$$

so that

$$\|w^h - w^h\|_{\mathcal{V}}^2 \leq \|D(w - \hat{w}^h)\|_{\mathcal{S}}\|\psi^h\|_{\psi}\ . \tag{3.7}$$

To finalize the estimate, the following hypotheses are needed

a) Continuity; $\|D\underline{u}\| \leq \Gamma\|\underline{u}\| \qquad \forall\ \underline{u} \in \mathcal{V} \qquad \Gamma \neq \Gamma(h)$

b) Stability of D_h; $\gamma^2\|\psi^h\| \leq \|D_h^*\psi^h\| \qquad \forall\ \psi^h \in \mathcal{S}^h$

where $\qquad 0 < \gamma^2 \neq \gamma^2(h)$

(actually, b) is already used in order to assert $\mathfrak{w}^h = \mathcal{R}(D_h^*)$). Then (3.7) and the triangle inequality give

$$\|w - w^h\| \leq K\|w - \hat{w}^h\| \qquad K \equiv 1 + \Gamma/\gamma^2$$

as the error estimate. Thus, the approximation is optimal in $\mathfrak{w}^h$. To show the optimality in $\mathcal{V}^h$, slightly more complex arguments are required, but the hypotheses remain sufficient [13].

It is essential to realize that in general, hypothesis b) is false, and will be true only for subspaces $\mathcal{V}^h$ and $\mathcal{S}^h$ related in particular ways. This is unlike the direct energy case, in which the analogous estimate follows from that of the continuous problem. The implied total or partial loss of stability can considerably complicate the question of "optimality" of the numerical approximation. For it is possible to obtain technical optimality by use of relatively unnatural norms, while at the same time having suboptimal accuracy (with a partial instability implied) in other more reasonable looking norms.

To show that b) is false in general, consider the square with $\mathcal{V}^h$ defined as above but with $\mathcal{S}^h$ defined to be piecewise constants defined this time on the same triangulation as $\mathcal{V}^h$. It is more convenient to work with the adjoint of condition b), which is the following:

$$\|D_h w^h\| \geq \gamma^2 \|w^h\| \qquad \forall\, w^h \in \mathcal{W}^h \qquad \gamma^2 \neq \gamma^2(h) > 0 . \tag{3.8}$$

This is the condition which will be shown to be impossible. The reason for this is that $\mathcal{W}^h \equiv \mathcal{V}^h$. This follows from the facts that in the present case $D_h\, \underline{u}^h \equiv \operatorname{div} \underline{u}^h$ and that the only solenoidal piecewise linear field vanishing on Γ is the zero field. Thus, Z^h contains only the zero element and $\mathcal{V}^h \equiv \mathcal{W}^h$. But (3.8) cannot then hold, for $\underline{w}^h$ can always be chosen to be approximately solenoidal, but of unit norm. Hence, in general b) or (3.8) will not hold. In general, the stability estimate is hard to prove, and at present appears to be grid dependent, [6].

3.5. As an example of the second type, consider the Poisson case, in which

$$c(\underline{u},\psi) \equiv \int_\Omega \psi \operatorname{div} \underline{u}.$$

The approximation problem is now

$$\max_{\mathcal{V}^h} - \int_\Omega \underline{u}^h \cdot \underline{u}^h \tag{3.9}$$

subject to

$$\int_\Omega \psi^h \mathrm{div}\, \underline{u}^h = \int_\Omega \psi f \qquad \forall\, \psi^h \in \mathcal{S}^h \subset L^2(\Omega).$$

Let $\mathbb{W}^h$ denote the linear manifold of solutions (each assumed unique) obtained by varying f in (3.9). $\mathbb{W}^h$ will play the same role as the space of the same name in the previous paragraph. The present (ad hoc) characterization is used because the earlier rigorous construction is not now easily available. If $\underline{w}$ and $\underline{w}^h$ denote the (unique) solutions of the exact and approximate problems respectively, then

$$\int_\Omega \psi \ \mathrm{div}\ \underline{w} = \int_\Omega \psi f \qquad \forall\ \psi \in L^2(\Omega) \tag{3.10}$$

and from (3.9) and (3.10)

$$\int_\Omega \psi^h \mathrm{div}(\underline{w} - \underline{w}^h) = 0 \qquad \forall\ \psi^h \in \mathcal{S}^h. \tag{3.11}$$

Let $\hat{\underline{w}}^h$ be arbitrary in $\mathbb{W}^h$. Then using (3.11) in the identity

$$\int_\Omega \psi \ \mathrm{div}(\hat{\underline{w}}^h - \underline{w}^h) = \int_\Omega \psi \ \mathrm{div}(\hat{\underline{w}}^h - \underline{w}) + \int_\Omega \psi \ \mathrm{div}(\underline{w} - \underline{w}^h)$$

gives

$$\begin{aligned}\int_\Omega \psi \ \mathrm{div}(\hat{w}^h - \underline{w}^h) &= \int_\Omega \psi \ \mathrm{div}(\hat{\underline{w}}^h - \underline{w}) \\ &\quad + \int_\Omega (\psi - \psi^h)\mathrm{div}(\underline{w} - w^h).\end{aligned} \tag{3.12}$$

At this point ψ and ψ^h are arbitrary. If ψ is chosen equal to ψ^h, the last term goes, and continuity and stability, if available, could be used to estimate respectively the right and left sides. This is essentially what was done in §3.4. In the absence of the continuity it is necessary to proceed a little differently.

Let $H^1_B(\Omega)$ denote all $\varphi \in L^2(\Omega) \cap H^1(\Omega)$ such that $\varphi|_{\Gamma_1} = 0$, and define

$$\|\varphi\|_{H_B^1(\Omega)} \equiv \|\nabla\varphi\|_{\vec{L}^2(\Omega)}$$

and let $\|\ \|_*$ denote the norm dual to $\|\ \|_{H_B^1(\Omega)}$. In (3.12) integrate by parts in the 1st term on the right, and apply Cauchy's inequality to the second. Then take the supremum over all ψ such that $\|\psi\|_{H_B^1(\Omega)} = 1$ to obtain

$$\|\text{div}(\hat{\underline{w}}^h - \underline{w}^h)\|_* \le \|\hat{\underline{w}}^h - \underline{w}\|_{\vec{L}^2(\Omega)} + \delta_h\|\text{div}(w - w^h)\|_{L^2(\Omega)} \quad (3.13)$$

where δ_h is the approximation theoretical quantity

$$\delta_h = \sup_{\|\varphi\|_{H_B^1(\Omega)}=1} \ \inf_{\psi^h \in \mathcal{S}^h} \|\varphi - \psi^h\|.$$

To obtain the desired estimates from (3.13), first of all the stability condition

$$\|\text{div}\ \tilde{\underline{w}}^h\|_* \ge \gamma^2\|\tilde{\underline{w}}^h\|_{\vec{L}^2(\Omega)} \qquad \forall\ \tilde{\underline{w}}^h \in W^h \qquad \gamma^2 \ne \gamma^2(h) > 0$$

is required. This is essentially similar to the estimate required for the continuous case and, as with the latter, in general will be false. It involves D rather than D_h however. The estimation of the last term on the right is the new feature. In (6), the size of this term was controlled by means of a new hypothesis, namely that $\mathcal{V}^h$ and $\mathcal{S}^h$ should satisfy

$$\text{div}\ \mathcal{V}^h \equiv \mathcal{S}^h. \quad (3.14)$$

If (3.14) holds, then it follows from the constraint in (3.9) that

$$\text{div}\ \underline{w}^h = \hat{f}^h \quad \text{(pointwise)}$$

where $\hat{f}^h = \Pi_{\mathcal{S}^h} f$. Since also (at least formally)

$$\text{div}\ w = f,$$

it follows that

$$\|\operatorname{div}(w-w^h)\|_{L^2(\Omega)} = \|f - \hat{f}^h\|_{L^2(\Omega)} .$$

In this way, the optimal (in $\vec{L}^2(\Omega)$) accuracy is preserved [6] without any assumption of continuity on the form $c(\cdot,\cdot)$. The condition (3.14) is essentially equivalent to the requirement that $Z^h \subset Z$ or that $D_h \equiv \operatorname{div}$, or in other terms, that a div free basis be used for Z^h. It is not clear that optimal $L^2(\Omega)$ accuracy can be achieved without some condition such as (3.14) to replace the continuity hypothesis.

The analysis of this paragraph may also be applied to the case where there is, in fact, a continuity condition satisfied. This is done in [7,8].

4. DISCRETE SYSTEM ALGORITHMS

4.1. The previous discretizations lead, in the most general case, to algebraic equation systems of the form

$$\begin{pmatrix} A & B \\ C & 0 \end{pmatrix} \begin{pmatrix} U \\ \Phi \end{pmatrix} = \begin{pmatrix} F \\ G \end{pmatrix} \tag{4.1}$$

where A has dimensions $dN{\times}dN$ (for a problem in d spatial dimensions), and B and C are respectively of dimensions $dN{\times}N$ and $N{\times}dN$. The stability conditions mentioned previously guarantee the uniform invertibility (w.r.t. h) of the systems (4.1). In this section, two broad classes of algorithms for solving (4.1) will be mentioned, one of them iterative, the other what will be called "semi-direct."

The first class of algorithms is the following [8]:

$$\Phi^{k+1} = C_k(\Phi^k,\Phi^{k-1}) + \alpha_k(CU^{(k)} - G) \tag{4.2}$$

$$U^{k+1} = T(U^k,\Phi^{k+1})$$

where $C_k(\cdot,\cdot)$ denotes a convex combination of the arguments, in general dependent on k, α_k is a scalar parameter and $T(\cdot,\cdot)$ is some algorithm (the "inner" algorithm) for getting a new U^{k+1} from the current U^k and Φ^{k+1}. (4.2)-(4.3) may be motivated by the Poisson case, in which A will be essentially an identity operator (actually a mass matrix). Suppose then $A \equiv I$. For (4.3), it is then natural to use

$$U^{k+1} = F - B\Phi^{k+1} \tag{4.4}$$

Substitution of (4.4) into (4.2), when $F = 0$ shows immediately (making the identifications $B \sim$ "grad," $C \sim$ "div") that for this special case, (4.2) becomes the general non-stationary second order iterative process. As such, it includes as special cases (by choosing $\mathcal{C}_k(\cdot,\cdot)$ and α_k) the conjugate gradient method [4,5], semi-iterative methods [10, 17] and the second order Richardson method [18]. In the case of a more general A, some more general technique than (4.4) is needed. Exactly what must be done is dependent on qualitative properties of A, but in all cases, the question of matching the "outer" and inner convergence rates becomes an issue. Rather than pursue this topic here, the second class of algorithms will be introduced.

<u>4.2.</u> Usually, any attempt at direct elimination of system (4.1) soon runs into bandwidth difficulties. For example, in the Stokes case, A will be some discretization of the (vector) Laplacian. Its bandwidth is intrinsic in a sense, and, given that some direct method is to be used, it seems reasonable that one should be willing at least to store and eliminate A in banded or similar appropriate form. However, the bandwidth (or appropriate measure) of the coefficient matrix in (4.1) is severely increased by the presence of the B and C sub-

matrices. The idea of the "semi-direct" method, is to reduce the elimination essentially to that of a matrix comparable in bandwidth to A. The price of this reduction will be some iterations, but it is very important to point out a qualitative difference between the iterations required by the semi-direct method, and those required by the iterative method described in (4.2)-(4.3). In the earlier case, the algorithm is intrinsically iterative, while in the present case the iterative part of the algorithm can have its rate of convergence arbitrarily high simply by choosing a parameter small enough.

The semi-direct method is defined by the following iteration:

$$A_1U^{k+1} + B\Phi^{k+1} = F - A_2U^k \tag{4.5}$$

$$CU^{k+1} + D\Phi^{k+1} = G + D\Phi^k \tag{4.6}$$

where D is an arbitrary $N\times N$ matrix, $A \equiv A_1 + A_2$, and U^0 and Φ^0 are given initial guesses. The motivation for this procedure is as follows. First, consider the case where A is not split, i.e., $A_2 \equiv 0$. Then if D was zero, the iteration would converge in one step to the exact solution of (4.1). It follows that D should be chosen to be as "small" as possible, and the smaller the better from the rate of convergence standpoint. The reason for introducing D is simply that (4.5)-(4.6) can be solved more economically when $D \neq 0$ than when $D \equiv 0$. In fact, multiplying (4.6) by BD^{-1} and subtracting from (4.5) gives the equation for U^{k+1}

$$(A_1 - BD^{-1}C)U^{k+1} = (F - A_2U^k - B\Phi^k) - BD^{-1}G. \tag{4.7}$$

Having solved (4.7) for U^{k+1}, Φ^{k+1} is given by (4.6) as

$$\Phi^{k+1} = \Phi^k - D^{-1}(CU^{k+1} - G) \tag{4.8}$$

Notice that (4.7)-(4.8) actually require only D^{-1} to be chosen. There is no need to choose D and then to invert it. In addition to being "large," D^{-1} can be chosen to control scaling, sparseness and, assuming (4.7) is to be solved directly, eliminability in $A_1 - BD^{-1}C$. For most of the problems considered in §1-§3, $B \equiv C^T$ and it makes sense to choose $D^{-1} = \epsilon^{-1}I$, and also $A_2 \equiv 0$. For the Navier-Stokes case, however, some low Re flows may be better computed with $A_2 \neq 0$. And for cases with $B \neq C^T$, some nonscalar choice of D^{-1} will be appropriate, although still with $\|D^{-1}\| = O(\epsilon^{-1})$.

How should ϵ be chosen? The limiting factor will be

$$\text{cond}(A_1 - BD^{-1}C) \tag{4.9}$$

It seems natural to require that (4.9) should be no worse in the order of magnitude sense than is the case with $D^{-1} = 0$. This means that for the Poisson, complementary elasticity theory and similar problems, $\epsilon = O(N^{-1})$ while for the various flow problems governed by the Navier-Stokes equations and its various linearizations, $\epsilon = O(1)$. In each of these cases, the expectation is that the required number of iterations of (4.5)-(4.6) is independent of N. This is discussed in a forthcoming report by the author which will also contain numerical results. Related solution techniques are discussed in [1], [2], and [15].

ACKNOWLEDGMENTS

I would like to thank Jim Boland, George Fix, and Max Gunzburger for direct and indirect help with the work reported above.

REFERENCES

1. Axelsson, O., "Preconditioning of Indefinite Problems by Regularization," SIAM J. Num. Anal. 1, 58 (1979).
2. Bercovier, M., "A Family of Finite Elements with Penalization for the Numerical Solution of Stokes and Navier-Stokes Equations," Proc. IFIP 1977, p. 97, North-Holland Publishing Co., (1977).
3. Brezzi, F., "On the Existence, Uniqueness and Approximation of Saddle Point Problems Arising from Lagrangian Multipliers," R.A.I.R.O. 8-R2, 129 (1974).
4. Chandra, R., Eisenstat, S. C., and Schultz, M. H., "Conjugate Gradient Methods for Partial Differential Equations," in "Advances in Computer Methods for Partial Differential Equations" (R. Vichnevetsky, ed.) p. 60, AICA, Bethlehem, Pennsylvania, (1975).
5. Concus, P., Golub, G. H., and O'Leary, D., "A Generalized Conjugate Gradient Method for the Numerical Solution of Elliptic Pole," in "Sparse Matrix Computations" (James R. Bunch and D. J. Rose, eds.) Academic Press, (1976).
6. Fix, G. J., M. D. Gunzburger, and Nicolaides, R. A., "On Mixed Finite Element Methods for First Order Elliptic Systems," Num. Math. (to appear).
7. Fix, G. J., Gunzburger, M. D., and Nicolaides, R. A., "Theory and Application of Mixed Finite Element Methods," in "Constructive Approaches to Mathematical Models" (G. J. Fix and C. V. Coffman, eds.) p. 375, Academic Press, Inc., (1979).
8. Fix, G. J., Gunzburger, M. D., and Nicolaides, R. A., "On Mixed Finite Element Methods for a Class of Nonlinear Boundary Value Problems," in "Computational Methods in Nonlinear Mechanics" (J. T. Oden, ed.) p. 245, North-Holland Publishing Co., (1980).
9. Gartling, D. K., Nickell, R. E., and Tanner, R. I., "A Finite Element Convergence Study for Accelerating Flow Problems," Int. J. Num. Meth. Eng. 11, 1155 (1977).
10. Golub, G. H., and Vargh, R. S., "Chebychev Semi-iterative Methods, Successive Overrelaxation Iterative Methods and Second Order Richardson Iterative Methods I and II," Num. Math. 3, 147 (1961).
11. Griffiths, D. F., "The Construction of Approximately Divergence Free Finite Elements, in "The Mathematics of Finite Elements and Its Applications III" (J. R. Whiteman, ed.) 237, Academic Press, Inc., (1979).
12. MacCamy, Journal Math. Anal. and Appl. (to appear).
13. Nicolaides, R. A., "Existence, Uniqueness and Approximation for Generalized Saddle Point Problems," SIAM Num. Anal. (to appear).

14. Raviart, P. A., and Girault, V., "Finite Element Approximations for the Navier-Stokes Equations," Lecture Notes in Math., Springer, (1979).
15. Segal, A., "On the Numerical Solution of the Stokes Equations Using the Finite Element Method," Comp. Meth. Appl. Mech. Eng. 19, 165 (1979).
16. Temam, R., Navier-Stokes Equations, North-Holland Pub. Co., (1977).
17. Varga, R. S., Matrix Iterative Analysis, Prentice-Hall, (1962).
18. Young, D. M., Iterative Solution of Large Linear Systems, Academic Press, Inc., (1971).

ELLPACK: PROGRESS AND PLANS

John R. Rice[1]

Mathematical Sciences
Purdue University
West Lafayette, Indiana

I. INTRODUCTION

This paper presents the origins and objectives of the ELLPACK cooperative effort, summarizes the progress made to the summer of 1980 and outlines the plans for future work. Two aspects of the effort are emphasized: the software performance evaluation results obtained and a new "structure" for ELLPACK which preserves the current facilities and facilitates a variety of extensions to these facilities.

ELLPACK started in 1975 when G. Birkhoff proposed a cooperative effort between people at Harvard (Birkhoff), Purdue (Lynch, Rice) and University of Texas (Kincaid, Young) to systematically test some ideas and eventually produce a modest, but useful, software facility for elliptic problems. This idea was developed and by the summer of 1976 J. Ortega organized a small meeting at ICASE to explore broadening this effort. About a dozen people[2] met for two days to discuss this and a cooperative effort was agreed upon with me as coordinator.

---

[1]Sponsored by the United States Army under Contract Nos. DAAG29-75-C-0024 and DAAG29-80-C-0041 and National Science Foundation Grant MCS77-01408.
[2]G. Birkhoff, A. Brandt, G. Fix, A. George, G. Golub, J. Ortega, J. Rice, M. Schultz, R. Sweet, R. Varga, O. Widlund and D. Young.

ISBN 0-12-632620-7

ELLPACK's long term objective was stated as follows: "provide a tool for research in the evaluation and development of numerical methods for solving elliptic PDE's. Various software components can be interchanged and the resulting accuracies, efficiencies, etc., measured. A person interested in one aspect of the problem can use other people's software for those aspects where he has less interest and expertise."

II. THE ELLPACK SYSTEM AND MODULES

The approach selected for ELLPACK was to provide an umbrella ELLPACK system to which modules could be contributed by members of the cooperative group - or others. ELLPACK includes a simple problem statement language for elliptic problems, see Figure 1. Such a system is more than a convenience for everyone involved, it provides an important facility to concisely and permanently identify runs made for performance evaluation. ELLPACK is designed to allow anyone to verify performance evaluation results on his own computer. Thus it is easily transported and has been used on a wide variety of computers; it is available on magnetic tape from IMSL, Inc. for a nomial handling charge. More details about this system are given in [Rice, 1977].

This umbrella system allows elliptic problem solving modules to be added with a minimum of effort. The current list of modules in ELLPACK is given in Table I. In addition there are numerous modules which have been put into ELLPACK and since replaced by better ones. All of these are in the area of solving linear systems of equations by direct or iterative methods. This retirement process is consistant with

```
*                    JULY TEST OF ELLPACK 78 NO. 2
*                    KASE 17 OF DOMAIN TEST SET
*           REGION FROM HALL,LUCZAK,SERDY - TOMS 2(1976) 257-274
*
*
EQUATION.   LAPLACE  $ UXX$ + UYY = 0  $  SELF-ADJOINT
*
OPTIONS.  TIME
DOMAIN. 1. TO 2.  LINE  0,0 TO  0,23 TO 10.,23.
    3. X =XDOM3(P)  $  Y = YDOM3(P)    $ P = 0 TO 19.
    4. LINE  34,13 TO  34,7
    5. X = XDOM5(P) $  Y = YDOM5(P)    $ P = 0 TO 20.
    6. X = 12.-P    $  Y = 0.0         $ P = 0 TO 12.
*
BOUND.    1. U = TRUE(X,Y)
          2. U = TRUE(X,Y)
          3. U = TRUE(X,Y)
          4. U = TRUE(X,Y)
          5. U = TRUE(X,Y)
          6. U = TRUE(X,Y)
GRID.   UNIFORM X = 13, 0.0 TO 34.  $  UNIFORM Y =  9,  0.0 TO 24.
*
OUTPUT.     TABLE-BOUNDARY $ TABLE-DOMAIN $ PLOT-DOMAIN
DIS.        5-POINT STAR
INDEX.      NATURAL
DIS(2).     2DEPEP(NTRI = 47, MEM = 6500)
SOLUTION.   LINPACK BAND
OUTPUT(2).         $ MAX-ERROR $ TABLE(8,8)-U
OPTIONS.        TIME  $  MEMORY
*
SEQUENCE.   OUTPUT
            STOP
            DIS  $ INDEX  $  SOL  $  OUT(2)  $  DIS(2)  $  OUT(2)
*
*          FORTRAN FUNCTIONS FOR KASE 17
F.
      FUNCTION XDOM3(T)
      XDOM3 = 15.+ T
      IF( T .GE. 5.)              RETURN
         P1730 = AMAX1(AMIN1(T,1000.),-2.)
         XDOM3 = 20.-10.*COS(3.141592654*P1730*.1)
         RETURN
      END
      FUNCTION YDOM3(T)
      YDOM3 = 13.
      IF( T .GE. 5. )                       RETURN
         P1730 = AMAX1(AMIN1(T,1000.),-2.)
         YDOM3 = 23.-10.*SIN(3.141592654*P1730*.1)
         RETURN
      END
      FUNCTION XDOM5(T)
      XDOM5 = 34. - T
      X = 349 - 7
      IF( T .LE. 15. )                      RETURN
         P1755 = AMAX1(AMIN1(T,75.),-25.)
         XDOM5 = 19.-7.*SIN(3.141592654*(P1755-15.)*.1
         RETURN
      END
      FUNCTION YDOM5(T)
      YDOM5 = 7.
      IF( T .LE. 15. )                      RETURN
         P1755 = AMAX1(AMIN1(T,75.),-25.)
         YDOM5 = 7.*COS(3.141592654*(P1755-15.)*.1)
         RETURN
      END
      FUNCTION TRUE(X,Y)
      TRUE = X*Y + Y -2.
      RETURN
      END
      FUNCTION D3EST(X,Y)
      D3EST=1.0
      RETURN
      END
END.
```

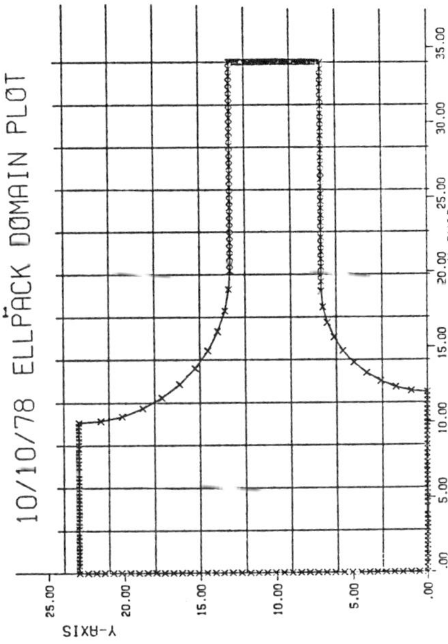

FIGURE 1. An ELLPACK language statement of an elliptic problem on a n x n -rectangular domain.

ELLPACK's objective of evaluating software, identifying weaknesses and developing improved software.

A special user interface has been prepared to allow ELLPACK to be used remotely via telephone connections. This network allows members of the cooperative group direct access to the latest versions of ELLPACK and its data base. The interface requires almost no knowledge of the Purdue operating system; it is also convenient for terminal use at Purdue.

TABLE I. The elliptic problem solving modules in the ELLPACK system as of June 1, 1980.

Name/Reference	Author	Brief Description
5-POINT STAR	R. Boisvert	DISCRETIZATION. The usual finite difference approximation for a general problem on a general domain
7-POINT 3D	ITPACK[a]	DISCRETIZATION. Finite difference approximation to self-adjoint problems on a rectangular 3D domain.
P3-C1 COLLOCATION	E. Houstis	DISCRETIZATION. Collocation with bicubic Hermite elements for a general problem on a rectangular domain.
P3-C1 GALERKIN	E. Houstis	DISCRETIZATION. Galerkin approximation with bicubic Hermite elements for a self-adjoint problem with homogeneous boundary conditions on a rectangular domain.
HODIE-HELMHOLTZ [Boisvert, 1979]	R. Boisvert	DISCRETIZATION. A 4th or 6th order difference approximation to $u_{xx} + u_{yy} + f(x,y)u = g(x,y)$ with Dirichlet conditions on a rectangular domain.
HODIE-ACF [Lynch and Rice, 1978]	R. Boisvert	DISCRETIZATION. A 4th or 6th order difference approximation to $au_{xx} + bu_{yy} + fu = g$ with Dirichlet conditions on a rectangular domain.

Name/Reference	Author	Brief Description
HODIE 27-POINT 3D [Lynch, 1977]	R. Lynch	DISCRETIZATION. A 6th order difference approximation to the Poisson equation with Dirichlet conditions on a cubical domain.
FFT 9-POINT [Houstis and Papatheodorou, 1979]	E. Houstis T. Paptheodorou	PROBLEM SOLVING. Uses 2nd, 4th, or 6th order difference approximation and FFT method for $au_{xx} + bu_{yy} + fu = g$ with Dirichlet conditions on a rectangular domain
2 DEPEP [Sewell, 1979]	G. Sewell	PROBLEM SOLVING. Galerkin approximation with 6-node triangular elements to self-adjoint problems on general domains.
MARCHING ALGORITHM [Bank, 1978]	R. Bank	PROBLEM SOLVING. Symmetric 5-Point difference approximation and generalized marching algorithm for self-adjoint, separable problems in rectangular domains.
DYAKANOV-CG [Bank, 1977]	R. Bank	PROBLEM SOLVING. Symmetric 5-Point difference approximation and conjugate gradient iteration for self-adjoint problem on a rectangular domain.
DYAKANOV-CG4 [Bank, 1977]	R. Bank	PROBLEM SOLVING. Richardson extrapolation of DYAKANOV-CG to obtain 4th order approximation.
NATURAL	R. Boisvert	INDEXING. Order of discretization module used.
RED-BLACK	ITPACK[a]	INDEXING. The red-black or checker board ordering for difference methods.
YALE MIN DEG [Eisenstat et al, 1977]	YALEPACK[b]	INDEXING. Applies minimal degree algorithm to $A + A^T$.
YALE RCM [Eisenstat et al, 1977]	YALEPACK[b]	INDEXING. Applies Reverse Cuthill-McKee algorithm.

TABLE I (Cont)

Name/Reference	Author	Brief Description
NESTED DISSECTION [Eisenstat et al, 1977]	YALEPACK[b]	INDEXING. The nested dissection ordering.
LINPACK BAND [Dongarra et al, 1979]	LINPACK C. Moler	SOLUTION. Gauss elimination with pivoting for a band matrix.
LINPACK SPD BAND [Dongarra et al, 1979]	LINPACK C. Moler	SOLUTION. Cholesky decomposition of a symmetric positive definite band matrix.
SPARSE GE-PIVOTING [Sherman, 1978]	A. Sherman	SOLUTION. Sparse Gauss elimination
YALE SPARSE [Eisenstat et al, 1977]	YALEPACK[b]	SOLUTION. Sparse factorization of a symmetric or general matrix, plus solution of triangular system.
YALE ENVELOPE [Eisenstat et al, 1977]	YALEPACK[b]	SOLUTION. Matrix is put in envelope form and a sparse LU factorization made.
SOR [Grimes et al, 1979]	ITPACK[a]	SOLUTION. Iterative solution of linear system by SOR methods with automatic parameter selection.
JACOBI SI [Grimes et al, 1979]	ITPACK[a]	SOLUTION. Iterative solution by Jacobi method with semi-iteration acceleration.
JACOBI CG [Grimes et al, 1979]	ITPACK[a]	SOLUTION. Iterative solution by Jacobi method with conjugate gradient acceleration.
SYMMETRIC SOR SI [Grimes et al, 1979]	ITPACK[a]	SOLUTION. Iterative solution by symmetric SOR with semi-iterative acceleration.
Symmetric SOR CG [Grimes et al, 1979]	ITPACK[a]	SOLUTION. Iterative solution by symmetric SOR with conjugate gradient acceleration.
REDUCED SYSTEM SI [Grimes et al, 1979]	ITPACK[a]	SOLUTION. Iterative solution by reduced system method with semi-iterative accleration.

REDUCED SYSTEM CG [Grimes et al, 1979]	ITPACK[a]	SOLUTION. Iterative solution by reduced system method with conjugate gradient acceleration.

[a]The ITPACK project principal authors are R. Grimes, D. Kincaid and D. Young.
[b]The Yale sparse matrix project principal authors are S. Eisenstat, M. Schultz and A. Sherman

III. PERFORMANCE EVALUATION METHODOLOGY

One of ELLPACK's goals is to provide scientifically sound performance evaluations. The computational measurement system must meet the basic principles of scientific experiments:

Explicitly stated assumptions and hypotheses.

Objective criteria and measurements.

Reproducibility of results.

Access to actual experimental data.

Complete description of the experimental apparatus.

The general methodology for implementing these principles is discussed in [Rice, 1980] and its application to the case of elliptic problem solvers is detailed in [Houstis and Rice, 1980].

It is not easy nor straight forward to use a sound methodology; almost all computational experiments with software fail to be satisfactory. The ELLPACK attempt uses a software system explicitly designed for this purpose, see [Boisvert, Rice and Houstis, 1979]. The basic organization is shown in Figure 2. Typical results produced by this system are seen in the discussion of the next section.

A cruical element of experimental evaluations is a satisfactory population of problems. The population given by [Rice, Houstis and Dyksen, 1980] has 56 PDE operators, most

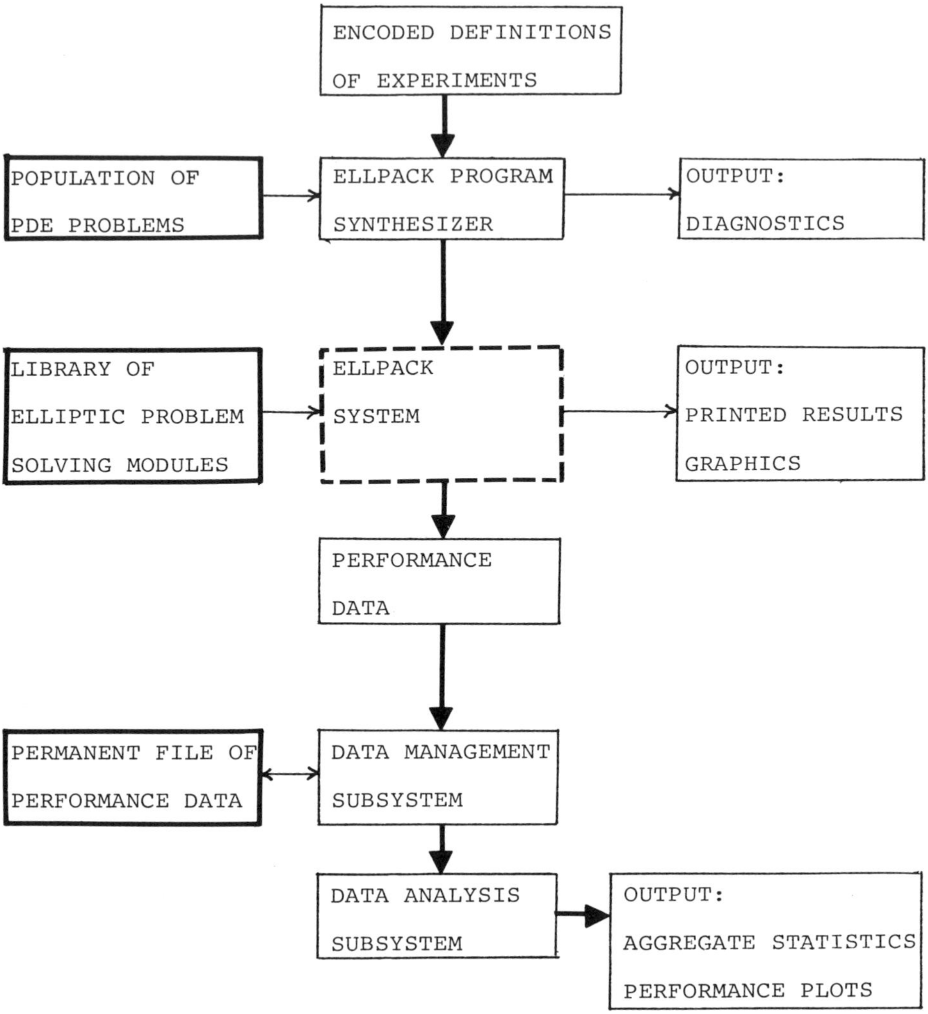

FIGURE 2. Block diagram of the ELLPACK system for performance evaluation of elliptic problem solvers.

have parameters, and the population in current use has 189 specific elliptic problems. An excerpt from this population is shown in Figure 3. A serious attempt was made to have this population represent the spectrum of elliptic problems: from easy to hard, from artificial to real world, from smooth to singular, from the Laplacian to general operators, etc. It was exhausting to locate or generate this number of PDE's to represent such a large domain. Even so, the population of

PROB 45 Nonlinear pth order reaction [20]

$$u_{xx} + u_{yy} - \alpha r^{\beta-1} u = 0$$

DOMAIN unit square
BC $u = g$
TRUE Unknown

Operator: Helmholtz type, homogeneous.
Right side: Zero
Boundary conditions: Dirichlet
Solution: Approximate solution given for $r = r(x,y)$ tabulated from a solution to the nonlinear problem; r should be u.
Parameters: α and β are physical parameters. Three cases are given: (1,2), (1,1000) and (2,2).

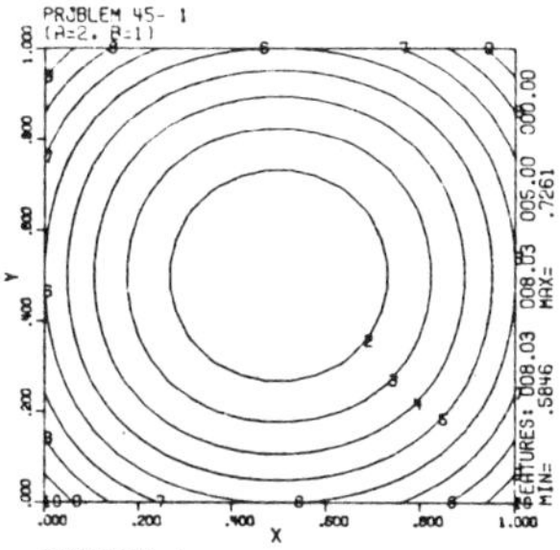

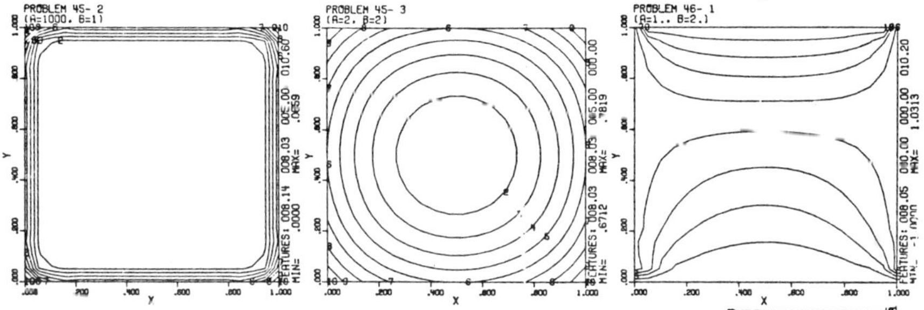

PROB 46 Magnetohydrodynamics [19]

$$u_{xx} + u_{yy} - \beta u_y = 0$$

DOMAIN $[0,\alpha] \times [0,1]$
BC $u=0$ for $x=0,\alpha$; $u=1$ for $y=1$; $u=-1$ for $y=0$
TRUE Unknown

Operator: Constant coefficients, homogeneous.
Right side: Zero
Boundary conditions: Dirichlet
Solution: Approximate solution given for 4 cases: $\alpha = 1$ and $\beta = 2$, $\alpha = 4$ and $\beta = 2$, $\alpha = 4$ and $\beta = 10$, $\alpha = 8$ and $\beta = 2$.

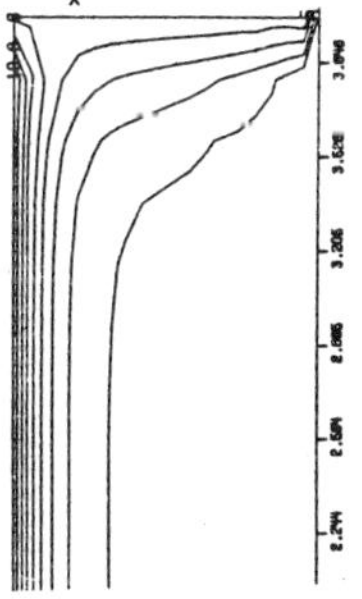

FIGURE 3. An excerpt of the PDE population for the ELLPACK system. The machine readable form of this population requires consists of well over 100 pages.

problems is not overly large. Statistical methods of analysis normally require a substantial amount of data and one of the most common weaknesses of computational experiments is that too little data is obtained to justify statistically significant conclusions.

IV. PERFORMANCE EVALUATIONS

This section discusses the application so far of ELLPACK to its principal objective: The comparative evaluation of elliptic problem solvers. Only summary or typical results are given due to space limitations.

A. Evaluation 1: Ordinary Finite Differences verses P3-C1 Collocation

The study [Houstis et al, 1978] concluded that collocation using bicubic Hermite polynomials (P3-C1 COLLOCATION) is superior to ordinary finite differences (5-POINT STAR). More specifically, it was concluded that 5-POINT STAR starts out at rough accuracy with better efficiency, then there is a crossover point where P3-C1 COLLOCATION becomes better. This crossover point is at low accuracy and thus 5-POINT STAR was more efficient only on problems requiring modest computational power. This study used a sample of 17 PDE of various types, including 5 with NON-rectangular domains.

We have repeated this evaluation with 48 PDEs on rectangular domains divided into two groups:

(1) smooth 2-1, 4-1, 5-2, 5-4, 10-1, 11-1, 12-3, 22-1, 25-4, 26-1, 30-2, 31-1, 32-1, 36-1, 38-1, 40-1, 42-1, 42-2, 45-1, 52-1

(2) not smooth 3-1, 7-1, 9-1, 9-2, 10-2, 10-3, 10-4, 10-7, 11-2, 11-3, 11-4, 15-1, 15-2, 17-1, 17-2, 20-1, 20-2, 30-4, 30-8, 34-1, 35-1, 36-2, 39-4, 44-2, 44-3, 47-2, 51-1, 54-1

Table II gives the average ranks in four measures of performance: slope of total time versus error, total time to achieve error levels of 5%, 0.5% and 0.05%. The confidence levels for the significance of the differences are given if they exceed the 90% level.

TABLE II: Average Ranks and Significance Levels for total time versus error at the nodes in comparing P3-C1 COLLOCATION (Col.) and 5-POINT STAR (FD)

Performance Measure	Group 1 Rank Col.	Group 1 Rank FD	Group 1 Sig. level	Group 2 Rank Col.	Group 2 Rank FD	Group 2 Sig. level	Combined Rank Col.	Combined Rank FD	Combined Sig level
Slope	1.05	1.95	99%	1.25	1.75	99%	1.17	1.83	99%
5% Error Level	1.67	1.33	-	1.79	1.21	99%	1.73	1.27	99%
0.5% Error Level	1.33	1.67	-	1.39	1.61	-	1.35	1.65	95%
0.05% Error Level	1.05	1.95	99%	1.29	1.71	95%	1.17	1.83	99%

This data support the conclusion of the earlier study with a high level of confidence (confidence levels are lower for Group 1 primarily because it has fewer PDEs).

B. Evaluation 2: The Hodie Method versus ordinary finite differences and collocation

The study in [Lynch and Rice, 1978a] compares the 9-point Hodie high order difference approximation using either Gauss elimination or FFT with ordinary finite differences and bicubic Hermite collocation. Thirty problems were used in that study, all of which are in the PDE population of the ELLPACK system. Thus we can apply statistical analysis to the data to quantitatively asses the comparative performance of these methods. The conclusion reached is that the Hodie method outperforms both ordinary finite differences and collocation. A different study [Boisvert, 1979] reaches similar conclusions.

The results for 9-point Hodie (HODIE-HELMHOLTZ) versus 5-POINT STAR and P3-C1 COLLOCATION are shown below. Eighteen

TABLE III. Average Ranks[a] and median values for total time versus error at the nodes.

Problems: 3-1, 3-2, 4-1, 6-1, 8-2, 9-2, 10-2, 10-3, 11-2, 11-4, 17-1, 17-3 20-2, 34-1, 35-1, 47-1, 47-2, 47-3

Method	Slope		5% error		0.5% error		0.05% error	
	Rank	Median	Rank	Median	Rank	Median	Rank	Median
5-POINT STAR	2.78	-.66	2.00	.036	2.56	1.1	2.61	35
P3-C1 COLLOCATION	1.94	-1.3	2.72	.1	2.22	.74	2.22	5.3
HODIE-HELMHOLTZ	1.28	-1.5	1.28	.0053	1.22	.058	1.17	.25

[a]Average rank differences are significant with 90% (99%) certainty if greater than 0.68 (0.97).

PDE's are used as the Hodie method is implemented in two forms in ELLPACK, depending on the nature of the operator.

C. Evaluation 3. Direct versus iterative methods for the Galerkin equations

It is widely believed (though not conclusively demonstrated) that iterative methods are eventually (if the grid is fine enough) more efficient in solving the linear system obtained from 5-POINT STAR. A study [Rice, 1980] made with the ELLPACK system shows that the same holds for solving the linear system obtained from P3-C1 GALERKIN. A typical performance plot is shown in Figure 4. This study was unusual in that not only were the results statistically significant with a high level of confidence, but, in every case, the slope of work versus accuracy was steeper for iterative methods. This implies that there is some level of accuracy (usually

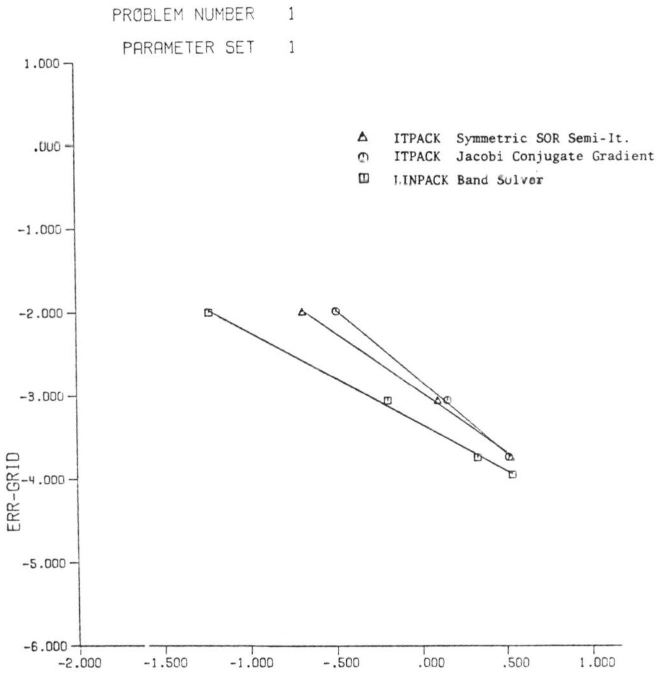

FIGURE 4. A typical performance plot comparing iterative and direct methods for the solution of the Galerkin equations. The plot is log of PDE error versus log of seconds to solve a linear system.

corresponding to about 300-500 unknowns) where iterative methods become more efficient. Of course, when the linear system gets very large then the sparse direct methods such as in YALEPACK should be used instead of the LINPACK band matrix solver.

D. Evaluation 4. High Order Methods versus Low Order Methods for Singular Problems

An extensive study [Houstis and Rice, 1980a] has been made to compare the effect of order on the performance of elliptic problem solvers. It is widely believed that high order methods are less effective for singular or wildly behaving problems because, in the cases of practical interest, the higher order contributes "instabilities" and larger errors rather than smaller errors. This study involves three modules (5-POINT STAR; FFT9, IORDER=2; DYAKANOV-CG) based on the same second order discretization, four modules (FFT9, IORDER=4; DYAKANOV-CG 4; HODIE-HELMHOLTZ; P3-C1 COLLOCATION) using various fourth order discretizations and two modules (FFT9, IORDER=6; HODIE-HELMHOLTZ, METHOD=60) using sixth order discretizations. Various comparisons were made among these methods to evaluate the effect of order. The conclusion is unequivocable: <u>higher order methods are more effective for solving singular or nearly singular linear, elliptic problems in 2-dimensional rectangular domains</u>.

E. Evaluation 5. Compiler and Machine Effects

The study [Rice, Dyksen and Ward, 1980] considers the effects of compilers and machines on the performance of ELLPACK software. While numerical analysts are primarily interested in the effect of using different methods, it is important to know the level of these other effects so as to set a level of difference in performance that is significant across a range of computing environments. Thus if changing compilers or machines can make an unpredictable 50% difference in the relative performance of software (see [Parlett and Wang, 1975] for examples of such large differences) then one can ignore somewhat an observed 20-30% difference in the relative performance of two methods. There is, of course, also an inherent interest in the relative performance of compilers and machines.

At the time of this writing this study is incomplete; a preliminary analysis suggests that:

A. Changing machines can have a significant effect on the relative performance of different modules. Extreme cases may be as large as 500%, even though the average effect is much smaller, perhaps 25%.

B. The CDC 6600 executes ELLPACK software at about 1.5 to 2 times the rate of the CDC 6500; the CDC 6500 executes ELLPACK software at about 1.5 the rate of the VAX (without speeded up floating point).

Table IV shows the ratios of execution times for a few modules on some runs for selected combinations of machines and compilers. These ratios are reasonably stable except those involving discretizations on the VAX; the UNIX Fortran is terrible in its evaluation of elementary functions and this work might or might not dominate the discretization time.

TABLE IV. Ratios of execution times (in parentheses after the module timed) for three combinations of machines and compilers.

CDC6500 + FTN/CDC6600 + FTN	
5-POINT STAR (.75)	YALE SPARSE (.63,.70,.71)
LINPACK BAND (.60,.66.,58)	P3-C1 GALERKIN (.47)
CDC6500 + FTN/CDC6500 + MNF	
5-POINT STAR (.84)	YALE SPARSE (.73,.71,.69)
LINPACK BAND (.69,.58,.66)	P3-C1 GALERKIN (.86)
VAX + UNIX/6500 + MNF	
5-POINT STAR (.81,.27,.35,.2,.32)	YALE SPARSE (.73,.65,.73)
LINPACK BAND (.68,.76,.66,.67,.62)	P3-C1 COLLOCATION (.55,.31,.33)
JACOBI CG (.55,.75.,78)	

F. Evaluation 6: Effect of Short Word Length Machines

The study [Dyksen and Rice, 1980] is to evaluate the effect of using a 32-bit machine to solve elliptic problems. The limiting accuracy in a 32-bit machine (such as IBM 360/370 or VAX single precision) is about 6 decimal digits so one could not expect to obtain more than 5 or 6 correct digits in solving a numerical problem. Most elliptic problems are solved to 2 or 3 decimal digits accuracy and the question is: Can one expect to obtain 3 correct decimal digits in solving elliptic problems on a 32-bit machine?

The experiment is simple, we just solve the problems on a 32-bit VAX and compare the results with a 60-bit computation on a CDC-6500. The preliminary results are striking: for about 10 problems and all ELLPACK modules, no significant round off magnification has been observed on the 32-bit machine. These experiments involve linear systems of order up to 3000 for direct methods and 65,000 for FFT methods. They suggest that there is no need to use a word length much longer than that needed to represent the final accuracy requirement in solving the PDE.

Figure 5 shows the results for three methods applied to three different problems. Case (a) shows no effect of the short word length for ordinary finite differences up to a 57 x 57 mesh. Case (b) shows the Hodie high order difference method quickly achieves the limiting accuracy and thereafter the round-off effects are magnified exponentially. Case (c) shows an FFT method that achieves the limiting accuracy with a 65 x 65 mesh and then little change takes place up to a 256 x 256 mesh.

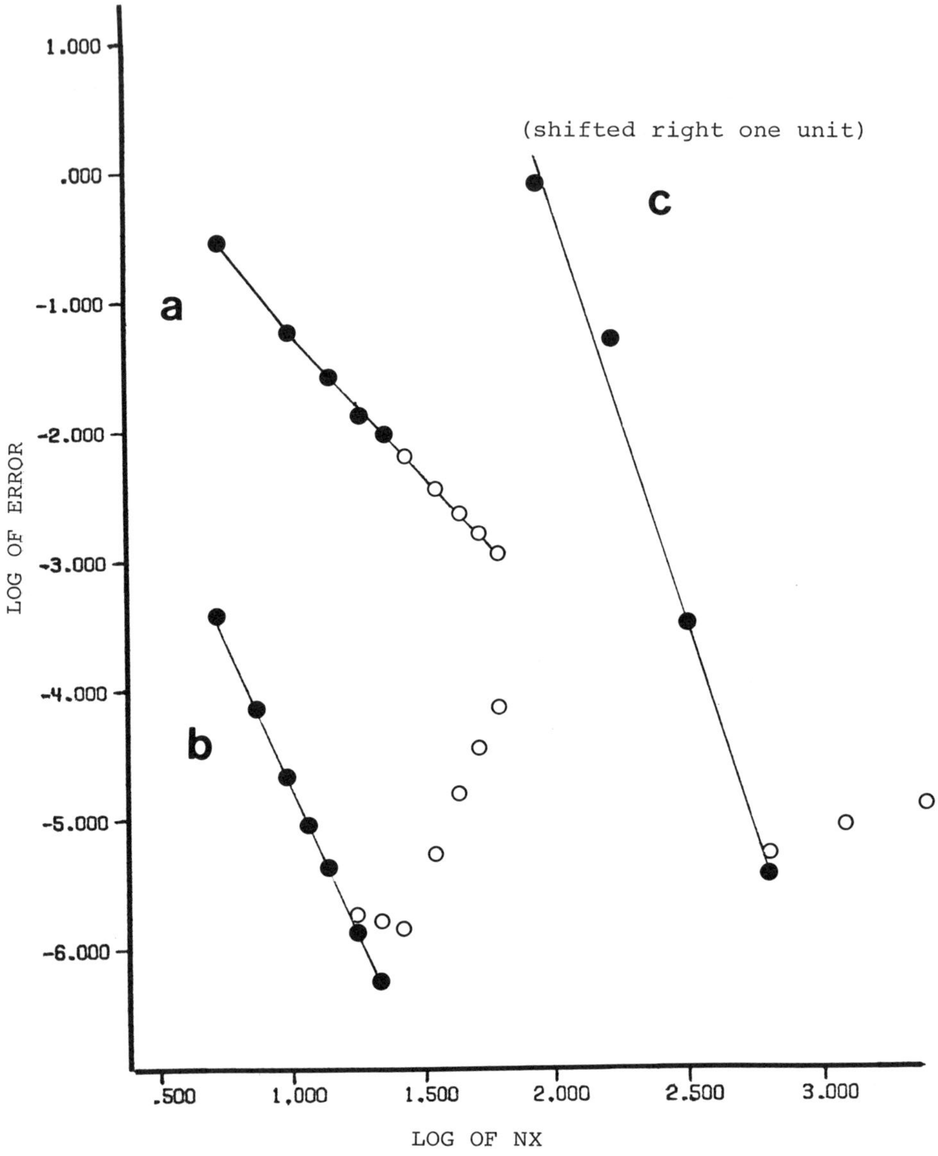

FIGURE 5. Comparison of results on a 32-bit machine (VAX - open circles) and a 60-bit machine (CDC6500 - solid circles) for (a) 5-POINT STAR + LINPACK BAND on PDE 6-1, (b) HODIE-HELMHOLTZ + LINPACK BAND on PDE 4-1 and (c) FFT 9-POINT (IORDER=6) on PDE 10-4.

IV. PLANS FOR A "CORE" ELLPACK

The ELLPACK group has decided to define a "core" ELLPACK that will be a stable system for the indefinite future. Stable means that facilities will remain available in later versions. This system will appear essentially the same as the present ELLPACK system from the outside. There may be some minor changes in options or similar items and a modest capability for non rectangular 3-dimensional geometry may be added. There will be comprehensive, semi-permanent documentation of this system for the user and for the researcher.

Internally, the new system will be quite different. The language processor will be rewritten to make it more transparent and to simplify the mechanism for adding modules to ELLPACK. Currently, adding a module requires adding 3-10 lines of FORTRAN to the preprocessor, recompiling several routines and substituting them into the ELLPACK system. It is quite straight forward, but still a tedious chore. The new language processor will be completely table driven as far as modules go. To add (or delete) a module, one just changes the tables and adds (or deletes) the corresponding module to the library. The entire library could be replaced without changing the language processor.

A second, stronger motivation for changing the internal structure of the system is to facilitate various extensions of ELLPACK. This structure is presented in detail in a later section. Various members of the group are interested in extensions like nonlinear problems, time dependent problems, simultaneous equations, multiple domain methods, etc. It is not feasible to continuously modify the system to accomodate more and more facilities. It is hoped that this approach will make it easy for interested parties to make the extensions they desire. The ELLPACK group does not take on the responsibility of making any of these extensions; it is possible that at some future date the successful extensions would be merged into a new system under the auspices of the ELLPACK group.

VI. THE FUTURE ELLPACK STRUCTURE

The basic principle of the ELLPACK design is to define interfaces between problem solving modules and make them the fixed parts of the structure. Then problem solving modules can do whatever they want between interfaces and even skip some of them. In computer science terms, the system consists of a standard set of data structures (the interfaces) and the numerical methods modules map one data structure to another.

The new system follows the same principle; it just envisages more data structures. A block diagram of this new structure is given in Figure 6. The current ELLPACK structure is separated into two parts: (A) fixed information about geometry and the PDE, (B) the current ELLPACK structure for solving a PDE problem on a given domain. In addition, there may be associated PDE problems, domains and solutions. These are called MESH and SOL in the diagram. The two new module types may be called STEP and MAP. Both modules transform a given MESH and SOL pair to another. STEP modules keep the underlying domain and grid fixed (e.g. as in time dependent problems or iteration methods) while MAP also changes the underlying geometry or grid. Two potential applications follow to illustrate the possibilities; these are quite

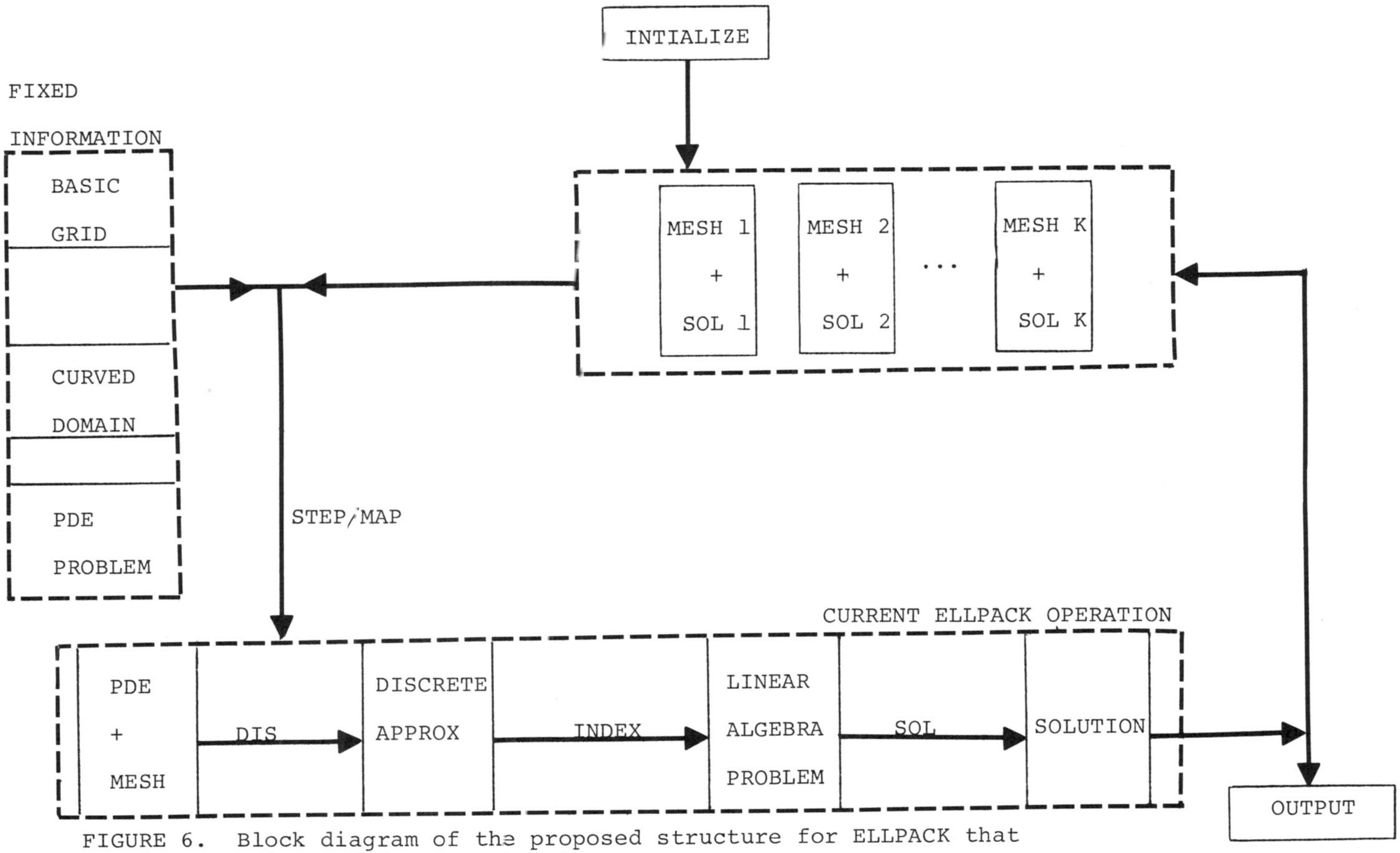

FIGURE 6. Block diagram of the proposed structure for ELLPACK that permits a variety of future extensions.

preliminary and should not be interpreted as specific plans.

The terms that appear in Figure 5 are defined as follows:

BASIC GRID	= Rectangular grid containing all other grids. ~~May also~~ have associated boundary conditions.
CURVED DOMAIN	= Curved boundary pieces plus associated boundary conditions.
PDE PROB	= Operator given by EQUATION.
MESH k	= A rectangular grid or a rectangular grid plus curved domain may have associated boundary conditions or PDE operator.
SOL k	= A "SOLUTION" associated with mesh k.
STEP	= A module type that carries out a step in: (a) Time, (b) A nonlinear method or (c) An iteration.
MAP	= A module type that transfers a "SOLUTION" from one mesh to another, creating (possibly) a new PDE.

Application 1: Nonlinear elliptic PDE. For simplicity 1-dimensional problem is considered.

```
EQUATION.     UXX + A(X,U)UX + B(X,U) = 0  $  NONLINEAR
BOUNDARY.     X = 0, U = BC(X,Y)  $  X = 1, U = BC(X,Y)
              Y = 0, U = BC(X,Y)  $  Y = 1, U = BC(X,Y)
INITIALIZE.   U = EXP(X,Y)        $   SOL1 = 1.05 * EXP(X,Y)
GRID.         UNIFORM X = 10

STEP.         SECANT
DIS.          5-POINT STAR
INDEX.        NATURAL
SOL.          LINPACK BAND

SEQUENCE.     LOOP = 100
              STEP
              DIS  $  INDEX  $  SOL
              TEST
```

Explanation of Module Actions:

```
Step: 1) Move  SOL1  →  SOL2,  U  →  SOL1

      2) Compute AU(X,U) = (A(SOL1,X) - A(SOL2,X))/(SOL1-SOL2)
                 BU(X,U) = (B(SOL1,X) - B(SOL2,X))/(SOL1-SOL2)

      3) Compute R(X) = residual of U = UXX + A(X,U)UX + B(X,U)

      4) Create PDE Problem:
      VXX + AU(X,U) * VX + (AU(X,U) * UX + BU(X,U)) * V = -R(X)

TEST: 1)  U = U + V

      2)  Stop if ||V|| is small
```

<u>Application 2:</u> Coordinate Transformation Methods

```
EQUATION.           UXX + UYY = F(X,Y)
DOMAIN.         1.  X=P      $  Y=0             $  P=0  to  1
                2.  X=1+P    $  Y=P             $  P=0  to  1
                3.  X=2-P    $  Y=1+P*(3.-P)    $  P=0  to  3
                4.  X=-1+P   $  Y=1-P           $  P=0  to  1
BOUNDARY.       1.  U=0
                2.  U=Y
                3.  U=1
                4.  U=Y
BOUNDARY(2).        X = -1, U=1  $  X=2,  U=1
                    Y = 0 , U=0  $  Y=3.5,U=1
GRID.               UNIFORM X=11,  UNIFORM Y=11

DIS.                5-POINT STAR
INDEX.              NATURAL
SOL.                LINPACK BAND
MAP(1).             PDE TRANSFORM
MAP(2).             INVERT U

SEQUENCE.           DOMAIN
                    MAP(1)
                    DIS  $  INDEX  $  SOL
                    MAP(2)
```

Explanation of Module Actions:

PDE TRANSFORM — Generates mapping of domain onto grid. Then makes this change of variable numerically in the PDE.

INVERT U — Takes solution defined on rectangular image and maps it back to original domain.

VII. FUTURE PERFORMANCE EVALUATION PLANS

The new core ELLPACK will be a stable system, but this does not mean that more modules cannot be added. In fact, several more modules are expected to be added in the near future and other modules are invited. Thus a continuous program of performance evaluation is foreseen. There are many investigations with the present set of modules yet to be made. Examples include

A. How do residuals behave, can accuracy be reliably estimated from a study of residuals?

B. How do first order terms affect various methods?

C. What methods are most effective for problems with oscillatory solutions?

D. What is the effect of highly non-rectangular geometry on the linear equation solution modules? (Gauss elimination, iterative methods or sparse matrix methods)

E. How do boundary conditions involving derivatives affect the performance of methods?

F. How radical is the change in relationship between linear equation solution methods as one goes from 2 to 3 dimensions?

The data base facility must be extended in some way to accomodate data from different machines. In addition to the obvious advantages of this, there is the fact that the CDC 6500 being used by the ELLPACK performance evaluation system is a disappearing breed. It is straight forward to change machines and start a new data base, but that wastes the previous data. A master data base needs to be created which contains all the data on an "equivalent" basis and yet allows the original raw data to be retrieved.

VIII. NON RESEARCH APPLICATIONS OF ELLPACK

There has been a deliberate effort by the ELLPACK group to avoid becoming obligated to producing a user oriented system. The reason for this is that good user oriented software requires a higher level of robustness, documentation, etc. than the group wants to commit itself to. On the other hand, it is obvious that the ELLPACK system provides a convenient tool to solve a variety of real problems. The study [Rice, 1979a] suggests a startling improvement in problem solving efficiency in using ELLPACK compared to normal FORTRAN programming. It is not clear that it is reasonable for a research oriented group to produce user oriented software systems (even though some instances of this exist).

Ignoring for the moment whether ELLPACK can or will produce a robust, user oriented system, we consider the modes of use one would expect for such a system.

1. Educational. Several people within the ELLPACK group have used it in conjunction with courses. It provides an easy means for students to try out methods without having to spend long hours on implementations. In such an environment it would be useful to have modules available for "simple" methods which might not be interesting for real problems.

2. Real Problem Solving. One usually hears most about the really tough problems, the ones that require many man years of analytical and programming effort and run for hours on hugh machines. Yet there are also a lot of more modest

problems that need to be solved. I believe there is a real need for software tools that can realiably solve modest problems with standard techniques. Thus the routine problems can be handled quickly and the scientist can concentrate his attention on the more difficult ones. ELLPACK could well provide such a tool for elliptic problems.

3. Pilot Studies. When one invests 10 or 20 man years in PDE software, one is usually frozen into the basic numerical method fairly early in the project. One can use ELLPACK to evaluate various methods for simpler problems similar to the actual application. This methodology is standard in many areas; it makes good sense in software development.

4. Starting Point for More Complex Problems. A variation on the pilot study approach is to pose a problem to ELLPACK that is as close as possible to the real problem and then use the FORTRAN program that ELLPACK generates as a starting point for the real problem. This approach has proved quite satisfactory in several instances (see [Gherson, Lykoudis and Lynch, 1978] for an example). This approach requires one to become involved in the details of FORTRAN programming and some aspects of the ELLPACK system, yet it has involved much less work than writing all the programs from scratch.

REFERENCES

Bank, R.E. (1978). Algorithm 527. A FORTRAN implementation of the generalized marching algorithm, ACM Trans. Math. Software, 4, pp. 165-176.

Boisvert, Ronald R. (1979). High order finite difference for elliptic boundary value problems, Ph.D. Thesis, Purdue University.

Boisvert, R.F., J.R. Rice and E.N. Houstis (1979). A system for performance evaluation of partial differential equations software, IEEE Trans. Software Eng., 5, pp. 418-425.

Dongarra, J.J., J.R. Bunch, C.B. Moler and G.W. Stewart (1979). LINPACK User's Guide, SIAM, Philadelphia, Pa.

Dyksen, W.R. and J.R. Rice (1980). Effect of short word length machines on elliptic PDE software, CSD-TRxxx, Computer Science, Purdue University.

Eisenstat, S.C., M.C. Gursky, M.H. Schultz and A.H. Sherman (1977). Yale Sparse matrix Package: I. The Symmetric Codes, Rpt. 112, II. The Nonsymmetric Codes, Rpt. 114, Computer Science, Yale University.

Gherson, P., P.S. Lykoudis and R.E. Lynch (1978). The use of ELLPACK 77 for solving the Laplace equation on a region with interior slits, application to a problem in magnetohydrodynamics, CSD-TR 275, Computer Science, Purdue University.

Grimes, R.G., D.R. Kincaid and D.M. Young (1979). ITPACK 2.0 Users Guide, Center for Numerical Analysis, University of Texas.

Hageman, L.A., and D.M. Young (1980). Applied Iterative Methods, Academic Press, to appear.

Houstis, E.N., R.E. Lynch, T.S. Papatheodorou and J.R. Rice (1978). Evaluation of numerical methods for elliptic partial differential equations, J. Comp. Physics, 27, pp. 323-350.

Houstis, E.N. and T.S. Papatheodorou (1979). Algorithm 543, FFT9, Fast solution of Helmholtz-type partial differential equations, ACM Trans. Math. Software, 5, pp. 490-493.

Houstis, E.N. and J.R. Rice (1980). An experimental design for the computational evaluation of elliptic partial differential equation solvers, in Production and Assessment of Numerical Software (M. Hennell and L. Delves, ed.) Academic Press, pp. 55-65.

Houstis, E.N. and J.R. Rice (1980a). The effectiveness of higher order methods for solving singular elliptic problems, CSD-TR 341, Computer Science, Purdue University.

Kincaid, D.R., and D.M. Young (1978), "Survey of Iterative Methods," CNA-135, Center for Numerical Analysis, University of Texas, Austin. Also in Encyclopedia of Computer Sciences and Technology, Vol. 13, J. Belzer, A.G. Holzman, and A. Kent (eds.), Dekker Inc., New York, 1979, pp. 353-590.

Lynch, R.E. (1977). $O(h^6)$ accurate finite difference approximation to solutions of the Poisson equation in three variables, CSD-TR 221, Computer Science Dept. Purdue University.

Lynch, R.E., and J.R. Rice (1978). High accuracy finite difference approximations to solutions of elliptic partial differential equations, Proc. Nat. Acad. Sci. 75, pp. 2541-2544.

Lynch, R.E. and J.R. Rice (1978a). The Hodie method and its performance for solving elliptic partial differential equations, in Recent Advances in Numerical Analysis (C. de Boor, ed.), Academic Press, pp. 143-175.

Parlett, B.N. and Y. Wang (1975). The influence of the compiler on the cost of mathematical software - in particular on the cost of triangular factorization, ACM Trans. Math. Software, 1, pp. 35-46.

Rice, John R. (1977). ELLPACK, A research tool for elliptic partial differential equations software, in Mathematical Software III (J. Rice, ed.) Academic Press, pp. 319-341.

Rice, J.R. (1979). Methodology for the algorithm selection problem, in Performance Evaluation of Numerical Software (L. Fosdick, ed.) North-Holland, pp. 301-307.

Rice, J.R. (1979a). Programming effort analysis of the ELLPACK language, SIGNUM Newsletter, 14, pp. 109-111.

Rice, J.R. (1980). The effectiveness of iteration for solving the Galerkin method equations, CSD-TRxxx, Computer Science, Purdue University.

Rice, J., Dyksen, W. and Ward, W. (1980). Machine and compiler effects on ELLPACK software performance, CSD-TRxxx, Computer Science, Purdue University.

Rice, J.R., E.N. Houstis and W.R. Dyksen (1980). A population of second order, linear elliptic partial differential equations on rectangular, two-dimensional domains, Part I: MRC Rpt. 2078, Part II: MRC Rpt. 2079, Mathematics Research Center, University of Wisconsin-Madison.

Sewell, G. (1979). A finite element program with automatic user controlled mesh grading in Advances in Computer Methods for Partial Differential Equations III (R. Vishnevetsky, ed.), Rutgers Univ., New Brunswick, NJ.

Sherman, A.H. (1978). Algorithm 533. NSPIV, A FORTRAN subroutine for sparse Gaussian elimination with partial pivoting, ACM Trans. Math. Software, 4, pp. 391-398.

THE ITPACK PACKAGE FOR LARGE SPARSE LINEAR SYSTEMS[1]

David M. Young
David R. Kincaid

Center for Numerical Analysis
University of Texas
Austin, Texas

I. INTRODUCTION AND BACKGROUND

For several years the Center for Numerical Analysis at The University of Texas at Austin has been involved in a project, known as ITPACK, which is concerned with the development and use of research-oriented programs for solving large sparse linear systems using iterative algorithms. A package of programs, known as ITPACK, has been completed which provides for the use of seven alternative procedures. It is the purpose of this paper to describe the aims, objectives, and current status of ITPACK and to indicate plans for the future.

Among the objectives of ITPACK are the following:

(a) to provide a research and teaching tool for iterative methods. In particular, to facilitate comparisons of various iterative methods and between direct and iterative methods.

(b) to provide linear equation modules for use with ELLPACK (see Rice [1981]) and for solving moderate-size industrial problems.

(c) to contribute to research on mathematical software methodology.

It is not intended that ITPACK software should be production-oriented.

---

[1]The work was supported in part by the National Science Foundation through Grant No. MCS-7919828 with The University of Texas at Austin.

ISBN 0-12-632620-7

Flexibility rather than maximum efficiency is stressed. While it is doubtful that the programs of ITPACK would be used to solve large industrial problems, their use for model problems might well provide a guide for the construction of large "tailor-made" programs.

The work on ITPACK is being carried on under the direction of the authors and with the support of the National Science Foundation. A major part of the programming for the current version of ITPACK was done by Roger Grimes. Others who contributed to the analysis and programming include Minna Chao, Koso Idigbe, Kang Jea, William MacGregor, and John Respess.

Many of the iterative algorithms used in ITPACK have been developed in collaboration with L. A. Hageman. Complete descriptions of these algorithms, including derivations, are given in the forthcoming book by Hageman and Young [1981].

The ITPACK project is an outgrowth of a collaborative effort proposed by Garrett Birkhoff in the summer of 1974 for the development of software for solving elliptic partial differential equations. Discussions were held with John Rice, Robert Lynch, and the authors, as well as with others. These discussions eventually led to the establishment of the ELLPACK project (in September 1976) and the ITPACK project. We believe it is fair to say that Garrett Birkhoff is the "godfather" of ELLPACK and of ITPACK. His interest, enthusiasm, encouragement, and advice have played an important part in the development of both projects.

This paper is intended to provide general information concerning the overall capabilities of ITPACK without going into great detail. It is assumed that the reader has some acquaintance with iterative methods. In sections II, III, and IV, we give a general description of the iterative algorithms used in ITPACK. The storage format used in ITPACK is described in section V. The programs of ITPACK are described in section VI. Sections VII and VIII are concerned with the use of ITPACK with ELLPACK and with numerical experiments. Future plans are discussed in section IX. The discussion of future plans is supplemented by the Appendix, which describes one of the generalized conjugate gradient procedures to be considered for accelerating nonsymmetrizable methods. A brief guide to ITPACK literature is given in section X.

II. ITERATIVE ALGORITHMS: BASIC ITERATIVE METHODS

The ITPACK package includes programs for solving, by several alternative iterative algorithms, the linear system

$$Au = b. \tag{1}$$

Here A is a given nonsingular square matrix of order N, b is a given column vector, and the column vector u is to be determined. It is assumed that the matrix A is large and sparse.

In all but one case, each algorithm of ITPACK consists of four components:

(a) a basic iterative method such as the Jacobi method, the RS method, or the SSOR method;

(b) an acceleration method such as Chebyshev acceleration or conjugate gradient acceleration;

(c) an automatic procedure for computing acceleration parameters where needed;

(d) a stopping procedure to determine whether the approximate solution is sufficiently close to the exact solution of (1).

A basic iterative method is a method of the form

$$u^{(n+1)} = Gu^{(n)} + k \tag{2}$$

where $u^{(0)}$ is arbitrary. To be "completely consistent" with (2) in the sense of Young [1971], such a method must correspond to a "splitting" of the original matrix A, i.e., to a representation of A in the form

$$A = Q - (Q - A)$$

where Q is a "splitting matrix." To derive the iterative method corresponding to Q we rewrite (1) in the form

$$Qu = (Q - A)u + b,$$

and we then write

$$Qu^{(n+1)} = (Q - A)u^{(n)} + b. \tag{3}$$

Evidently (2) is equivalent to (3) with

$$\begin{cases} G = I - Q^{-1}A \\ k = Q^{-1}b \end{cases}$$

For the more well-known iterative methods we have the following splitting matrices:

Method	Q
Richardson	I
Jacobi	D
Gauss-Seidel	$D - C_L$
SOR	$\frac{1}{\omega}D - C_L$
SSOR	$\frac{\omega}{2-\omega}\left(\frac{1}{\omega}C - C_L\right)D^{-1}\left(\frac{1}{\omega}D - C_U\right)$.

Here D is the diagonal matrix with the same diagonal elements as A, C_L is the strictly lower triangular matrix, and C_U is a strictly upper triangular matrix such that

$$A = D - C_L - C_U.$$

Thus for the matrix

$$A = \begin{bmatrix} 4 & -1 & -1 & 0 \\ -1 & 4 & 0 & -1 \\ -1 & 0 & 4 & -1 \\ 0 & -1 & -1 & 4 \end{bmatrix}$$

we have

$$D = \begin{bmatrix} 4 & 0 & 0 & 0 \\ 0 & 4 & 0 & 0 \\ 0 & 0 & 4 & 0 \\ 0 & 0 & 0 & 4 \end{bmatrix}, \quad C_L = \begin{bmatrix} 0 & 0 & 0 & 0 \\ 1 & 0 & 0 & 0 \\ 1 & 0 & 0 & 0 \\ 0 & 1 & 1 & 0 \end{bmatrix}, \quad C_U = \begin{bmatrix} 0 & 1 & 1 & 0 \\ 0 & 0 & 0 & 1 \\ 0 & 0 & 0 & 1 \\ 0 & 0 & 0 & 0 \end{bmatrix}$$

One of the basic iterative methods included in ITPACK is the "RS method," which is Richardson's method applied to the "reduced system." We assume here that the matrix A is a "red/black matrix" in the sense that A has the form

$$A = \begin{bmatrix} D_R & H \\ K & D_B \end{bmatrix}$$

where D_R and D_B are diagonal matrices. We thus can write (1) in the form

$$\begin{bmatrix} D_R & H \\ K & D_B \end{bmatrix} \begin{bmatrix} u_R \\ u_B \end{bmatrix} = \begin{bmatrix} b_R \\ b_B \end{bmatrix} \qquad (4)$$

We let the unknowns corresponding to u_R and u_B be "red" and "black," respectively. The above form can be obtained from a 5-point discretization of a partial differential equation on a rectangular mesh by labelling the mesh points according to a checkerboard on "red/black" ordering. If we eliminate u_R from (4) we get the <u>reduced system</u>

$$(I - F_B F_R) u_B = F_B c_R + c_B$$

where

$$\begin{cases} F_R = D_R^{-1} H, & F_B = D_B^{-1} K \\ c_R = D_R^{-1} b_R, & c_B = D_B^{-1} b_B \end{cases}$$

If we apply Richardson's method to the reduced system we get the <u>RS method</u> defined by

$$u_B^{(n+1)} = F_B F_R u_B^{(n)} + F_B c_R + c_B.$$

Evidently, the eigenvalues of the iteration matrix $F_B F_R$ are nonnegative and $M(F_B F_R) = M(B^2) < 1$.[2] Thus the RS method converges about twice as fast as the Jacobi method.

As an example, consider the usual 5-point finite difference discretization of the Poisson equation $u_{xx} + u_{yy} = -1$ in the unit square with mesh size $h = 1/3$ and with zero boundary values. With the interior mesh points numbered as indicated,

---

[2]Throughout this paper M(H) denotes the largest eigenvalue of matrix H.

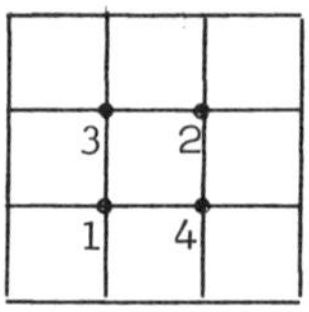

we obtain, after multiplying by $-h^2$, the red/black system

$$\left[\begin{array}{cc|cc} 4 & 0 & -1 & -1 \\ 0 & 4 & -1 & -1 \\ \hline -1 & -1 & 4 & 0 \\ -1 & -1 & 0 & 4 \end{array}\right] \left[\begin{array}{c} u_1 \\ u_2 \\ \hline u_3 \\ u_4 \end{array}\right] = \frac{1}{9} \left[\begin{array}{c} 1 \\ 1 \\ \hline 1 \\ 1 \end{array}\right]$$

the reduced system is easily determined and the RS method is defined by

$$\begin{pmatrix} u_3^{(n+1)} \\ u_4^{(n+1)} \end{pmatrix} = \begin{pmatrix} \frac{1}{8} & \frac{1}{8} \\ \frac{1}{8} & \frac{1}{8} \end{pmatrix} \begin{pmatrix} u_3^{(n)} \\ u_4^{(n)} \end{pmatrix} + \begin{pmatrix} \frac{27}{8} \\ \frac{27}{8} \end{pmatrix} .$$

III. ACCELERATION PROCEDURES

Many iterative methods can be speeded up by the use of a polynomial acceleration procedure such as Chebyshev acceleration or conjugate gradient acceleration. For this to be possible it is sufficient for the iterative method to be symmetrizable, i.e., that the matrix $W(I-G)W^{-1}$ is symmetric and positive definite (SPD) for some nonsingular matrix W. If A is SPD, then the Richardson, Jacobi, and SSOR methods are symmetrizable (assuming $0 < \omega < 2$ for the SSOR method). If an iterative method is symmetrizable, then the eigenvalues of the iteration matrix G are real and less than unity. In a typical case, both A and Q are SPD and W can be taken to be $Q^{\frac{1}{2}}$ or $A^{\frac{1}{2}}$. It should be noted that all necessary formulas can be written in terms of W^TW instead of W so that one does not need to compute $Q^{\frac{1}{2}}$ or $A^{\frac{1}{2}}$.

Both Chebyshev and conjugate gradient acceleration of the iterative method (2) can be written in the form

$$u^{(n+1)} = \rho_{n+1}\{\gamma_{n+1}(Gu^{(n)} + k) + (1 - \gamma_{n+1})u^{(n)}\} + (1 - \rho_{n+1})u^{(n-1)} \tag{5}$$

where $\rho_1 = 1$. In the case of Chebyshev acceleration we have

$$\gamma_1 = \gamma_2 = \cdots = \frac{2}{2-M-m}$$

$$\rho_{n+1} = \begin{cases} 1, & \text{if } n = 0 \\ \left(1 - \frac{1}{2}\sigma^2\right)^{-1}, & \text{if } n = 1 \\ \left(1 - \frac{1}{4}\sigma^2\rho_n\right)^{-1}, & \text{if } n \geq 2 \end{cases} \tag{6}$$

$$\sigma = \frac{M-m}{2-M-m}$$

Here $M = M(G)$ and $m = m(G)$ are the largest and smallest eigenvalues of G, respectively. If M and m are not known, estimates must be used.

In conjugate gradient acceleration we have, with A and Q SPD and with $W = Q^{\frac{1}{2}}$,

$$\gamma_{n+1} = \left(1 - \frac{(\delta^{(n)}, QG\delta^{(n)})}{(\delta^{(n)}, Q\delta^{(n)})}\right)^{-1}$$

$$\rho_{n+1} = \left[1 - \frac{\gamma_{n+1}}{\gamma_n}\,\frac{(\delta^{(n)}, Q\delta^{(n)})}{(\delta^{(n-1)}, Q\delta^{(n-1)})}\,\frac{1}{\rho_n}\right]^{-1}, \quad n \geq 2 \tag{7}$$

where the <u>pseudo-residual</u> vector $\delta^{(n)}$ is given by

$$\delta^{(n)} = Gu^{(n)} + k - u^{(n)}. \tag{8}$$

To illustrate the effectiveness of Chebyshev acceleration, consider the model problem $u_{xx} + u_{yy} = -1$, $0 \leq x \leq 1$, $0 \leq y \leq 1$ with $u = 0$ on the boundary. Listed below are the approximate number of iterations needed to reduce the error, measured in the ordinary norm, to 10^{-6} times the initial error:

	$h = \frac{1}{20}$	$h = \frac{1}{40}$
Jacobi	1120	4480
Jacobi-Chebyshev	92	184

Thus, when the mesh size is halved, the accelerated Jacobi method takes twice as many iterations whereas the unaccelerated procedure takes four times as many.

Unfortunately, the behavior of the Chebyshev acceleration process is very sensitive to the estimate M_E for $M(G)$. Thus, if $M(G) = .99990$ and if the (apparently accurate) estimate $M_E = .99989$ were used, the rate of convergence would be reduced by 30%. Except in special cases it would not appear possible to estimate $M(G)$ to sufficient accuracy in advance. However, it is possible to use an adaptive procedure as we now describe. For simplicity let us assume that a reasonably good lower bound $\underline{m}$ is known for the eigenvalues of G. We let $m_E = \underline{m}$ and do not change m_E. We choose $M_E^{(1)}$ such that $\underline{m} \leq M_E^{(1)} < M(G)$ and iterate. We measure the convergence rate based on the norms of the pseudo-residual vectors $\{\delta^{(n)}\}$. If the convergence rate is substantially less than it would be if $M(G)$ were equal to $M_E^{(1)}$, we modify our estimate to obtain $M_E^{(2)}$. The process is repeated, with new estimates $M_E^{(3)}$, $M_E^{(4)}$, ... being obtained, until convergence (for $u^{(n)}$ but not necessarily for the $M_E^{(n)}$) is achieved. Further details are given in Hageman and Young [1981] and Grimes, et al. [1979].

The use of conjugate gradient acceleration yields a convergence rate which is nearly always faster than Chebyshev acceleration and often considerably faster. (It is always faster if the error is measured in a suitable norm.) Moreover, no estimates of $m(G)$ and $M(G)$ are required. On the other hand, the formulas for conjugate gradient acceleration are somewhat more complicated. By a modification of the formulas for conjugate gradient acceleration the number of extra operations required can be minimized; see Hageman and Young [1981] and Grimes, et al. [1979].

For the case of the SSOR method, even with conjugate gradient acceleration the parameter ω must be estimated. An adaptive procedure for doing this is included in ITPACK; see Hayes and Young [1977] and Grimes, et al. [1979] for details.

IV. STOPPING PROCEDURES

The problem of determining when an approximate solution $u^{(n)}$ is close enough to the true solution $\bar{u}$ of (1) is one which is often treated lightly. In many cases the iterative process is stopped when $u^{(n+1)} - u^{(n)}$, measured in some norm, is sufficiently small. Alternatively, the process may be stopped when the residual vector $r^{(n)} = b - Au^{(n)}$ or the pseudo-residual vector $\delta^{(n)} = Gu^{(n)} + k - u^{(n)}$ is sufficiently small. Unfortunately, the error vector $\epsilon^{(n)} = u^{(n)} - \bar{u}$ may be quite large even though the above quantities are small.

Let us consider the use of an acceleration process for the basic iterative method (2). We continue to assume that the method is symmetrizable and that $W(I-G)W^{-1}$ is SPD for some matrix W. We propose the following theoretical stopping test, namely,

$$\frac{\|u^{(n)} - \bar{u}\|_W}{\|\bar{u}\|_W} \leq \zeta \tag{9}$$

where ζ is a prescribed small number, say 10^{-6}. Here for any vector v we let $\|v\|_W = \|Wv\|_2 = (Wv,Wv)^{\frac{1}{2}}$. In order to estimate $u^{(n)} - \bar{u}$ we have by (2)

$$\epsilon^{(n)} = (G-I)^{-1}\delta^{(n)}$$

since $\bar{u} = G\bar{u} + k$, and hence

$$\|u^{(n)} - \bar{u}\|_W \leq \|(G-I)^{-1}\|_W\|\delta^{(n)}\|_W = \|W(G-I)^{-1}W^{-1}\|_2\|\delta^{(n)}\|_W$$

$$= M((G-I)^{-1})\,\|\delta^{(n)}\|_W = \frac{\|\delta^{(n)}\|_W}{1 - M(G)}$$

since $\|H\|_W = \|WHW^{-1}\|_2$ for any matrix H. We note that since $W(G-I)^{-1}W^{-1}$ is symmetric, then $\|W(G-I)^{-1}W^{-1}\| = M(W(G-I)^{-1}W^{-1}) = M((G-I)^{-1})$. Thus we take for our stopping test

$$\frac{\|\delta^{(n)}\|_W}{(1-M(G))\,\|u^{(n)}\|_W} \leq \zeta. \tag{10}$$

If we are using the adaptive Chebyshev procedure, we replace $M(G)$ by the latest estimate M_E. If we are using conjugate gradient acceleration, then we can estimate $M(G)$ as the largest eigenvalue of a certain tridiagonal matrix involving previously computed $\{\rho_i\}$ and $\{\gamma_i\}$; see Hageman and Young [1981] and Grimes, et al. [1979].

The estimate (10) is very accurate for the adaptive Chebyshev procedure. For many numerical experiments the number of iterations required for (10) to be satisfied was, within one or two, the same as the number required for (9) to be satisfied. For nonadaptive Chebyshev acceleration, where the optimum value of $M(G)$ is used at the outset, the agreement was not as close but was still within 10-15%. For conjugate gradient acceleration a similar agreement was obtained.

One should be careful not to specify too small a value of ζ, especially with a machine of limited word length. If ζ is too small the process may never converge because of roundoff errors. Thus, one cannot expect the pseudo-residual $\delta^{(n)}$ in (10) to be less than several times the basic machine rounding unit ϵ. If $M(G)$ is very close to unity, then the left member of (10) may be many times ϵ. For example, if $M(G) = .999$, then a factor of 1000 would be involved. If ϵ is of the order of 10^{-8}, then one might be wise to choose ζ no smaller than 10^{-4} or possibly 10^{-3}.

Another problem which may arise if a large value of ζ such as 10^{-1} or 10^{-2} is used is "premature convergence." Suppose for small n that one has a very good initial estimate $u^{(0)}$ so that $u^{(n)} - \bar{u}$ is small and $M_E^{(1)} = 0$. Then the value of $M(G)$ used in (10) may be very small, since the adaptive process has not yet increased the estimate of $M(G)$. Thus the left member of (10) may be much smaller than it should be. This difficulty can be partly overcome if the user can provide a better estimate of $M(G)$. However, it should be noted that for the adaptive process to be effective $M_E^{(1)}$ must be less than $M(G)$.

V. SPARSE STORAGE FORMAT

The programs of ITPACK 2.0 use the symmetric sparse storage format as in the Yale sparse matrix package to store the matrix A; see Eisenstat, Gursky, Schultz, and Sherman [1979, 1979a]. A modified version of ITPACK 2.0, to be called ITPACK 2A, will also allow the use of the nonsymmetric sparse storage format. With the nonsymmetric format all nonzero elements of A are stored while with the symmetric format only the nonzero elements on or above the main diagonal are stored. In this section we give a brief description of the nonsymmetric format.

For the nonsymmetric sparse storage format the NZ nonzero elements of the matrix A are stored in an array A(1), A(2), ..., A(NZ). The nonzero elements are stored row-wise, but for each row the elements need not be stored by increasing column numbers. The array JA(1), JA(2), ..., JA(NZ) gives the column numbers. Thus, JA(K) gives the column number for A(K). The array IA(1), IA(2), ..., IA(N), IA(N+1) is used as follows. IA(1) = 1 and IA(L+1) - IA(L) indicates the number of nonzero elements in the L-th row of the matrix A. Thus, if IA(L+1) > IA(L), then A(IA(L)) is an element of the L-th row of the matrix A which appears in the array A(1), A(2), ..., A(NZ).

As an example, consider the matrix

$$A = \begin{pmatrix} 0 & 0 & 0 \\ 0 & 3 & 4 \\ 0 & 5 & 0 \end{pmatrix}.$$

(Of course, a matrix with zero diagonal elements could not be used by ITPACK.) Here N = 3 and NZ = 3. For the array A(1), A(2), A(3) we let A(1) = 4, A(2) = 3, A(3) = 5. Note that the elements A(1) and A(2) are not stored by increasing column number. Evidently, JA(1) = 3, JA(2) = 2, JA(3) = 2. Also, IA(1) = 1. Since there are no nonzero elements in the first row, IA(2) = 1. Since there are two nonzero elements in the second row, IA(3) = 3. Since there is one nonzero element in the third row, we have IA(4) = 4.

The use of the sparse storage format provides much greater flexibility. However, for many linear systems arising from discretization applied to partial differential equations, the matrix is "structured" as well as sparse. In many cases substantial savings in machine time and storage could be achieved by taking advantage of the structured property; see the discussion of Hageman [1975]. Another disadvantage for really huge problems in which the coefficients elements can be easily generated when needed is the explicit storage scheme in ITPACK. For these reasons it is not expected that general-purpose programs like ITPACK would be appropriate for large industrial problems.

VI. PROGRAMS OF ITPACK

The current ITPACK package, which is referred to as "ITPACK 2.0," contains the following seven methods:

Abbreviation	Basic Method	Acceleration
J-SI	Jacobi	Chebyshev
J-CG	Jacobi	conjugate gradient
SOR	SOR	--
SSOR-SI	SSOR	Chebyshev
SSOR-CG	SSOR	conjugate gradient
RS-SI	RS	Chebyshev
RS-CG	RS	conjugate gradient

For each case involving Chebyshev acceleration the user can specify that the adaptive process should be used or, alternatively, he can specify that a fixed estimate of M(G) be used. For the SSOR and RS methods the estimate m_E for the smallest eigenvalue m(G) is taken to be zero. However, for the Jacobi method the user must choose a value m_E which is a lower bound for m(G).[3] In many problems arising from elliptic partial differential equations it is known that $m(G) \geq -1$ for the Jacobi method.

For the SSOR method the user has the option of specifying ω or letting the program determine ω adaptively as described in Hayes and Young [1977] and Grimes, et al. [1979].

The SOR method is based on an algorithm contributed by L. A. Hageman and does not use an acceleration procedure. The routine uses an adaptive procedure to determine the relaxation factor ω; see Hageman and Young [1981] and Grimes, et al. [1979].

In order to use a particular program of ITPACK, one needs to provide as input the number of equations N, the number of nonzero elements NZ of A, the arrays A, JA, and IA, and the array RHS which corresponds to the right-hand side of (1). Such items as an initial guess $u^{(0)}$, a stopping number ζ, and initial values for iteration parameters may be preset by the user or automatically generated when adaptive procedures are used.[4] There is also a "strategy parameter" F for the adaptive process which could be specified; however, the default option F = .75 has proven satisfactory in all cases which we have run.

VII. THE USE OF ITPACK WITH ELLPACK

The ELLPACK project, see, e.g., Rice [1981], is concerned with the development of a package of programs for solving a class of elliptic partial differential equations by various methods. A typical problem is solved in several stages including: input; domain processing; discretization or equation formulation; equation indexing; equation solution; and output. Rigid interfaces are provided between each stage so that the

---

[3] This may not be required in a later version of ITPACK. A procedure for estimating m(G) is given in Hageman and Young [1981], Chapter 6. Another scheme has been suggested by T. Manteuffel, private communication.

[4] Procedures for obtaining good starting vectors are described by Gordon [1981].

the user can loop through certain stages without having to repeat previous ones. ITPACK routines can be used in the equation-solution stage. Alternatively, modules based on direct methods can be used. A high-level language is provided to enable the user to easily define the problem and to specify the solution procedure in a convenient manner. This high-level language allows all information needed for ITPACK to be generated and available when needed.

VIII. NUMERICAL EXPERIMENTS

The ITPACK programs have been used successfully to solve a large number of sparse linear systems both in conjunction with ELLPACK and independently. Generally, the performance of the program has been quite satisfactory; at least we have received very few reports of difficulties.

In this section we describe some numerical experiments designed to test the effectiveness of the algorithms used and to compare the different ITPACK routines. We describe two types of experiments. The first type of experiments is designed to test the effectiveness of the algorithms. We refer to these as "simulation studies." The second type of experiments involves the use of ITPACK to solve linear systems arising from partial differential equations.

A. Simulation Studies

The effectiveness of adaptive and nonadaptive Chebyshev acceleration procedures and of conjugate gradient acceleration procedures can be studied by the use of "simulation." Thus, given the basic iterative method (1) and a symmetrization matrix W such that $W(I-G)W^{-1}$ is SPD, we let P be any orthogonal matrix such that $\Lambda = PWGW^{-1}P^{-1}$ is a real diagonal matrix (with diagonal elements less than unity). From (5)

$$\hat{u}^{(n+1)} = \rho_{n+1}\{\gamma_{n+1}(\Lambda\hat{u}^{(n)} + \hat{k}) + (1 - \gamma_{n+1})\hat{u}^{(n)}\} + (1 - \rho_{n+1})\hat{u}^{(n-1)}$$

where $\hat{u}^{(n)} = Pu^{(n)}$, $\hat{k} = Pk = (\Lambda - I)\hat{\bar{u}}$. Also,

$$\|\delta^{(n)}\|_W = \|\hat{\delta}^{(n)}\|$$

where $\delta^{(n)}$ is given by (8) and

$$\hat{\delta}^{(n)} = \Lambda\hat{u}^{(n)} + \hat{k} - \hat{u}^{(n)}$$

and

$$(W\delta^{(n)}, WG\delta^{(n)}) = (\hat{\delta}^{(n)}, \Lambda\hat{\delta}^{(n)}) .$$

Thus, by (6) and (7), we can carry out either the Chebyshev or the conjugate gradient acceleration procedure given any Λ, $\hat{u}^{(0)}$, and $\hat{\bar{u}}$. The stopping test (9) can be written in the form

$$\frac{\|\hat{u}^{(n)} - \hat{\bar{u}}\|}{\|\hat{\bar{u}}\|} \leq \zeta.$$

Except for roundoff, the behavior of a simulated problem is the same as for the original problem. However, there is a great saving of machine time and storage since the iteration matrix for the simulated program is diagonal. Also, one can easily study the effect of various distributions of the eigenvalues of G and of the "Fourier coefficients" of the initial error vector.

The results of a number of simulation studies are described in Hageman and Young [1981]. The following conclusions were obtained:

(a) With $\zeta = 10^{-6}$ and with $u^{(0)} = 0$, the number of iterations using adaptive Chebyshev acceleration was 20-40% more than for optimal Chebyshev acceleration no matter how bad the initial choice of M(G). The choice of F = .75 for the strategy parameter was always satisfactory. It should be noted that if one were to use a larger value of ζ or, equivalently, a value of $u^{(0)}$ such that $\|u^{(0)} - \bar{u}\| \ll \|\bar{u}\|$, then the adaptive process might be somewhat less effective.

(b) Conjugate gradient acceleration usually required considerably fewer iterations than Chebyshev acceleration. The behavior of

Chebyshev acceleration depended almost entirely on the eigenvalue range while the behavior of conjugate gradient acceleration depended on the number and distribution of the eigenvalues in the range. If the density of eigenvalues in part of the range is low, the number of iterations might be greatly reduced.

(c) The stopping tests were accurate to within one or two iterations with adaptive Chebyshev acceleration and to within 10-15% for optimal nonadaptive Chebyshev acceleration and for conjugate gradient acceleration.

B. The Solution of Partial Differential Equations

A number of numerical experiments have been carried out on linear systems arising from partial differential equations. Some of these experiments are described in Kincaid, Grimes, Young, and MacGregor [1979], Kincaid, Grimes, and Young [1979], and Eisenstat, et al. [1979]. The problems which were solved were relatively small--on the order of 1000-2000 unknowns. The results suggest the following tentative conclusions.

1. For red/black problems, the RS-SI and RS-CG methods are best for problems of moderate size. It should be noted that by use of a suitable grouping of the unknowns (such as correspond to lines of mesh points), it may be possible to obtain a block red/black problem. This capability may be included in a future version of ITPACK.
2. For problems for which it is not convenient or not possible to use the red/black form, the J-CG method is best for moderate-size problems.
3. For problems with a very small mesh size where the coefficients of the differential equation are very well-behaved, the SSOR-SI or SSOR-CG methods seem to be best. It is believed that in some cases, for example piecewise constant coefficients, the use of SSOR with

variable ω or the use of an approximate factorization similar to the SSOR splitting (see, e.g., Gustafsson [1979]) would be quite effective.

IX. FUTURE PLANS

As stated earlier, by the end of the summer of 1980 we expect to complete work on ITPACK 2A. The programs of ITPACK 2A will allow not only the symmetric storage format to be used but also the nonsymmetric storage format described in section V. ITPACK 2A will also use a slightly different form of the conjugate gradient algorithm (for the J-CG, RS-CG, and SSOR-CG methods) which are equivalent to the existing ones in the symmetric case but has been found to work much better for nonsymmetric problems. These modifications are described in Kincaid and Young [1981]; see also the Appendix.

Other short-range modifications being considered include block methods and the determination, where possible, of a diagonal matrix P such that PAP^{-1} is symmetric and positive definite. This would be very useful for the conjugate gradient algorithms for certain nonsymmetric problems.

Aside from short-range changes as described above, we plan to leave the existing version of ITPACK alone for at least a year or two. We have, however, begun work on a new package with a completely different structure. Instead of having 7 complete programs, which essentially allow for three basic iterative methods and two acceleration procedures, as in the present version, we plan to introduce a modular structure. The new package will include a number of subroutines--some for basic iterative methods, some for acceleration procedures, and some for stopping procedures. In order that a particular algorithm can be used, a "driver" program will have to be written which will call the appropriate subroutines as needed. The feasibility of developing a control program to permit the use of a high-level language, as in ELLPACK, will be investigated.

The acceleration methods will include the adaptive Chebyshev and conjugate gradient procedures now included in ITPACK. These procedures are designed for symmetric problems but also work well for some mildly nonsymmetric problems. The new procedures to be added will be designed to handle nonsymmetric problems. Included will be

(a) three generalizations of conjugate gradient acceleration, ORTHODIR, ORTHOMIN, and ORTHORES described in Young and Jea [1979, 1981]; some of these procedures have been considered by Axelsson [1978], Vinsome [1976], and Eisenstat, Elman, and Schultz [1979]. Each of these procedures has a number of variants and each reduces to conjugate gradient acceleration in the symmetric case. We describe one of the methods in more detail in the Appendix.

(b) adaptive Chebyshev acceleration for problems where the eigenvalues of the iteration matrix are complex; see Manteuffel [1977].

(c) the Lanczos method; see Lanczos [1950, 1952] and Young and Jea [1979, 1981].

Among the basic iterative methods to be included will be the following:

(a) Richardson's method

(b) Jacobi method

(c) SSOR method

(d) Peaceman-Rachford [1955] alternating direction implicit method

(e) a method proposed by Concus and Golub [1976] and by Widlund [1978], for nonsymmetric systems. The splitting matrix is $\frac{1}{2}(A + A^T)$, which is assumed to be positive definite.

(f) methods based on the use of approximate factorization as described by Varga [1960], Stone, [1968], Dupont, Kendall, and Rachford [1968], Meijerink and van der Vorst [1977], Manteuffel [1978], Gustafsson [1979], and others.

Included in some of the algorithms, such as the SSOR method where the basic method itself involves an iteration parameter, will be a scheme to modify this parameter based on the observed convergence behavior.

X. ITPACK INFORMATION

The ITPACK package is available for distribution through the International Mathematical and Statistical Libraries (IMSL) Inc. A source tape and documentation can be ordered from IMSL at the following address: Sixth Floor-NBC Building, 7500 Bellarie Blvd., Houston, Texas 77036. Questions concerning the performance of the software and requests for documentation only should be directed to the authors. Descriptions of the algorithms used in ITPACK are given in the "ITPACK 2.0 User's Guide"; see Grimes, Kincaid, and Young [1979]. A description of the package is given in the forthcoming TOMS article by Grimes, Kincaid, and Young [1981]. As stated earlier, more complete descriptions of the algorithms, with derivations, are given in the book by Hageman and Young [1981].

APPENDIX
ORTHORES: A GENERALIZED CONJUGATE GRADIENT ACCELERATION PROCEDURE

We illustrate the use of a generalized conjugate gradient acceleration procedure which we refer to as "ORTHORES." We consider the case where the splitting matrix Q for the basic iterative method (2) can be written in the form

$$Q = Q_L Q_D Q_U$$

where Q_D is a diagonal matrix with positive diagonal elements and where Q_L and Q_U are unit lower triangular and unit upper triangular matrices, respectively.

To apply ORTHORES one must choose a positive integer s. The formulas are given by (see Young and Jea [1981])

$$u^{(n+1)} = \gamma_{n+1} f_{n+1,n} \delta^{(n)} + f_{n+1,n} u^{(n)} + \dots + f_{n+1,n-s} u^{(n-s)}$$

where

$$\begin{cases} \gamma_{n+1} = \sigma_{n+1,n}^{-1} \\ \sigma_{n+1,i} = \begin{cases} \dfrac{(YQ^{-1}A\delta^{(n)}, \delta^{(i)})}{(Y\delta^{(i)}, \delta^{(i)})}, & i = n-s,\dots,n \text{ and } i \geq 0 \\ 0 \quad \text{if } i < 0 \end{cases} \\ f_{n+1,n} = \left[1 + \gamma_{n+1} \sum_{i=n-s}^{n-1} \sigma_{n+1,i} \right]^{-1} \\ f_{n+1,i} = \sigma_{n+1,i} \gamma_{n+i} f_{n+1,n}, \qquad i = n-s,\dots,n-1 \\ \delta^{(n)} = Gu^{(n)} + k - u^{(n)} \end{cases}$$

Here we let

$$Y = Q_U^T Q_D Q_U$$

If Q and A are SPD, then $Y = Q$ and $YQ^{-1}A = A$; thus Y and $YQ^{-1}A$ are SPD and we have the symmetrizable case.[5] It then turns out that $f_{n+1,i} = 0$ for $i < n-1$. Letting $f_{n+1,n} = \rho_{n+1}$, we get $f_{n+1,n-1} = 1 - \rho_{n+1}$ and

$$u^{(n+1)} = \rho_{n+1}\{\gamma_{n+1}\delta^{(n)} + u^{(n)}\} + (1-\rho_{n+1}) u^{(n-1)}$$

where

$$\gamma_{n+1} = \frac{(Q\delta^{(n)}, \delta^{(n)})}{(Q\delta^{(n)}, (I-G)\delta^{(n)})} = \left[1 - \frac{(\delta^{(n)}, QG\delta^{(n)})}{(\delta^{(n)}, Q\delta^{(n)})} \right]^{-1} \tag{11}$$

$$\rho_{n+1} = \begin{cases} 1, & \text{if } n = 0 \\ \left[1 + \gamma_{n+1} \dfrac{(QG\delta^{(n)}, \delta^{(n-1)})}{(Q\delta^{(n-1)}, \delta^{(n-1)})} \right]^{-1}, & n \geq 1 \end{cases} \tag{12}$$

---

[5] Note that a basic iterative method is symmetrizable if $YQ^{-1}A$ is SPD for some SPD matrix Y.

After some algebraic manipulation one can obtain (7). While the formulas (11) and (12) are equivalent to (7) in the symmetrizable case, they work much better in those nonsymmetrizable cases which we have tried.

In the nonsymmetrizable case, it is shown by Young and Jea [1981] that if $YQ^{-1}A$ is positive real, then the "idealized" version of ORTHORES where $s = \infty$ will not break down and indeed will converge in at most N steps. (Of course this version of the method is not practical in general since one must store all previous vectors.) It seems reasonable to expect, or at least hope, that for many iterative methods which are "nearly" symmetrizable, in some sense, ORTHORES with some small positive value of s will be at least as effective as Chebyshev acceleration. This has indeed been the case in some preliminary experiments which we have run. One would expect that the greater the degree of nonsymmetry the larger the value of s which will be required. Clearly, a great deal of theoretical analysis and numerical experimentation are needed to give an accurate evaluation of ORTHORES and related methods.

REFERENCES

Axelsson, O. [1978]. "On Preconditioned Conjugate Gradient Methods," CNA-143, Center for Numerical Analysis, University of Texas, Austin.

Concus, Paul, and Golub, Gene H. [1976]. "A Generalized Conjugate Gradient Method for Nonsymmetric Systems of Linear Equations," Report Stan-CS-76-535, Computer Science Department, Stanford University, California.

Dupont, T., Kendall, R., and Rachford, H. H., Jr. [1968]. "An Approximate Factorization Procedure for Solving Self-adjoint Elliptic Difference Equations," SIAM J. Numer. Anal. 5, 559-573.

Eisenstat, S. C., Elman, H., and Schultz, M. H. [1979]. "Variational Iterative Methods for Nonsymmetric Systems of Linear Equations," unpublished manuscript.

Eisenstat, S. C., Gursky, M. C., Schultz, M. H., and Sherman, A. H. [1977]. "Yale Sparse Matrix Package, I, The Symmetric Codes,"Research Report No. 112, Department of Computer Science, Yale University, New Haven, Conn.

Eisenstat, S. C., Gursky, M. C., Schultz, M. H., and Sherman, A. H. [1977a]. "Yale Sparse Matrix Package, II, The Nonsymmetric Codes," Research Report No. 114, Department of Computer Science, Yale University, New Haven, Conn.

Eisenstat, S. C., George, A., Grimes, R. G., Kincaid, D. R., and Sherman, A. H. [1979]. "Some Comparisons of Software Packages for Large Sparse Linear Systems," in Advances in Computer Methods for Partial Differential Equations III (R. Vichnevetsky and R. S. Stepleman, eds.), publ. IMACS, Dept. Computer Science, Rutgers Univ., New Brunswick, N. J.

Grimes, R. G., Kincaid, D. R., and Young, D. M. [1979]. "ITPACK 2.0 User's Guide," CNA-150, Center for Numerical Analysis, University of Texas, Austin.

Grimes, R. G., Kincaid, D. R., and Young, D. M. [1981]. "ITPACK 2.0: A FORTRAN Implementation of Adaptive Accelerated Point Iterative Methods for Solving Symmetric Positive Definite Systems ," to appear in the Transactions of Mathematical Software.

Gordon, William J. [1981]. "Initial Approximations for the Iterative Solution of Boundary Value Problems," this volume.

Gustafsson, I. [1979]. "Stability and Rate of Convergence of Modified Incomplete Cholesky Factorization Methods," Report 79.02R (Ph.D. thesis), Department of Computer Sciences, Chalmers University of Technology, Göteborg, Sweden.

Hageman, L. A. [1955]. "The Solution of Linear Equations Resulting from Finite Element Discretizations of Multi-dimensional Boundary Value Problems," Report WAPD-TM-1209, Bettis Atomic Power Laboratory, Westinghouse Electric Corp., West Mifflin, Pa.

Hageman, L. A., and Young, D. M. [1981]. "Applied Iterative Methods," Academic Press (to appear).

Hayes, L. J., and Young, D. M. [1977]. "The Accelerated SSOR Method for Solving Large Linear Systems," CNA-123, Center for Numerical Analysis, University of Texas, Austin.

Kincaid, D. R., and Young, D. M. [1981]. "Adapting Iterative Algorithms Developed for Symmetric Systems to Nonsymmetric Systems," this volume.

Kincaid, D. R., Grimes, R. G., Young, D. M., and MacGregor, W. I. [1979]. "ITPACK - Adaptive Iterative Algorithms Using Symmetric Sparse Storage," Proceed. of Fifth Symp. on Reservoir Simulation, Society of Petroleum Engineers, Denver, Colorado.

Kincaid, D. R., Grimes, R. G., and Young, D. M. [1979]. "The Use of Iterative Methods for Solving Large Sparse PDE-related Linear Systems," in Advances in Computer Methods for Partial Differential Equations III (R. Vichnevetsky and R. S. Stepleman, eds.), publ. IMACS, Dept. Computer Science, Rutgers Univ., New Brunswick, N. J.

Lanczos, C. [1950]. "An Iteration Method for the Solution of the Eigenvalue Problem of Linear Differential and Integral Operators," J. Research NBS 45.

Lanczos, C. [1952]. "Solution of Systems of Linear Equations by Minimized Iterations," J. Research NBS 49.

Manteuffel, T. A. [1977]. "The Tchebyshev Iteration for Nonsymmetric Linear Systems," Numer. Math. 28.

Manteuffel, T. A. [1978]. "The Shifted Incomplete Cholesky Factorization," Report SAND 78-8226, Sandia Laboratories, Livermore, Calif.

Meijerink, J. A., and van der Vorst, H. A. [1977]. "An Iterative Solution Method for Linear Systems of Which the Coefficient Matrix Is a Symmetric M-Matrix," Math. of Comput. 31.

Peaceman, D. W., and Rachford, H. H., [1955]. "The numerical solution of parabolic and elliptic differential equations," J. Soc. Indus. Appl. Math. 3.

Rice, John R., [1981]. "ELLPACK: Progress and Plans," this volume.

Stone, H. L. [1968]. "Iterative Solutions of Implicit Approximations of Multidimensional Partial Differential Equations," SIAM J. Numer. Anal. 5.

Widlund, O. [1978]. "A Lanczos Method for a Class of Nonsymmetric Systems of Linear Equations," SIAM J. Numer. Anal. 15.
Varga, R. S. [1960]. "Factorization and normalized iterative methods," in Boundary Problems in Differential Equations (R. E. Langer, ed.), Univ. Wisconsin Press, Madison, Wisconsin.
Vinsome, P. K. W. [1976]. "Orthmin, an Iterative Method for Solving Sparse Sets of Simultaneous Linear Equations," Proceed. Fourth Symposium of Numerical Simulation of Reservoir Performance, Soc. of Pet. Engrs. of AIME, Los Angeles, California.
Young, D. M., and Jea, K. C. [1979]. "Some Generalizations of Conjugate Gradient Acceleration for Nonsymmetrizable Iterative Methods (Preliminary Report)," CNA-149, Center for Numerical Analysis, University of Texas, Austin.
Young, D. M., and Jea, K. C. [1981]. "Generalized conjugate gradient acceleration of nonsymmetrizable iterative methods," to appear in Special Issue on Large Scale Matrix Problems, SIAM Journal on Linear Algebra and Its Applications.

EFFICIENT FORTRAN SUBPROGRAMS FOR THE SOLUTION OF ELLIPTIC PARTIAL DIFFERENTIAL EQUATIONS

John C. Adams
Paul N. Swarztrauber
Roland A. Sweet

National Center for Atmospheric Research
Boulder, Colorado

I. INTRODUCTION

FISHPAK (Version 3) is a package of Fortran subroutines that has been developed at the National Center for Atmospheric Research. The first published edition was extensively described in (1) and summaried in (2). Version 3 has approximately 20,000 lines of FORTRAN, including comments, and contains modifications of earlier versions plus extensive additions.

The package provides a basic capability to automatically produce finite difference approximations to Helmholtz's equation defined on a rectangle in a particular co-ordinate system. Also, there are routines for the more general separable elliptic equation and three-dimensional Helmholtz equation in cartesian coordinates.

II. THE PACKAGE

The package is comprised of a set of drivers, a set of solvers, and an extensive sub-package for computing Fourier transforms.

ISBN 0-12-632620-7

A driver is a subroutine that determines a second-order finite difference approximation to a Helmholz equation in a specific co-ordinate system. The routine sets up the finite difference equations, incorporates the given boundary data, adjusts the right side of the resultant linear system when it is singular, and calls the appropriate solver for solution.

Drivers are available on two different types of finite difference grids; namely, centered and staggered. In a centered grid the physical boundaries coincide with the grid lines, whereas in a staggered grid the physical boundaries are located at one-half a grid spacing away from the grid lines. All drivers use a uniform or equally spaced grids in the given coordinate system. Non-uniform grids can be handled by using the solvers. (See the example for subroutine BLKTRI.)

The drivers treat two types of singularities; namely, operator and coordinate. A singular operator results from the specification of certain combinations of boundary conditions, e.g., normal derivative of the solution zero on the boundaries. The linear system that results from the finite difference approximation is then also singular. This singularity imposes a restriction on the right side of the system or the data of the problem. The data is perturbed so that the restriction is satisfied. The solution of this perturbed system, is, in fact, a weighted least squares solution of the original problem. More detail on this perturbation technique may be found in (1). Coordinate singularities may occur in non cartesian co-ordinate representations of the Laplacian. (For example, in polar coordinates the term $\frac{1}{r} \frac{\partial}{\partial r} \left(r \frac{\partial u}{\partial r}\right)$

vanishes at the origin r = o). Special formulas are required to approximate singular terms.

There are thirteen derivers in the package. Ten of them result from approximately a Helmholtz equation on either a centered or staggered grid in one of the five two-dimensional co-ordinate systems:

1. cartesian
2. polar
3. cylindrical
4. spherical surface
5. spherical cross-section.

One driver provides an approximation on a centered grid to the Helmholtz equation in three-dimensional cartesian co-ordinates. The remaining two drivers are for the general separable elliptic equations and have the option to provide fourth-order finite difference approximations via deferred corrections.

A solver is a subroutine designed to solve a specific type of linear system, i.e., those resulting from approximations to separable elliptic equations. Five of the solvers implement generalizations of the cyclid odd-even reduction and the sixth is based on the use of the last Fourier transform.

An extensive sub-package of Fourier transform routines has been developed to facilitate the solution of elliptic equations. In addition to the general complex transform there are routines to compute the transform of real periodic, sine, cosine, and quarter-wave vectors. These are particularly useful for treating a variety of boundary conditions, and as such, may be used to solve constant coefficient elliptic equations subject to a variety of exiting boundary conditions.

The package is completely documented on comment cards.

REFERENCES

Swarztrauber, P., and Sweet, R., "Efficient FORTRAN subprograms for the solution of elliptic partial differential equations," NCAR Technical Note TN/IA-109, 1975. (1)

. . . ., Algorithm 541: "Efficient FORTRAN subprograms for the solution of Separable Elliptic Partial Differentail Equations," TOMS 5(1979), pp352-364. (2)

ITERATIVE METHODS FOR FINITE ELEMENT EQUATIONS

O. Axelsson
University of Nijmegen, The Netherlands

Some preconditioning techniques for the efficient solution of symmetric finite element equations by a preconditioned conjugate gradient method are described.

1. INTRODUCTION

For second order elliptic problems, in particular for general three dimensional problems, there is a need for fast solvers. One effective class of methods which is also fairly easily implemented into a software package, is based on the by now classical preconditioned conjugate gradient method (see for instance [1], [2],[6]).

In the most common and successful preconditioning method, one calculates an approximate factorization LU of the given matrix A and then applies the conjugate gradient method on for instance the matrix $L^{-1}AU^{-1}$. For simplicity, we consider here only symmetric problems and then $U = L^T$.

We survey some preconditioning techniques like 'chopping off methods', modified and standard incomplete factorizations with a priori and/or a posteriori chosen positions of accepted fill in.

In particular, we discuss an adaptive incomplete factorization technique, which utilizes available storage fully, in the respect that the incomplete factorization is made more or less accurate during the factorization process, depending on the relative amount of storage, which is currently available.

The above methods are numerically stable for M-matrices, but may be instable for other classes of matrices. Higher order Lagrangian finite element matrices are in general not M-matrices. For such problems we present a method based on spectral equivalence, where we show that, without degrading the asymptotic rate of convergence, the incomplete factorization may as well be based on the linear finite element matrix (for which it is stable).

ISBN 0-12-632620-7

2. SOME PRECONDITIONING TECHNIQUES, IN PARTICULAR FOR FINITE ELEMENT EQUATIONS

a. "Chopping off" methods

When performing a complete LU-decomposition, one often finds that most of the fill-in entries are quite small, in particular well away from the non-zero entries of A. Hence an obvious method to preserve sparsity during the solution process, would be to neglect all entries in the triangular factors L, U, for which the relative size of the absolute value is small enough. This naturally demands a full factorization to begin with, but since it is done only once, it needs only a few data transfers from backing store. One finds for elliptic problems that often only 5 - 10 % of the fill-in has to be kept along for a satisfactorily accurate approximate factorization, so each iteration step can be performed much faster than in an iterative refinement method. On the other hand, quite a number of iterations have to be performed, so the total CPU-time is not smaller. Hence the advantage with this method is its ability to stay in core during the solution process even for quite large problems.

However, although, most entries in the full factorization will not be used, they have still to be calculated, and this makes the method impractical, for instance on large three-dimensional elliptic problems. The method has been used, among others by [6].

b. Incomplete factorization (IC)

A method to avoid a full factorization is described in a particular caseby [9] and in a more general context by Meijerink, van der Vorst [7], and Beauwens [4]. In this, certain fill-in entries are neglected during the factorization. Only fill-in in certain in advance chosen positions is accepted and neglection of fill-in in each pivot-row is done prior to the elimination. Hence neglected fill-in entries will not cause any fill-in at later stages and the extent of fill-in is controlled.

A problem with incomplete factorization is that the factorization may be unstable, since the pivot entries may become small or negative, even if A itself is positive definite. For M-matrices, Meijerink and van der Vorst prove however, that the factorization is always stable. See also [10] for a more general result.

One easily finds that for discretized elliptic problems, the asymptotic rate of convergence of this method is the same as for the unpreconditioned conjugate gradient method.

c. Modified incomplete factorization (MIC)

In order to get a faster asymptotic rate of convergence, one can easily modify the IC-methods, by adding neglected fill-in entries to the diagonal (or to an off-diagonal positive entry in the same row, see [8]). This method has been examined in [5] and one also finds that if fill-in is accepted in certain in advance determined positions the total execution time decreases as more accurate incomplete factorizations are used but eventually starts to increase when about 100 % or more fill-in (compared to A) is accepted.

To consider the stability of the algorithm, let's define the rowsum of a matrix row as the sum of all entries in the row. It is easily seen that the MIC type methods leave the rowsum of all rows unchanged. This means in particular that if the rowsums of A are non-negative and if A is a positive definite M-matrix, i.e. having all off-diagonal entries non-positive, then the modification step preserves both the positive definiteness, and the M-matrix properties.

However for positive definite matrices, which are not M-matrices, one may loose positive definiteness due to the modification step.

d. Fixed space incomplete factorization

In addition to specify a pattern for the non-zeros in the incomplete factorization we may allow fill-ins subject to a numerical criterion.

In this we move (or just drop) entries if they are fill-ins and if their numerical values relative to the corresponding diagonal elements are less than a relative drop tolerance. In the k'th pivot step we drop $a_{ij}^{(k+1)}$ if

$$|a_{ij}^{(k+1)}| < c \sqrt{a_{ii}^{(k)} a_{jj}^{(k)}} .$$

The amount of fill-in is determined by the size of c. If c is close to zero we obtain almost a complete factorization, while c = 1 corresponds to a factorization which only includes the specified non-zero pathern.

A major problem in factorizing sparse matrices is to determine the amount of storage needed to hold the factorized form. This problem can be overcome by, if necessary, increasing c during the factorization so that we never need more space than the user allocates for the arrays holding the matrix. In [8] it is described how c may be changed automatically during the factorization in order to utilize available storage in a nearly optimal way.

3. SPECTRAL EQUIVALENCE

Let $K_h^{(i)}$, $i = 1,2$ be two discretization matrices corresponding to a self adjoint elliptic differential operator of second order on a region Ω. Let $N = N(h)$ be the number of Lagrangian finite element nodes for $K_h^{(1)}$ and $K_h^{(2)}$ (the same number), where h is a discretization parameter.

Definition. $K_h^{(1)}$, $K_h^{(2)}$ are said to be *spectrally equivalent* if

$$0 < c \le (K_h^{(1)} \underset{\sim}{u}, \underset{\sim}{u}) \,/\, (K_h^{(2)} \underset{\sim}{u}, \underset{\sim}{u}) \le C,$$

$\forall \underset{\sim}{u} \in \mathbb{R}^N$, $\underset{\sim}{u} \neq \underset{\sim}{0}$, where c, C are independent on h.

Consider a finite element mesh consisting of T elements and N Lagrangian nodes. Let element number $\ell(e_1)$ have n_ℓ nodes, $\ell = 1,2,\ldots T$, and let the global matrices be $K_h^{(1)}, K_h^{(2)}$. The matrices local to e_ℓ are $K_{h,\ell}^{(i)}$, $i = 1,2$. These latter are positive semidefinite and all entries corresponding to nodes outside e_ℓ are zero. We have

$$K_h^{(i)} = \sum_{\ell=1}^{T} K_{h,\ell}^{(i)}.$$

Theorem 3.1. Assume that $K_{h,\ell}^{(i)}$, $i = 1,2$, have the same nullspace, that is

$$N_\ell = N(K_{h,\ell}^{(i)}) = N(K_{h,\ell}^{(2)}),\ \ell = 1,2,\ldots,T$$

Let $\mu_{\ell,j}^{(i)}$, $j = 1,2$ be the extreme *positive* eigenvalues of $K_{h,\ell}^{(i)}$, $i = 1,2$ and assume that

(3.1) $$\max_{1\le\ell\le T} \mu_{\ell,2}^{(2)} / \mu_{\ell,1}^{(1)}$$

and

(3.2) $$\max_{1\le\ell\le T} \mu_{\ell,2}^{(1)} / \mu_{\ell,1}^{(2)}$$

are independent of the number of elements N(h). Then $K_h^{(1)}$ and $K_h^{(2)}$ are spectrally equivalent.

Proof. A proof of this theorem can be found in [3].

The finite element matrices $K_h^{(i)}$ have in fact only the constant vector as nullvector.

Let us construct the finite element mesh by a uniform meshdivision starting from a given coarse mesh.

Hence all angles of $\cup e_\ell$ are given by the original mesh and thus independent of the meshparameter h. This implies that the quotients (3.1), (3.2) are independent on h, so Theorem 3.1 implies that $K_h^{(1)}$, $K_h^{(2)}$ are spectrally equivalent.

Let now $K_k^{(1)}$ correspond to piecewise linear approximations and $K_h^{(2)}$ to a piecewise approximation of higher degree. Assuming that all angles in Ue_ℓ are not larger that $\pi/2$, $K_h^{(1)}$ is a positive definite M-matrix. Further, let C_h be a MIC-preconditioning of $K_h^{(1)}$. Hence it is stable. If we use C_h as a preconditioning for $K_h^{(2)}$ we then get an effective spectral condition number

$$H_0 = H(C_h^{-\frac{1}{2}} K_h^{(2)} C_h^{-\frac{1}{2}}) \leq H(C_h^{-\frac{1}{2}} K_h^{(1)} C_h^{-\frac{1}{2}}) H(K_h^{(1)^{-\frac{1}{2}}} K_h^{(2)} K_h^{(1)^{-\frac{1}{2}}}) = O(h^{-1}),$$

$h \to 0$.

Hence, it follows from [2] that the number of iterations for any finite element matrix problem of the above kind, is bounded by $O(h^{-\frac{1}{2}}) \ln h^{-1}$, $h \to 0$.

REFERENCES

1. O. Axelsson (1972), A generalized SSOR method, *BIT 12*, 443-467.
2. O. Axelsson (1976), A class of iterative methods for finite element equations, *Comp. Meth. Appl. Mech. Eng. 9*, 123-137.
3. O. Axelsson, I. Gustafsson (1980), A preconditioned conjugate gradient method for finite element equations, which is stable for rounding errors, IFIP Proceedings 80, to appear.
4. R. Beauwens, L. Quenon (1976), Existence criteria for partial matrix factorizations in iterative methods, *SIAM J. Numer. Anal. 13*, 615-643.
5. I. Gustafsson (1979), Stability and rate of convergence of modified incomplete Cholesky factorization methods, 7902R, Department of Computer Sciences, Chalmers University of Technology, Gothenburg, Sweden.
6. A. Jennings, G.A. Malik (1977), Partial elimination, *J. Inst. Maths. Applics. 20*, 307-316.
7. J.A. Meijerink, H.A. van der Vorst (1977), An iterative solution method for linear systems of which the coefficient matrix is a symmetric M-matrix, *Math. Comp. 31*, 148-162.
8. N. Munksgaard, O. Axelsson (1980), Analysis of incomplete factorization with fixed storage allocation, internal report.
9. R.S. Varga (1960), Factorization and normalized iterative methods, in Boundary Problems in Differential Equations, (R.E. Langer, ed.). University of Wisconsin Press, Madison, 1960, pp. 121-142.
10. R.S. Varga (1980), Incomplete factorizations of matrices and connections with H-matrices, this proceedings.

PREDICTOR-CORRECTOR METHODS FOR THE SOLUTION OF TIME-DEPENDENT PARABOLIC PROBLEMS ON PARALLEL PROCESSORS

Marsha Berger[1]
Joe Oliger[1]

Department of Computer Science
Stanford University
Stanford, California

Garry Rodrigue[2]

Computer Research
Lawrence Livermore National Laboratory
Livermore, California

I. INTRODUCTION

We consider predictor-corrector (P/C) methods for parabolic partial differential equations. Our original motivation for considering these methods was to develop methods which would be easy to implement in an efficient manner on distributed and highly parallel computers. Another objective was to have algorithms which could be easily used for computations which use Lagrangian coordinates over portions of the computational domain and Eulerian coordinates over other portions of the computational domain. The latter problem was considered by Hickman(1) and was our point of departure. His method, and obvious extensions of it, leads to block methods - using either direct or iterative methods on these blocks. These algorithms have a predictor-corrector nature on the "borders" of the

[1]Supported in part by ONR Contract No. N00014-75-C-1132.
[2]Supported in part by DOE/OBES Contract No. W-7405-Eng-48.

ISBN 0-12-632620-7

blocks. This led us to consider P/C methods applied at all points and our preliminary results indicate that the resultant algorithms are both efficient and implementable on scalar and parallel computers.

To begin, consider an implicit method for a parabolic equation, e.g., Crank-Nicholson (CN) on the heat equation. We propose to solve the implicit linear system at each time step by an iterative method and attempt to reduce the iterations by providing a good guess of the solution at the next time level - we use an explicit method, the forward Euler (FE), e.g., for this purpose. Finally, we recognize that we do not need to carry the iteration to convergence. The exact solution of the difference equations is only locally accurate to $O(\Delta t(\Delta t^2 + \Delta x^2))$, say, and so we can stop our iteration when we have approximated this solution to one additional order of accuracy, $O(\Delta t^2(\Delta t^2 + \Delta x^2))$.

Since we are not solving the unconditionally stable difference equations exactly, it is not clear what the stability properties of our proposed P/C methods will be. If we use the FE method as a predictor, then local errors are $O(\Delta t(\Delta t + \Delta x^2))$ and it can be shown that one Jacobi (J) iteration on the CN corrector will improve this by $O(\Delta t)$, which yields a local error of $O(\Delta t^2(\Delta t^2 + \Delta x^2))$ where we can consider stopping. Similar arguments hold for other P/C methods.

II. STABILITY ANALYSIS

The minimum number of iterations needed in solving the implicit corrector equations can be determined by a stability analysis of the model problem $v_t = v_{xx}$. Let $\lambda = \Delta t/\Delta x^2$ and consider the FE method as the predictor and the CN method as the corrector. One J iteration yields the difference scheme

$$(1 + \lambda)\ v_j^{n+1[1]} = v_j^n + \lambda/2\delta^2 v_j^n + \lambda/2\ (v_{j+1}^{n+2[0]} + v_{j-1}^{n+1[0]})$$

where "[•]" denotes iteration count. The magnitude of the amplification factor for this method is <1 for $\lambda<1$. Thus, while the predictor has a stability limit of $\lambda<1/2$, one J iteration doubles the time-step. If we iterate to convergence, the method is of course unconditionally stable. Table I lists the stability limit for λ as a function of the number of J iterations for this particular P/C method.

TABLE I.

No./Iterations	λ	No./Iterations	λ
1	1.0	6	1.8
2	1.0	7	3.3
3	1.8	8	2.2
4	1.4	9	3.9
5	2.6	10	2.6

A similar analysis can be done using Gauss-Siedel (GS) iteration for the above P/C method. A loose bound on the amplification matrix of this method yields stability for $\lambda \leq 1.1$ for one GS iteration, a small improvement over that of using a J iteration. In actual fact, $\lambda < 1.7$ yields a stable method. Other P/C combinations are being studied.

III. EXPERIMENTS

The heat equation $U_t - U_{xx}$ given in [3, p. 5] was solved with $\Delta x = .05$, $\Delta t = .0025$. Table II records the relative L_2 - error at $t = .05$ for predictors given by the previous solution (PS), Dufort-Frankel (DF), and FE, with a CN corrector, and J and GS iterative methods (here GSn and Jn denote n iterations). The last column of the table indicates the number of iterations to achieve full convergence for the corrector. If CN were used alone as a method, an L_2 error of 9.43 E-4 would be achieved at $t = .05$.

TABLE II.

	GS1	GS2	J1	J2	It/Conv
PS	1.77 E-1	5.42 E-2	2.62 E-1	1.23 E-1	10
DF	4.85 E-3	1.00 E-3	4.29 E-3	1.34 E-3	6
FE	2.06 E-3	1.63 E-4	1.97 E-3	4.86 E-4	6

Table III uses the second-order BE as a corrector where an L_2-error of 2.4 E-4 would be attained if this method were used alone.

TABLE III.

	GS1	GS2	J2	It/Conv
PS	2.74 E-1	1.11 E-1	2.11 E-1	11
DF	7.42 E-3	3.43 E-3	4.00 E-3	7
FE	1.94 E-3	1.09 E-3	2.17 E-3	6

The P/C method was used to solve the 2-D heat diffusion equation $\phi_t = \nabla \cdot \Lambda \nabla \phi$ given in [2]. The predictor is the FE method and the corrector is the Backward Euler (BE) method with the point-Jacobi conjugate gradient method as the iterative method. The problem is solved to steady-state with the time-step controlled so as to allow only a 15% change in ϕ (with an upper limit of $\Delta t = 10$). At each time level, the corrector is iterated to convergence (i.e., normalized L_2 residual in $<10^{-4}$) and the number of required iterations using predictors PS and FE are recorded. The first column indicates the time level.

TABLE IV.

n	Δt_n	itns/PS	itns/FE
10	1.5×10^{-2}	3	1
60	1.4	19	11
100	10.	42	28
140	10.	35	13
170	10.	30	13

Table V uses the CN method as a corrector.

TABLE V.

n	Δt_n	itns./PS	itns./FE
10	3.8×10^{-1}	5	4
60	10.	38	20
100	10.	30	17
140	10.	26	17

REFERENCES

1. Hickman, R. B., Lawrence Livermore National Laboratory Report UCRL-52471 (Thesis).
2. Rodrigue, G., Hendrickson, C., Pratt, M., Lawrence Livermore National Laboratory Report UCRL-82930 (Rev. 1).
3. Richtmyer, R. D., Martin, K. W., "Difference Methods for Initial-Value Problems".

EFFICIENT SOLUTION OF THE BIHARMONIC EQUATION

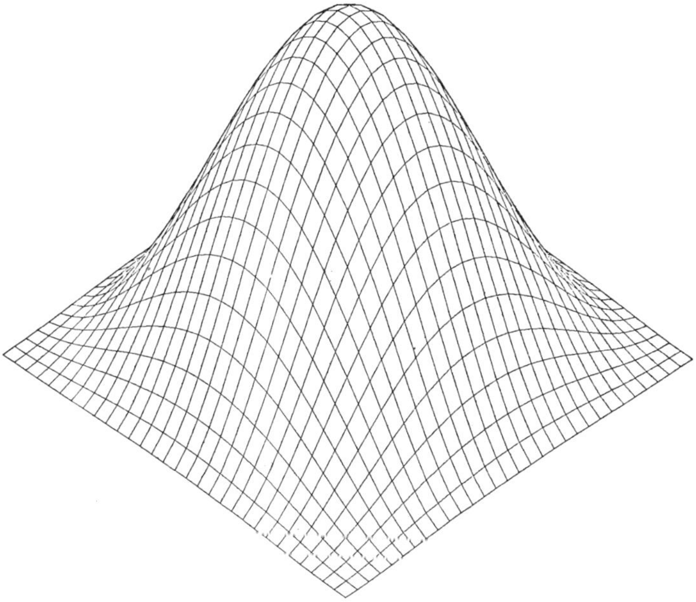

Petter Bjørstad[1]

Computer Science Department
Stanford University
Stanford, California

A new method for the numerical solution of the first biharmonic problem in a rectangular region is outlined. The theoretical complexity of the method is $N^2 + O(N)$ storage and $O(N^2)$ arithmetic operations. (In order to achieve a prescribed accuracy on an N by N grid.) Numerical results from a computer code that requires

---

[1]Supported in part by the Department of Energy Contract DE-AC03-76SF00326-PA #30.

$aN^2 + bN^2 \log N + O(N)$ operations with $b << a$, are presented using both a scalar and a vector computer. Extensions and some applications of the method for solving eigenvalue problems and certain nonlinear problems are mentioned.

I. INTRODUCTION

Consider the (first) Dirichlet problem for the biharmonic operator in a square S :

$$\begin{aligned} \Delta^2 u(x,y) &= f(x,y) & (x,y) &\in S \\ u(x,y) &= g(x,y) & (x,y) &\in \partial S \\ u_n(x,y) &= h(x,y) & (x,y) &\in \partial S . \end{aligned} \tag{1}$$

Here u_n denotes the normal derivative of u with respect to the exterior normal. Let S be covered by a regular grid with N^2 interior gridpoints. The biharmonic operator is approximated using the 13-point stencil

$$\begin{array}{ccccc} & & 1 & & \\ & 2 & -8 & 2 & \\ 1 & -8 & 20 & -8 & 1 \\ & 2 & -8 & 2 & \\ & & 1 & & \end{array}$$

combined with quadratic extrapolation near the boundary. This scheme has truncation error $||u = u_h||$ of order h^2 where $h = 1/(N+1)$. The discrete problem can be written as

$$\frac{1}{h^4} A\, u_h = b$$

where A is defined by the discretization defined above and b is a function of the data f, g and h .

Many methods for the numerical solution of the above linear system have been proposed. They can roughly be classified as follows:

i) Iterative methods working on the matrix A .

ii) Direct methods working on the matrix A .

iii) Iterative methods based on reducing the biharmonic problem to a coupled system of two second order equations involving the Laplace operator.

iv) Direct methods taking advantage of the fact that A can be split into $L^2 + V$, where L is the discrete Laplace operator and V has low rank.

The first approach i) can be found in many early papers on the subject (6), (13). Its main disadvantage is related to the fact that A has condition number proportional to N^4 resulting in slow convergence of the iterative techniques.

Approach ii) is discussed in (2), (12) and (15). The theoretical complexity of a direct method is $O(N^3)$ arithmetic operations and $O(N^2 \log N)$ storage locations using for example nested dissection. This and other sparse matrix methods for the problem were studied and compared in (15). This study indicates that the constants in the above estimates are quite large and that a regular band solver is competitive even when the number of unknowns approach one thousand. A block elimination scheme is discussed in (2), while a band solver is being used in (12). Both these methods require a prohibitive amount of storage if N is large and they have a typical running time proportional to N^4 , unacceptable for fine grid calculations.

The third and fourth approach is essentially two different ways of looking at the same underlying problem. A method based on iii) above was introduced by Smith in (16). It had a running time of $O(N^3)$. This was later improved to $O(N^{5/2})$ by Ehrlich (9) and Smith (17), (18). A drawback is the need to estimate iteration parameters.

The last approach iv) was pioneered by Golub (11) and a refined implementation is given by Buzbee and Dorr (5). This implementation, which is a direct method, requires $O(N^3)$ arithmetic operations. Despite being an $O(N^3)$ method it proved very competitive with the $O(N^{5/2})$ methods on realistic problems because those methods have an actual cost of $cN^{5/2}$ with c rather large.

Based on the above results it was concluded in (14) that the solution of the first biharmonic problem was an order of magnitude more difficult than the solution of Poisson's equation, even on parallel computers. The results of this paper show that the problems have the same complexity when a prescribed accuracy is desired.

II. OUTLINE OF THEORY

A brief description of the theory behind the new method will now be given. For a more detailed analysis with proofs and computer algorithms see (3) and (4).

Define the elementary matrices

$$R = \text{Tridiagonal } [-1,2,-1]_{N\times N}$$

and

$$T = \text{Diagonal } [1,0,\ldots 0,1]_{N\times N} .$$

The matrix A can be written

$$A = [(I \otimes R) + (R \otimes I)]^2 + 2(T \otimes I) + 2(I \otimes T)$$

where I is an N by N identity matrix and standard tensor product notation is used. Notice that the matrix

$$L = (I \otimes R) + (R \otimes I)$$

is the standard 5-point difference approximation to the Laplace operator.

Let

$$B = L^2 + 2(T \otimes I) .$$

Solving linear systems involving this matrix is no more difficult than solving Poisson's equations. Several $O(N^2)$ methods for this problem are known (1), (8) and such a method must be used in order to obtain an $O(N^2)$ method for the present problem. (Alternatively, an $O(N^2 \log N)$ method can be used, resulting in an overall operation count of $O(N^2 \log N)$

The matrix $2(I \otimes T)$ can be written

$$2(I \otimes T) = 2\, U\, U^T$$

where U is an N^2 by 2N matrix.

Using this and the generalization of the Sherman-Morrison formula (7) gives

$$A^{-1} = B^{-1}(I - 2\, U\, C^{-1}\, U^T\, B^{-1})$$

where

$$C_{2N\times 2N} = I + 2\, U^T\, B^{-1}\, U .$$

It is clear that the original problem can be solved efficiently if there is a way to solve the linear system of equations involving the matrix C in no more than (N^2) operations.

Define a real normalized sine transform of a vector $x \in R^n$ by the relation

$$y = Q_N x ,$$

and note that

$$Q_N = Q_N^{-1} = Q_N^T .$$

Let S_k be block number k in the block diagonal matrix

$$R(I \otimes Q_N) B (I \otimes Q_N) R^T$$

where R is the permutation matrix defined by

$$R(D \otimes E)R^T = E \otimes D .$$

Similarly, define $\tilde{S}_k$ to be the corresponding block when B is replaced by L^2. Let $Px \equiv \begin{pmatrix} x_1 \\ x_2 \end{pmatrix}$ and $Py \equiv \begin{pmatrix} y_1 \\ y_2 \end{pmatrix}$ where x_1 contains the odd numbered components of x, while x_2 contains the even numbered components.

The following theorems have important implications for the construction of efficient numerical methods.

Theorem 1.

Solving the linear system $Cx = y$ is equivalent to solving the two linear systems

$$T_1(x_1 + x_2) = y_1 + y_2$$

$$T_2(x_1 - x_2) = y_1 - y_2$$

where

$$T_1 = I + \frac{8}{N+1} \sum_{k=1,3\ldots}^{N} \sin^2 \frac{k\pi}{N+1} S_k^{-1}$$

$$T_2 = I + \frac{8}{N+1} \sum_{k=2,4\ldots}^{N} \sin^2 \frac{k\pi}{N+1} S_k^{-1}$$

Theorem 2.

Let T_i denote T_1 or T_2 and let $\tilde{T}_i$ be the corresponding matrix defined using $\tilde{S}_k$ instead of S_k. Then

$$P(Q\tilde{T}_i Q)^{-\frac{1}{2}} Q T_i Q (Q \tilde{T}_i Q)^{-\frac{1}{2}} P^T = I - \begin{bmatrix} (F^{i1})^T F^{i1} & O \\ O & (F^{i2})^T F^{i2} \end{bmatrix}$$

where all the singular values of F^{i1} and F^{i2} are in the interval

$$0 \leq \sigma < 0.8 .$$

Theorem 1 and 2 are proved in (3).

Since linear systems involving $\tilde{T}_i$ can be solved easily and since a matrix-vector product $T_i x$ can be formed at a cost of $O(N^2)$ arithmetic operations, it follows that the conjugate gradient method can be used to solve linear systems having coefficient matrix T_i . (Steepest descent would also work.) Moreover, the number of iterations required to achieve a prescribed accuracy is constant, independent of the grid size N .

III. COMPUTER IMPLEMENTATIONS

A large number of different computer implementations of the above ideas are possible. Which one is best will depend on the computer used and also on the underlying application. For a discussion of different alternatives see (3).

	N = 255		N = 511	
	SOLV	FFT	SOLV	FFT
IBM 370/168	7109	4324	-	-
CRAY-1 OFF=v	1804	882	-	-
CRAY-1 ON=v	251	548	878	2148

FIGURE 1. Time in milliseconds to solve the biharmonic equation on an NxN grid.

It is possible to design implementations that are well suited for parallel and vector computers. In order to illustrate this, a particular computer implementation was used on an IBM 370/168 and also on the CRAY-1 computer. Figure 1 displays some preliminary timing results.

Remarks:

i) The total solution time is the sum of the time spent in a fast Fourier transform routine (FFT) and in the remainder of the code (SOLV).

ii) The FORTRAN H(OPT=3) compiler was used on the 168 while the CFT compiler on the CRAY-1 was used with and without the vectorization option. (ON=v and OFF=v)

iii) The same FORTRAN code was used in all three cases with the single exception that a special vector innerproduct routine (written by Oscar Buneman) was used in the vectorized run.

iv) The iterative part of the algorithm was terminated when the 2-norm of the residual fell below 10^{-10}. (This typically takes less than 10 iterations.)

v) No attempt was made to optimize the Fourier transform part of the code. (The time spent in this part of the code can be substantially reduced, by doing the Fourier transforms in parallel.) The table should illustrate the practical behavior of the new algorithm; it does not represent a fully optimized implementation.

vi) It is estimated that the N=255 case would take at least 50 times longer using any of the earlier methods mentioned (aside from the difficulties of running previous algorithms in core for this grid size).

IV. EXTENSIONS AND APPLICATIONS

The handling of $N \times M$ grids $(N \neq M)$ is straightforward. It is also possible to solve more general fourth order problems having separable lower order terms in addition to the biharmonic operator. In particular, it is useful to be able to solve the following generalization:

$$\begin{aligned} \Delta^2 u + \alpha \Delta u + \beta u &= f \quad \text{in } R \\ u &= g \quad \text{on } \partial R \\ u_n &= h \quad \text{on } \partial R \end{aligned} \tag{2}$$

where α and β are two scalar parameters, and R is a rectangle. Formulation (2) was used (with $\alpha = 0$) in a Rayleigh iteration to compute a few of the eigenvalues of the biharmonic operator. Due to the cubic rate of convergence, this method is more satisfactory than the method reported in (2). Figure 2 shows the lowest eigenfunction near a corner. The entire plot is inside the 4 cells closest to the corner in the picture shown at the beginning of this paper. This reveals the existence of nodal lines near the corner and also gives a good description of the eigenfunction in this area. The presence of this phenomenon has been suspected, but due to the gridsize limitations of earlier algorithms, it has not been clearly demonstrated before. The theory outlined in Part II can be used to analyze this behavior also in the continuous case.

The existence of an efficient biharmonic solver makes it a useful computational tool in the construction of numerical methods for more complicated fourth order problems much in the same way as fast Poisson solvers have been used in the past ten years. As an illustration consider the driven cavity

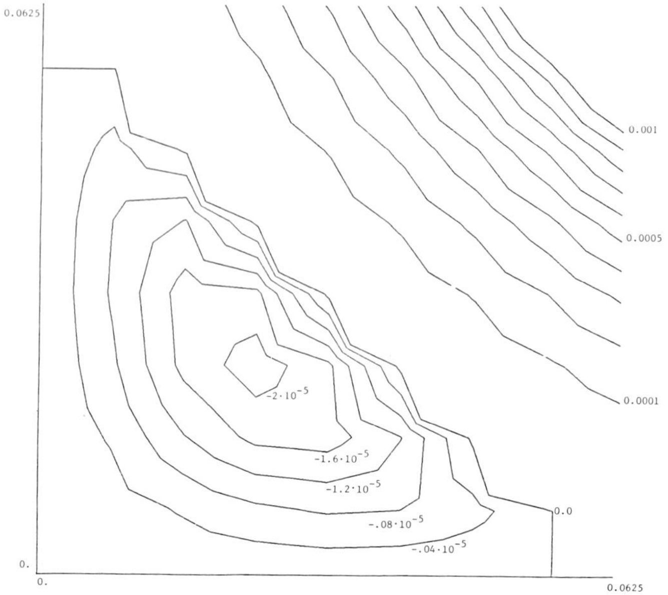

N = 127

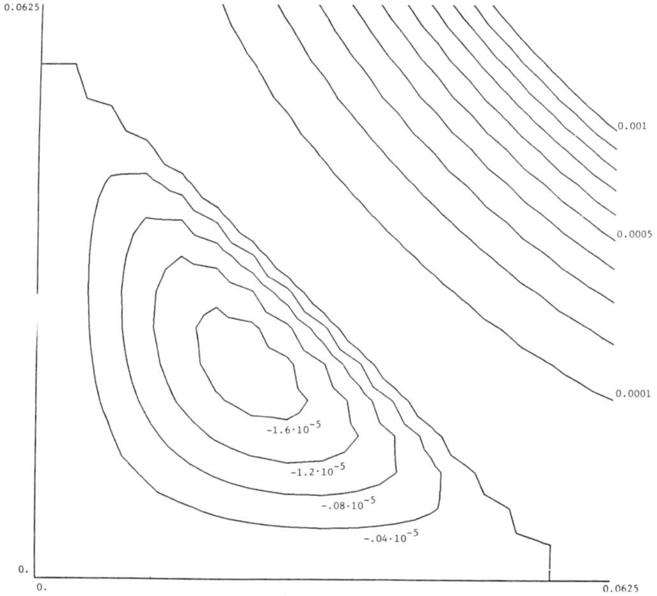

N = 255

FIGURE 2. Contour plots of the first biharmonic eigenfunction near a corner.

model problem for the nonlinear, time dependent Navier Stokes equation (19). Introducing a stream function Ψ in the usual way, the equation was solved using the following scheme:

$$\Delta^2 \Psi_{k+1} - \frac{R}{\Delta_t} \Delta \Psi_{k+1} = R(\Psi_y \Delta \Psi_x - \Psi_x \Delta \Psi_y)_k - \frac{R}{\Delta_t} \Delta \Psi_k$$

where k denotes the current time level. Notice that this again is a special case of (2) (this time with $\beta = 0$). This problem was solved in a square region with Reynold's number $R = 200$ and boundary conditions $\Psi = 0$ and $\Psi_n = 0$ except at the side $y = 1$ where

$$\Psi_n = \begin{cases} -\sin t & 0 \leqslant t \leqslant \pi/2 \\ -1 & t \geqslant \pi/2 . \end{cases}$$

This corresponds to an acceleration of the moving wall up to the standard velocity used in the stationary case. A 31×31 grid was used and 500 timesteps each of length 0.01 were used. (This is smaller than required for stability with this Reynold's number.) The run required approximately one minute on an IBM 370/168 and the velocity fields are shown in Figure 3a at two different time levels. It should be mentioned that the above scheme is unsatisfactory for large Reynold's number, due to the fact that the nonlinear term is fully explicit. The flow is not stationary at time 5, but changes very slowly into a final state after time equal 20 with a main vortex $\Psi = .105$ at coordinates (.41, .66) in agreement with stationary calculations.

ACKNOWLEDGMENTS

The author wishes to express his appreciation to Gene Golub for suggesting this problem and for contributing in many interesting discussions. Thanks are also due to Professor

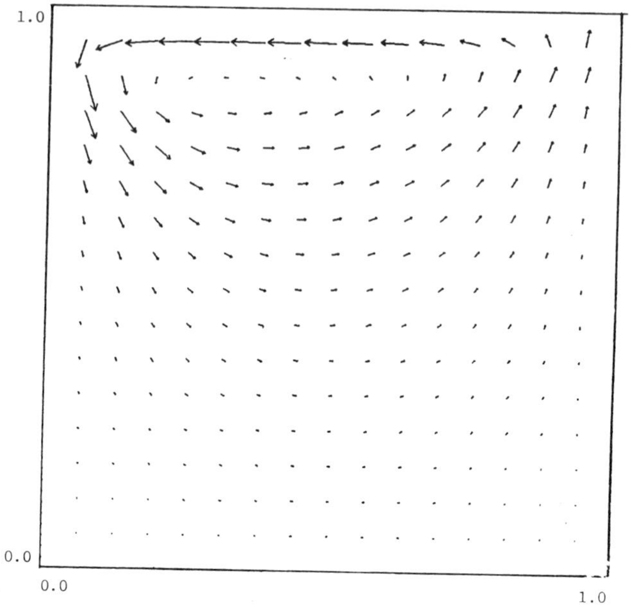

Velocity field t = 1.5

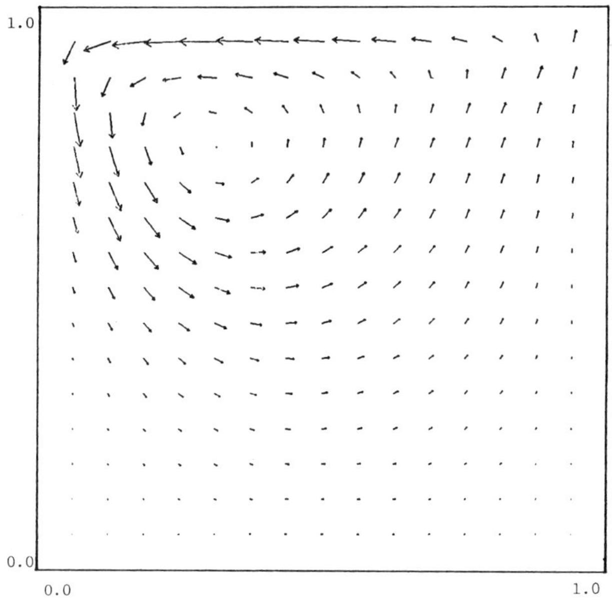

Velocity field t = 5.0

FIGURE 3a. Discrete solution of the time dependent Navier-Stokes equation at Reynold's number 200.

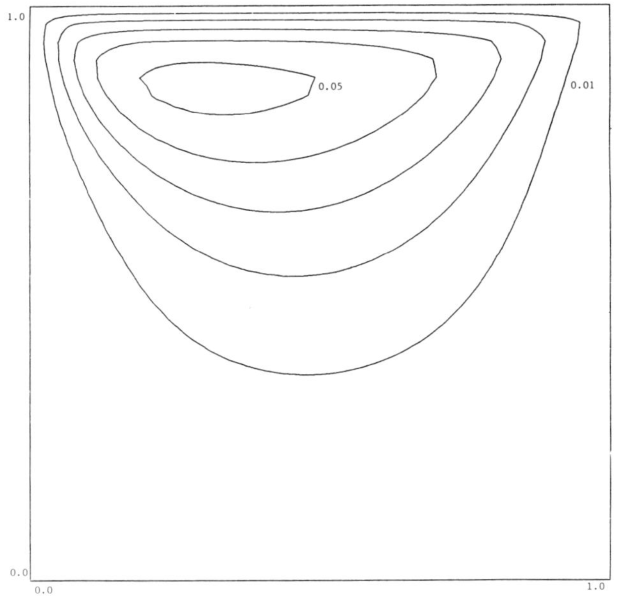

Stream function t = 1.5

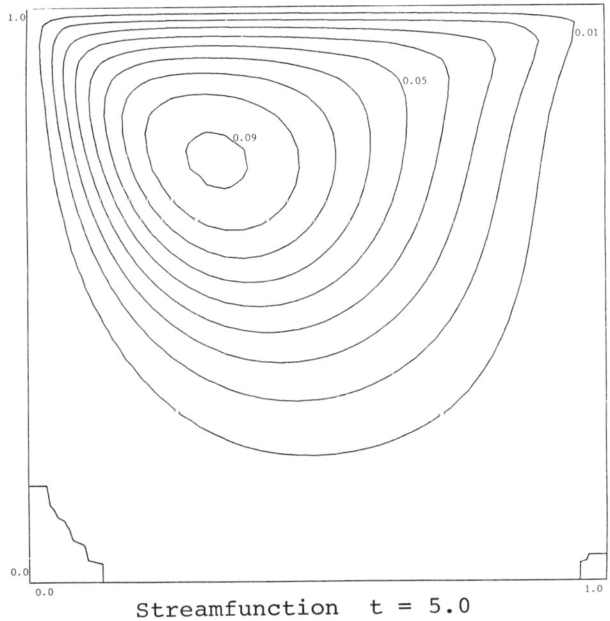

Streamfunction t = 5.0

FIGURE 3b. Discrete solution of the time dependent Navier-Stokes equation at Reynold's number 200.

Oscar Buneman of Stanford University for making the experiments on the CRAY-1 possible, and the Stanford Linear Accelerator for providing good computing facilities.

REFERENCES

1. Bank, R. E., and Rose, D. J. (1977). Marching algorithms for elliptic boundary value problems. I: The constant coefficient case. SIAM J. Numer. Anal. 14, pp. 792-829.
2. Bauer, L., and Reiss, E. L. (1972). Block five diagonal matrices and the fast numerical solution of the biharmonic equation. Math. Comp. 26, pp. 311-326.
3. Bjørstad, P. E. (1980). Numerical solution of the biharmonic equation, Ph.D. Thesis, Stanford University.
4. Bjørstad, P. E. Fast numerical solution of the biharmonic equation. To appear.
5. Buzbee, B. L., and Dorr, F. W. (1974). The discrete solution of the biharmonic equation on rectangular regions and the Poisson equation on irregular regions. SIAM J. Numer. Anal. 11, pp. 753-763.
6. Conte, S. D., and Dames, R. T. (1960). On an alternating direction method for solving the plate problem with mixed boundary conditions. J. Assoc. Comput. Mach. 7, pp. 264-273.
7. Dahlquist, G., and Björck, Å. (1974). Numerical methods. Prentice-Hall, Inc., New Jersey.
8. Detyna, E. (1979). Point cyclic reductions for elliptic boundary value problems. I: The constant coefficient case. J. Comp. Phys. 33, pp. 204-216.

9. Ehrlich, L. W. (1971). Solving the biharmonic equation as coupled finite difference equations. SIAM J. Numer. Anal. 8, pp. 278-287.

10. Ehrlich, L. W. (1973). Solving the biharmonic equation in a square: A direct versus a semidirect method. Comm. ACM, 16, pp. 711-714.

11. Golub, G. H. An algorithm for the discrete biharmonic equation. Unpublished. See Appendix of (10).

12. Gupta, M. M., and Manohar, R. P. (1979). Direct solution of the biharmonic equation using non-coupled approach. J. Comp. Phys. 33, pp. 236-248.

13. Parter, S. V. (1959). On two line iterative methods for the Laplace and biharmonic difference equations. Num. Mat. 1, pp. 240-252.

14. Sameh, A. H., Chen, S. C., and Kuck, O. J. (1976). Parallel Poisson and biharmonic solvers. Computing 17, pp. 219-230.

15. Sherman, A. H. (1975). On the efficient solution of sparse systems of linear and nonlinear equations. Ph.D. Thesis. Dept. of Comp. Sci., Yale Univ.

16. Smith, J. (1968). The coupled equation approach to the numerical solution of the biharmonic equation by finite differences, I. SIAM J. Numer. Anal. 5, pp. 323-339.

17. Smith, J. (1970). The coupled equation approach to the numerical solution of the biharmonic equation by finite differences, II. SIAM J. Numer. Anal. 7, pp. 104-111.

18. Smith, J. (1973). On the approximate solution of the first boundary value problem for $\nabla^4 u = f$. SIAM J. Numer. Anal. 10, pp. 967-982.

19. Teman, R. (1977). Navier Stokes Equations, North Holland, Amsterdam.

ATTAINABLE ACCURACY OF COMPACT DISCRETIZATIONS OF THE POISSON EQUATION

Ronald F. Boisvert

Center for Applied Mathematics
National Bureau of Standards
Washington, DC

I. INTRODUCTION

In this paper I outline results on attainable order of truncation error of finite difference discretizations of Poisson's equation, $\nabla^2 u = g$, defined on a bounded two-dimensional domain R with boundary ∂R and boundary data ϕ. Details are presented elsewhere [2], [3].

Let G denote the intersections of a rectangular grid placed over R. With each point $p = (x_p, y_p) \in G \cap R$ associate a grid cell $R_p = S_p \cap R$ and a boundary segment $\partial R_p = S_p \cap \partial R$, where S_p is the grid rectangle centered at p. We consider discretizations of the form

$$(1/h^2)\ L_p U = I_p g + (1/h^2)\ J_p \phi \qquad p \in G \cap R \tag{1}$$

$$L_p U \equiv \sum_{q \in D_p} \alpha_{pq} U_q, \quad I_p g \equiv \sum_{q \in E_p} \beta_{pq} g_q, \quad J_p \phi \equiv \sum_{q \in B_p} \gamma_{pq} \phi_q$$

where $D_p = R_p \cap G$ (discretization points), and $E_p \subset R_p \cup \partial R_p$ and $B_p \subset \partial R_p$ are additional discrete points sets (evaluation points and boundary evaluation points, respectively) that need not be grid intersections. Unless p is near ∂R, B_p is empty and D_p is a 3×3 grid of mesh points. These compact as

ISBN 0-12-632620-7

possible formulas are called HODIE methods (for High Order Differences with Identity Expansion) [6].

Traditionally, finite difference methods are derived by computing the Taylor series expansion of the local truncation error $T_p u \equiv I_p \nabla^2 u + J_p u - L_p u$ about p and determining coefficients α_{pq}, β_{pq}, γ_{pq} so that its leading terms are zero. The power of h in the first non-zero term is the order of the local error. Equivalently, we can determine the coefficients so that $T_p v \equiv 0$ for all $v \in P_n$, the space of polynomials of degree n or less. A formula of the form (1) which is exact on P_{n+1} for Poisson's equation has local error $O(h^n)$ for $u \in C^{n+2}$. One usually seeks discretizations of positive type [4] so that the local error is easily related to the global discretization error $e = ||U-u||$, which is $O(h^n)$ in this case.

II. OPTIMALITY RESULTS

One might believe that arbitrarily high order methods could be constructed by taking a sufficient number of evaluation points in (1) (this is the case for ordinary differential equations [5]). Unfortunately, we have

Theorem 2.1 Let k be the number of points in $D_p \cup B_p$. Suppose there exist harmonic polynomials $\psi_1, \ldots, \psi_k$ in P_{n+1} such that the matrix with elements $a_{ij} = \psi_i(x_j, y_j)$, $(x_j, y_j) \in D_p \cup B_p$ is non-singular. Then no formula of the form (1) can have truncation error smaller than $O(h^{n-1})$.

This result is independent of the number of evaluations of g. The conditions of Theorem 2.1 are sufficient to show that $\alpha_{pq} = 0$ for all $q \in D_p$. One can argue that no useful formula exists if simply $\alpha_{pp} = 0$. This yields

Theorem 2.2 Suppose there exists a harmonic polynomial $\psi \varepsilon P_{n+1}$ such that $\psi(x,y) = 0$ for each $(x,y) \varepsilon D_p \cup B_p$ except for the central point $p = (x_p, y_p)$. Then no formula of the form (1) can determine u_p with truncation error smaller than $O(h^{n-1})$.

These ideas have been used by several authors to obtain optimality results for p in the interior of R. In [1] Birkhoff and Gulati show that optimal order accuracy for 5-point formulas is $O(h^2)$ for square R_p and $O(h)$ for rectangular R_p. In [6] Lynch and Rice present an $O(h^6)$ 9-point discretization and show that it attains optimal order accuracy for square grids. For rectangular grids they show that only $O(h^4)$ is possible; a class of such discretizations is given in [2]. In the remainder of this paper I will indicate results I have obtained for the case in which p is near ∂R.

It is sufficient to consider the case $p = (0,0)$. We assume a square grid and first treat the case where ∂R_p is a curve passing through $(-h,0)$, $(0,-\theta h)$, $(h,-h)$, $0<\theta\leq 1$ (Fig. 1a). If we limit the set of boundary discretization points to these three, then using Theorem 2.2 one obtains

Theorem 2.3 Let R_p be the irregular cell of Figure 1a and $B_p = \{(-h,0), (0,-\theta h), (h,-h)\}$. Then no discretization of the form (1) can have truncation error smaller than $O(h^2)$.

However, (1) allows an arbitrarily large number of evaluations of the boundary data ϕ along ∂R_p; and indeed, this leads to higher order formulas, although limitations remain even in the simplest of cases. If ∂R_p is the line segment connecting $(-h,0)$ and $(h,-h)$, (Fig. 1b), then a fifth degree harmonic polynomial can be found which is zero on $D_p \cup \partial R_p - \{(0,0)\}$. A

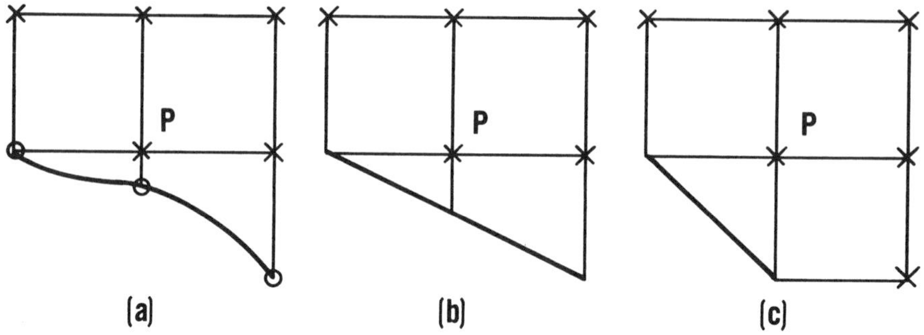

FIGURE 1. Irregular cells (X = discretization point).

seventh degree harmonic polynomial with similar properties can be constructed when ∂R_p is the line segment connecting (-h,0) and (0,-h), (Fig. 1c). This proves

Theorem 2.4 No formula of the form (1) can have more than $O(h^3)$ local accuracy when R_p is the irregular cell of Figure 1b or more than $O(h^5)$ when R_p is that of Figure 1c.

Discretizations attaining the optimum values are given in [3].

III. THE AFFECT ON GLOBAL ERROR

It is well known [4] that to achieve $O(h^n)$ global error we normally need only use formulas with local error $O(h^{n-2})$ at irregular points. Hence we may still obtain sixth order convergence when irregular cells of the form 1c are present, but no better than fifth order when cells 1b are present.

We give one example of an $O(h^4)$ method that uses formulas with $O(h^2)$ local error for its irregular points (Table 1) to indicate how little effect these can have even on coarse grids, and use it to solve $\nabla^2 u = 2\exp(x+y)$ on the trapezoid connecting (0,1), (1,1), (1,0), (0,1/2). See Table 2.

TABLE 1. Discretizations for Example Problem

Cell	Formula*
Square $O(h^4)$	$[-20u(0,0) + 4(u(h,0) + u(0,h) + u(-h,0) + u(0,-h)) + u(h,h) + u(-h,h) + u(-h,-h) + u(h,-h)]/6h^2 = g(0,0)/3 + [g(\eta,\eta) + g(-\eta,\eta) + g(-\eta,-\eta) + g(\eta,-\eta)]/6$
Fig 1b $O(h^2)$	$[-34u(0,0) + 5u(h,0) + 6u(0,h) + u(h,h) + u(-h,h)]/2h^2 = 2g(0,0) + g(0,\eta) - [16\phi(0,-\eta) + 5\phi(-h,0)]/2h^2$
Fig 1c $O(h^2)$	$[-4u(0,0) + u(h,0) + u(0,h)]/h^2 = g(0,0) - [\phi(-h,0) + \phi(0,-h)]/h^2$

$*\eta = h/2$

TABLE 2. Results for Example on CDC 6500

H	Max-error	Rate of convergence	Percentage of irregular points
1/8	4.6 E-5	--	19
1/10	2.0 E-5	3.8	15
1/12	9.7 E-6	3.9	12
1/14	5.3 E-6	3.9	10
1/16	3.2 E-6	3.9	9
1/18	2.0 E-6	3.9	8
1/20	1.3 E-6	4.0	7

REFERENCES

1. G. Birkhoff and S. Gulati, Optimal few-point discretizations of linear source problems, SIAM J. Num. Anal., 11 (1974), pp. 700-728.

2. R. F. Boisvert, High Order Finite Difference Methods for Elliptic Boundary Value Problems, Ph.D. Thesis, Purdue University, 1979.

3. R. F. Boisvert, Optimal order compact discretizations of Poisson's equation, in preparation.

4. J. H. Bramble and B. E. Hubbard, A theorem on error estimation for finite difference analogues of the Dirichlet problem for elliptic equations, Contr. to Diff. Eqns., 2, John Wiley & Sons, NY (1963) pp. 319-340.

5. R. E. Lynch, and J. R. Rice, A high order difference method for differential equations, Math. Comp., 34 (1980), pp. 333-372.

6. R. E. Lynch and J. R. Rice, High accurate finite difference approximations to solutions of elliptic partial differential equations, Proc. Nat. Acad. Sci., 75 (1979), pp. 2541-2544.

THE CONCEPT OF RIGIDITY AND ITS IMPLEMENTATION

Alain Bossavit

Electricité de France
Clamart, France

I. INTRODUCTION

"Rigidities" are abstract types corresponding to boundary operators.

Program objects fall into definite classes, or "types": INTEGERS, REALS, etc. We speak of "abstract types" when dealing with functional properties of these objects, the ones which should be verified in any language.

Given a domain and a function w on its boundary, the boundary operator is the one which associates with w the normal boundary derivative of the harmonic lifting of w in the domain.

These operators play a key role in a large class of elliptic PDE problems. In fact, for the class of problems we have in mind, the algorithms to solve them can be described entirely in terms of boundary operators and a few related concepts. In this way the description, which relies on the concept of "substructuration," can be kept direct and simple.

But, as is well known, it is not so easy to implement such algorithms, because abstract mathematical objects, conceptually simple as they may be, cannot be directly associated with actual program objects.

The package we present here has been made to overcome this

ISBN 0-12-632620-7

difficulty. It is a set of subroutines which allow the programmer to manipulate rigidities and other related types as easily as integers, arrays, etc. Given an algorithm for a problem of the target class, one can write, by using the package, a Fortran code structurally similar to the description of the algorithm in natural language.

II. CONCEPTS AND ASSOCIATED TYPES

We shall consider domains, their boundaries, and parts of these. The corresponding program objects are SUPPORTS. (Capitals will refer to types or Fortran subroutines.) Supports are implemented as sets of labels for the variables. Therefore, the correspondence boundary--support involves a discretization. It is done automatically by the system once supports have been associated to domains or to parts of boundaries.

Just as boundary parts support functions (traces and normal derivatives of the solution), SUPPORTS will support TRACES. Traces can be added together and locally modified (see MODTRA below) on specified parts of their support.

To each pair: domain Ω--part Γ of its boundary, corresponds an operator $R \in L(H^{1/2}(\Gamma), H^{-1/2}(\Gamma))$, called RIGIDITY as a program object, implemented of course as a matrix. R is the operator associated with the bilinear form

$$(Rw, w) = \inf \{ \int_\Omega \nabla^2\phi \mid \phi \in H^1(\Omega), \phi_{|\Gamma} = w\}.$$

It can be alternatively defined by $Rw = \partial u/\partial n_{|\Gamma}$, where u is the solution of the elliptic problem

$$- \Delta u = 0, u_{|\Gamma} = w, \partial u/\partial n = 0 \text{ on } \partial\Omega - \Gamma.$$

Rigidities can be assembled (the well-known finite-element concept) by the ASSBLY routine. The "condensation" (subroutine CNDENS) is this: Let γ be a part of Γ; the corresponding rigidity (w.r. to the same domain Ω) is the operator r of the form

$$(ru, u) = \inf \{(Rw, w) \mid w_{|\gamma} = u\}$$

and the operation required to get r is called "condensation of R with respect to γ". Algebraically, if R is the matrix $(\{R_{11}, R_{12}\}, \{R_{21}, R_{22}\})$, r is $R_{11} - R_{12}R_{22}^{-1}R_{21}$. The routine to compute r, corner-stone of the package, is a "selective Cholesky factorization" as described in Ref. (1).

III. DESIGN PRINCIPLES AND SCOPE

The system implements the sub-structure approach. Suppose we want to solve the stationary heat equation in Ω (Fig. 1) as a means to compute the flux through Γ_2 or Γ_6, where Dirichlet conditions u_2 and u_6 are imposed, Neuman conditions being assumed elsewhere. For some reason, it is desirable, or may be easier, to treat Ω_1 and Ω_2 separately. How can we do it?

The procedure is easy to describe, with the above concepts:

- associate supports to Σ and to the Γ_i, to $\partial\Omega_1$ and $\partial\Omega_2$,
- compute rigidities for the pairs $(\Omega_i, \partial\Omega_i)$, i = 1, 2,
- by an assembly, obtain R associated to $(\Omega, \partial\Omega \cup \Sigma)$,
- merge Γ_6 and Γ_2 into Γ_{26},
- condense R with respect to Γ_{26}.

The result r is the operator which gives $\partial u/\partial n$ on Γ_2 and Γ_6 in terms of u_2 and u_6. Therefore it remains to

- create a trace u equal to u_i, i = 2, 6, and a trace v

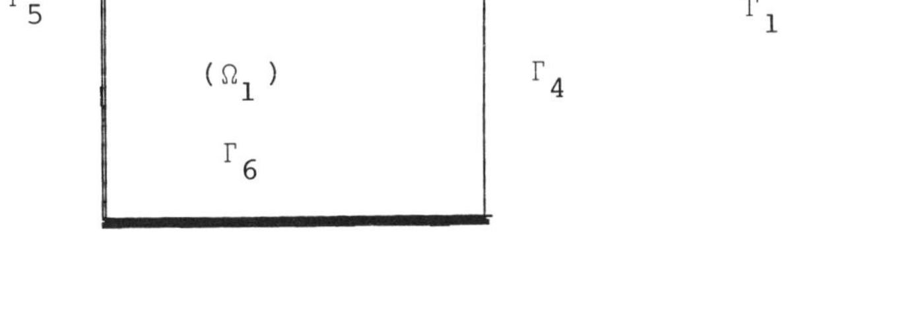

FIGURE 1. A simple example of substructuration

```
C                         SAMPLE PROGRAM
C
C     COMPUTE THE FLUX THROUGH AN L-SHAPED BODY, USING SUB-
C     STRUCTURATION.
C          SUB-REGIONS 1 AND 2 HAVE DIFFUSIVITIES DIF1 AND DIF2.
C
      INTEGER S(7), P(7), OR(8)
      INTEGER P1TO4, P4TO7, R1, R2, S26, T
      REAL DIF1, DIF2
      DATA DIF1 /5./, DIF2 /2.4/
C
C     List the vertices of the subdomains (POINTS declarations)
           CALL DCLPNT (P(1), 0., 0.)
           CALL DCLPNT (P(2), 2., 0.)
           CALL DCLPNT (P(3), 2., 1.)
           CALL DCLPNT (P(4), -1., 1.)
           CALL DCLPNT (P(5), -1., -3.)
           CALL DCLPNT (P(6), 0., -3.)
           P(7) = P(1)
C               (Two different labels for the same point; this
C                saves a few lines of code)
C     Define a family of supports  (to be associated with sides later)
           CALL DCLSPT (S(1), 15)
           CALL DCLSPT (S(2), 10)
           CALL DCLSPT (S(3), 20)
           CALL DCLSPT (S(4), 15)
           CALL DCLSPT (S(5), 25)
           CALL DCLSPT (S(6), 10)
           CALL DCLSPT (S(1), 18)
           S(8) = S(4)
C               (The same trick again)
C     Give an orientation to the sides
           DO 1 I = 1, 7
1               OR(I) = 1
           OR(8) = -1
C     Associate sides and supports. Create the rigidities
           CALL RIGFEM (4, P, OR, S, DIF1, P1TO4, R1)
           CALL RIGFEM (4, P(4), OR(4), S(5), DIF2, P4TO7, R2)
C     Assembly of the two rigidities; creation of a new support;
C     then condensation with respect to this support
           CALL ASSBLY (R1, R1, R2)
           CALL MERGER (S26, S(2), S(6))
           CALL CNDENS (R1, R1, S26)
C     Create a suitable trace on this support, and get the flux
           CALL DCLTRA (T, S26)
           CALL MODTRA (T, T, S2, 1.)
           CALL ENERGY (E, R1, T, T)
           PRINT 2, E
2          FORMAT (10X,13H THE FLUX IS , F8.3)
      STOP
      END
```

FIGURE 2. A Fortran code to compute the flux through the L-shaped body of Fig. 1.

equal to 1 on Γ_2 and 0 on Γ_6, then compute (ru, v), which is the desired flux. This is done using the subroutine ENERGY.

The guiding principle in the design was to allow the programmer to code the algorithm above just as described, apart from syntactical particularities. A functional specification of the abstract types and of the basic operations was established (see Ref. (2)) in order to check consistency and orthogonality of all the concepts. This specification is still open for extensions, but as it now stands, it reduces the scope of the system to a precise class of problems: linear elliptic operators, piece-wise constant coefficients, interest centered on boundary integrals of u or $\partial u/\partial n$, but not on values of u.

This target class of problems is large enough to include many engineering problems, but does not contain evolution equations, for instance. Other restrictions were made at the implementation level (dimension 2 only, polygonal domains, Laplace operator only) but could be suppressed in future evolutions of the present prototype.

Let us now read the sample code in Fig. 2. Integers like S, P, R1, etc., are labels for various objects. They contain addresses of memory locations where the information about the objects can be found. About RIGFEM: this routine associates supports S with the sides of a polygon of vertices P and computes the rigidity. It is recursive: the polygon is dissected into triangles, and these into smaller ones, RIGFEM being recursively invoked each time, down to the elementary triangles of a finite-element mesh, for which the rigidities are easily computed 3 by 3 matrices.

REFERENCES

(1) Bossavit, A., in "3rd IMACS Symposium on Computer Methods for Partial Differential Equations" (R. Vichnevetsky and R.S. Stepleman, ed.), p. 287. IMACS, New Brunswick, (1979).

(2) Bossavit, A., in "New Developments in Boundary Element Methods" (C. Brebbia, ed.), p. 380. CML Publications, Southampton, (1980).

THEOREMS OF STEIN-ROSENBERG TYPE II. OPTIMAL PATHS OF RELAXATION IN THE COMPLEX PLANE

John J. Buoni[1]

Department of Mathematics
Youngstown State University
Youngstown, Ohio

Richard S. Varga[2]

Department of Mathematics
Kent State University
Kent, Ohio

1. INTRODUCTION

In [2], results in ways similar to the classical Stein-Rosenberg Theorem (cf. [1], [4], and [5]) have been obtained for arbitrary splittings, and without the usual positivity assumptions. The main purpose of this note is to extend the results of [2] by obtaining <u>paths of optimal relaxation</u>, for small complex relaxation factors $\omega = re^{i\theta}$, for the extrapolated Jacobi iterative method (JOR) and for the successive overrelaxation iterative method (SOR).

By way of notation, let $\mathbb{C}^{n,n}$ denote the collection of all n x n complex matrices $B = [b_{i,j}]$. If $B \in \mathbb{C}^{n,n}$, its spectrum, $\sigma(B)$, is defined as usual by

$$\sigma(B) := \{\lambda : \det(\lambda I - B) = 0\},$$

and its spectral radius, $\rho(B)$, is defined as

$$\rho(B) := \max\{|\lambda| : \lambda \in \sigma(B)\}.$$

---

[1]Research supported by the Air Force Office of Scientific Research at Kent State University during the summers of 1979 and 1980.

[2]Research supported by the Air Force Office of Scientific Research and the Department of Energy.

To review the relevant parts of [2], suppose that $A \in \mathbb{C}^{n,n}$ admits the splitting

$$A = D - L - U. \tag{1.1}$$

Here, it is assumed only that D is nonsingular, with D, L, and U in $\mathbb{C}^{n,n}$ satisfying (1.1). Associated with this splitting are the extrapolated Jacobi matrix, J_ω, defined by

$$J_\omega := I - \omega D^{-1}A, \tag{1.2}$$

and the successive overrelaxation matrix, $\mathcal{L}_\omega$, defined by

$$\mathcal{L}_\omega := (D - \omega L)^{-1}\{(1 - \omega)D + \omega U\}. \tag{1.3}$$

Note that J_ω is defined for all complex numbers ω, while $\mathcal{L}_\omega$ is defined for all complex ω for which $D - \omega L$ is nonsingular. We further define

$$\Omega_J := \{\omega \in \mathbb{C}: \rho(J_\omega) < 1\}, \tag{1.4}$$

$$\Omega_{\mathcal{L}} := \{\omega \in \mathbb{C}: \rho(\mathcal{L}_\omega) < 1\}, \tag{1.5}$$

and

$$K(D^{-1}A) := \text{closed convex hull of } \sigma(D^{-1}A). \tag{1.6}$$

We next state some of the conclusions of [2, Thms. 3.1 and 3.4].

Theorem A. For the splitting of A in (1.1), assume that D is nonsingular. Then,

$$\Omega_J \cap \Omega_{\mathcal{L}} \neq \emptyset \quad \text{iff } 0 \notin K(D^{-1}A). \tag{1.7}$$

Moreover, if $0 \notin K(D^{-1}A)$, then there exist a real $\hat{\theta}$ and an $r_0 > 0$ for which

$$\Omega_J \cap \Omega_{\mathcal{L}} \supset \{\omega = re^{i\hat{\theta}}: \text{ for all } 0 < r < r_0\}. \tag{1.8}$$

Note that the results of (1.7) and (1.8) give the simultaneous convergence of J_ω and $\mathcal{L}_\omega$ (i.e., $\rho(J_\omega) < 1$ and $\rho(\mathcal{L}_\omega) < 1$) for all $\omega = re^{i\hat{\theta}}$ with $0 < r < r_0$, which is reminiscent of the well-known Stein-Rosenberg Theorem [4].

2. OPTIMAL PATHS OF RELAXATION FOR J_ω

Assuming $0 \notin K(D^{-1}A)$, we see from (1.8) that

$$\min_{0\leq\theta\leq 2\pi} \rho(J_{re^{i\theta}}) < 1, \text{ and } \min_{0\leq\theta\leq 2\pi} \rho(\mathcal{L}_{re^{i\theta}}) < 1, \tag{2.1}$$

for each $r > 0$ sufficiently small. The first goal of this section is i) to show that there exists a unique $\tilde{\theta}(r)$, for each r sufficiently small, such that

$$\rho(J_{re^{i\tilde{\theta}(r)}}) = \min_{0\leq\theta\leq 2\pi} \rho(J_{re^{i\theta}}), \tag{2.2}$$

and ii) to explicitly determine $\tilde{\theta}(r)$ asymptotically, as $r \to 0$, solely from the geometrical description of $K(D^{-1}A)$. To our knowledge, the existence of such an optimal path of relaxation and its geometrical description has not been discussed in this generality in the literature.

Since $0 \notin K(D^{-1}A)$, then

$$\tau := \min\{|\xi| : \xi \in K(D^{-1}A)\} > 0, \tag{2.3}$$

and there exists a unique point $\hat{z} \in \partial K(D^{-1}A)$ for which $\hat{z} = \tau e^{i\psi}$ (where $\partial K(D^{-1}A)$ denotes the boundary of $K(D^{-1}A)$). Two cases arise:

Case 1. The circle $|z| = \tau$ intersects $\partial K(D^{-1}A)$ in a vertex of $K(D^{-1}A)$, which implies that $\tau e^{i\psi}$ is an eigenvalue of $D^{-1}A$, and that all other eigenvalues ξ of $D^{-1}A$ satisfy

$$\min[\operatorname{Re} \xi e^{-i\psi} : \xi \in \sigma(D^{-1}A)\setminus\{\tau e^{i\psi}\}] > \tau, \tag{2.4}$$

as pictured in Figure 1 below.

Case 2. The circle $|z| = \tau$ intersects $\partial K(D^{-1}A)$ in a point of a line segment of $\partial K(D^{-1}A)$ which is perpendicular to the ray $\{z = re^{i\psi} : r \geq 0\}$, as shown in Figure 2 below. In this case, if we set

$$E := \{\xi \in \sigma(D^{-1}A) : \operatorname{Re} \xi e^{-i\psi} = \tau\}, \tag{2.5}$$

then E contains at least two eigenvalues of $D^{-1}A$. Moreover, we set

$$\mu_1 := \max\{\operatorname{Im} \xi e^{-i\psi} : \xi \in E\}; \quad \mu_2 := \min\{\operatorname{Im} \xi e^{-i\psi} : \xi \in E\}, \tag{2.6}$$

and it is evident in this case that $\mu_1 \geq 0$, $\mu_2 \leq 0$, with $|\mu_1| + |\mu_2| > 0$, and that $\xi_j := (\tau + i\mu_j)e^{i\psi}$ are in $\sigma(D^{-1}A)$ for $j = 1, 2$.

Next, it is convenient to define $E := \{\tau e^{i\psi}\}$ for Case 1, so that $\mu_1 = \mu_2 = 0$ for Case 1. In either case, we further set

$$\tilde{\theta}(r) := -\psi + \arcsin(-r(\mu_1 + \mu_2)/2) \text{ for all } 0 < r < 2/(|\mu_1| + |\mu_2|), \tag{2.7}$$

where $-\frac{\pi}{2} < \tilde{\theta}(r) + \psi < \frac{\pi}{2}$, and where the upper bound on r is infinite if $|\mu_1| + |\mu_2| = 0$.

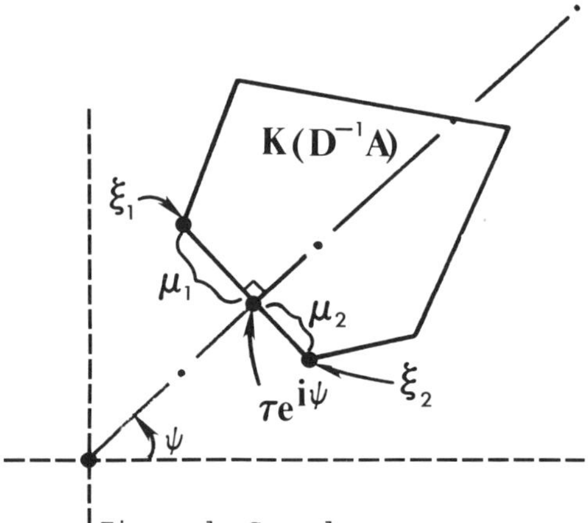

Figure 1: Case 1.

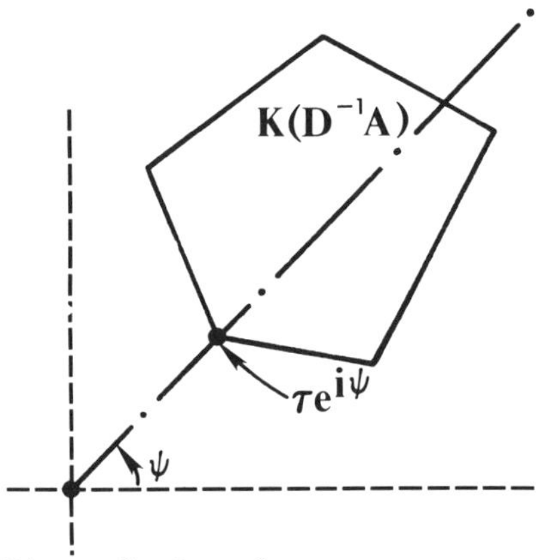

Figure 2: Case 2.

With these definitions, we establish

Theorem 2.1. *Assume that* $A = D - L - U$ *with* D *nonsingular, that* $0 \notin K(D^{-1}A)$, *and that* $\tau e^{i\psi}$ *is the point of* $K(D^{-1}A)$ *closest to the origin. Then, there exists a positive constant* m *such that, on each circle* $|\omega| = r$ *with* $0 < r < m$, *there is a unique* $\tilde{\theta}(r)$, *given by* (2.7), *for which*

$$\rho(J_{re^{i\tilde{\theta}(r)}}) = \min_{0 \leq \theta \leq 2\pi} \rho(J_{re^{i\theta}}) < 1. \tag{2.8}$$

Moreover, (cf. (2.6)),

$$\tilde{\theta}(r) = -\psi - r(\mu_1 + \mu_2)/2 - r^3(\mu_1 + \mu_2)^3/48 + \mathcal{O}(r^5), \text{ as } r \to 0, \tag{2.9}$$

so that

$$\lim_{r\to 0} \tilde{\theta}(r) = -\psi, \tag{2.10}$$

and

$$\rho(J_{re^{i\tilde{\theta}(r)}}) = 1-r\tau - \frac{r^2}{2}(\mu_1\mu_2) + \frac{r^3\tau}{4}(\mu_1^2+\mu_2^2) + \mathcal{O}(r^4), \text{ as } r \to 0. \tag{2.11}$$

Remark. Note that (2.10) gives us that the uniquely defined optimal path of relaxation, for J_ω with $|\omega|$ small, is tangential to the ray $\{re^{-i\psi} : r > 0\}$, as $|\omega| \to 0$.

Proof. Because of rotations, there is no loss of generality in assuming in the proof that $\psi = 0$. On considering Case 1 with $\psi = 0$, then τ is an eigenvalue of $D^{-1}A$, and

$$|1-re^{i\theta}\tau|^2 = 1-2r\tau\cos\theta + r^2\tau^2 > (1-r\tau)^2 \text{ for all } 0<\theta<2\pi. \tag{2.12}$$

On defining $m_1 := 2/M_1$ where

$$M_1 := \max\{\operatorname{Re}\xi + \tau + \frac{(\operatorname{Im}\xi)^2}{\operatorname{Re}\xi - \tau} : \xi \in \sigma(D^{-1}A)\setminus\{\tau\}\},$$

it can be verified that if $0 < r < m_1$, then $(1 - r\tau)^2 \geq |1 - r\xi|^2$ for all $\xi \in \sigma(D^{-1}A)$. Thus, with (2.12), $\rho(J_{re^{i\theta}}) > 1 - r\tau$ for all $0 < \theta < 2\pi$ and for all $0 < r < m_1$, while $\rho(J_r) = 1 - r\tau$. Thus, $\tilde{\theta}(r) = 0$ is the unique value of θ for which (2.2) holds, for all $0 < r < m_1$, under our normalization $\psi = 0$. Further, as a result of our convention that $\mu_1 = \mu_2 = 0$ for Case 1, $\tilde{\theta}(r)$, as defined in (2.7), is zero, and (2.9)-(2.11) follow immediately.

Assuming Case 2 with $\psi = 0$, it can be verified from the definitions of (2.5) and (2.7) that

$$\max\{|1-re^{i\tilde{\theta}}\xi| : \xi \in E\} > |1-re^{i\tilde{\theta}(r)}\xi_1| = |1-re^{i\tilde{\theta}(r)}\xi_2| \tag{2.13}$$

for all $\theta \neq \tilde{\theta}(r) \pmod{2\pi}$, and that there exists an $m_2 > 0$ such that

$$|1-re^{i\tilde{\theta}(r)}\xi_1| = |1-re^{i\tilde{\theta}(r)}\xi_2| \geq |1-re^{i\tilde{\theta}(r)}\xi| \text{ for all } \xi \in \sigma(D^{-1}A)\setminus E, \tag{2.14}$$

for all $0 < r < m_2$. From the above two inequalities, (2.8) follows. Then, from (2.7), (2.9)-(2.11) follow by direct calculation. ∎

If we assume that $D^{-1}A$ is a real matrix with $0 \notin K(D^{-1}A)$, then, from the fact that the nonreal eigenvalues of $D^{-1}A$ necessarily occur in conjugate complex pairs, it is evident that either $\psi = 0$ or $\psi = \pi$, and that (cf. (2.6)) $\mu_1 + \mu_2 = 0$. Consequently, from (2.7), $\tilde{\theta}(r) = 0$ or $\tilde{\theta}(r) = +\pi$ for all $r > 0$, and we immediately have from Theorem 2.1 the result of

Corollary 2.2. Assume that $A = D - L - U$ with D nonsingular, and that $D^{-1}A$ is a real matrix. If $\operatorname{Re} \xi > 0$ for all $\xi \in \sigma(D^{-1}A)$, then

$$\min_{0 \le \theta \le 2\pi} \rho(J_{re^{i\theta}}) = \rho(J_r) \tag{2.15}$$

for all $r > 0$, while if $\operatorname{Re} \xi < 0$ for all $\xi \in \sigma(D^{-1}A)$, then

$$\min_{0 \le \theta \le 2\pi} \rho(J_{re^{i\theta}}) = \rho(J_{-r}) \tag{2.16}$$

for all $r > 0$. In either case,

$$\begin{aligned}\min_{0 \le \theta \le 2\pi} \rho(J_{re^{i\theta}}) &= \{(1 - r\tau)^2 + r^2\mu_1^2\}^{1/2} = \\ 1 - r\tau + \frac{r^2\mu_1^2}{2} &+ \frac{r^3\tau\mu_1^2}{2} + \mathcal{O}(r^4), \text{ as } r \to 0.\end{aligned} \tag{2.17}$$

3. OPTIMAL PATHS OF RELAXATION FOR $\mathcal{L}_\omega$

We next obtain an analogue of Theorem 2.1 for the matrix $\mathcal{L}_\omega$ when $|\omega|$ is small. On defining

$$Q(\omega) := D^{-1}A + \omega D^{-1}L(I - \omega D^{-1}L)^{-1}D^{-1}A, \text{ for all } |\omega| \text{ small,} \tag{3.1}$$

we see from (3.1) that

$$\mathcal{L}_\omega = I - \omega Q(\omega), \quad \text{for all } |\omega| \text{ small.} \tag{3.2}$$

Next, if we set

$$K(Q(\omega)) := \text{closed convex hull of } \sigma(Q(\omega)), \tag{3.3}$$

then, since $Q(\omega) \to D^{-1}A$ as $|\omega| \to 0$ from (3.1), it follows from the continuity of the eigenvalues involved that

$$K(Q(\omega)) \to K(D^{-1}A), \quad \text{as } |\omega| \to 0. \tag{3.4}$$

Now, because the optimal path of relaxation for $|\omega|$ small for $J_\omega = I - \omega D^{-1}A$, from Theorem 2.1, depends only on the geometry of $K(D^{-1}A)$, it is not surprising from (3.4) that the same might be true for $\mathcal{L}_\omega$. More precisely, we have

Theorem 3.1. *Assume that* $A = D - L - U$ *with* D *nonsingular, that* $0 \notin K(D^{-1}A)$, *and that* $\tau e^{i\psi}$ *is the point of* $K(D^{-1}A)$ *closest to the origin. Then, there exists a positive constant* m' *such that, on each circle* $|\omega|=r$ *with* $0 < r < m'$, *there is a* $\hat{\theta}(r)$ *for which*

$$\rho(\mathcal{L}_{re^{i\hat{\theta}(r)}}) = \min_{0\leq\theta\leq 2\pi} \rho(\mathcal{L}_{re^{i\theta}}) < 1. \tag{3.5}$$

Moreover,

$$\lim_{r\to 0} \hat{\theta}(r) = -\psi, \tag{3.6}$$

and

$$\rho(\mathcal{L}_{re^{i\hat{\theta}(r)}}) = 1 - r\tau + \mathcal{O}(r^{1+1/n}), \text{ as } r \to 0. \tag{3.7}$$

Remark. As in Theorem 2.1, an optimal path of relaxation for $\mathcal{L}_\omega$, with $|\omega|$ small, is *tangential* to the ray $\{re^{-i\psi} : r > 0\}$ as $|\omega| \to 0$.

Proof. We shall establish this result only for Case 1, since Case 2 is similarly treated. That a $\hat{\theta}(r)$ exists on each circle $|\omega| = r$ for which (3.5) holds, follows of course by continuity. Now, in Case 1, we know that $\tau e^{i\psi} \in \sigma(D^{-1}A)$. Then, applying Ostrowski's classical result [3, p. 334] on the perturbation of eigenvalues of a matrix, there is at least one eigenvalue $\xi(\omega) \in \sigma(Q(\omega))$ for which

$$|\xi(\omega) - \tau e^{i\psi}| = \mathcal{O}(|\omega|^{1/n}), \text{ as } |\omega| \to 0, \tag{3.8}$$

where n is the order of A. Thus,

$$\eta(r, \theta) := 1 - re^{i\theta}\xi(re^{i\theta}) \tag{3.9}$$

is an eigenvalue of $\mathcal{L}_{re^{i\theta}}$ from (3.2) for each such $\xi(\omega)$. The modulus of each such $\eta(r, \theta)$ is, from (3.8), just

$$|\eta(r, \theta)| = 1 - r\tau \cos(\theta + \psi) + \mathcal{O}(r^{1+1/n}), \text{ as } r \to 0, \tag{3.10}$$

uniformly in θ. For any fixed ϵ with $0 < \epsilon < 1$, suppose that $\cos(\theta + \psi) \leq 1 - \epsilon$, so that from (3.10),

$$|\eta(r, \theta)| \geq 1 - r\tau + \epsilon r\tau + \mathcal{O}(r^{1+1/n}), \text{ as } r \to 0.$$

On the other hand, the choice of $\theta = -\psi$ in (3.10) gives the asymptotically smaller quantity $1 - r\tau + \mathcal{O}(r^{1+1/n})$ as $r \to 0$. Thus, we see that, for each fixed $r > 0$ sufficiently small, if $\tilde{\theta}(r)$ minimizes $|\eta(r, \theta)|$ over $0 \leq \theta \leq 2\pi$ and over all such eigenvalues $\eta(r, \theta)$, then $\tilde{\theta}(r)$ cannot satisfy $\cos(\tilde{\theta}(r) + \psi) \leq 1 - \epsilon$ for all $r \to 0$. Next, as in the proof of Theorem 2.1 in Case 1, it turns out that $\min_{0 \leq \theta \leq 2\pi} \rho(\mathcal{L}_{re^{i\theta}})$, for r small, is governed by the behavior of such eigenvalues $\eta(r, \theta)$, so that

$$\min_{0 \leq \theta \leq 2\pi} \rho(\mathcal{L}_{re^{i\theta}}) = 1 - r\tau + \mathcal{O}(r^{1+1/n}), \text{ as } r \to 0, \tag{3.11}$$

which gives the desired results of (3.5) and (3.7). The previous discussion concerning $\cos(\theta + \psi) \leq 1 - \epsilon$ also shows that $\lim_{r \to 0} \tilde{\theta}(r) = -\psi$. ∎

On comparing (2.11) of Theorem 2.1 and (3.7) of Theorem 3.1, we see that the spectral radii of the optimized (in θ) extrapolated Jacobi matrix and optimized successive overrelaxation matrix have the same asymptotic expansions through first order terms in r, as $r \to 0$. Because of the crude estimation of perturbation effects in $Q(\omega)$ for $|\omega|$ small, which resulted in the term $\mathcal{O}(r^{1+1/n})$ in (3.7), it is not possible to say which of these optimized procedures is <u>faster</u> as $r \to 0$, unlike the classical Stein-Rosenberg Theorem. Nevertheless, the following interesting observations can be deduced.

If the term $\mathcal{O}(r^{1+1/n})$ in (3.7) actually behaves asymptotically as

μr^{σ}, as $r \to 0$, where $|\mu| > 0$ and where $1 < \sigma < 2$, then it is evident from (2.11) and (3.7) that the optimized extrapolated Jacobi procedure is asymptotically faster than the optimized successive overrelaxation procedure for all $r > 0$, sufficiently small. This means, for example, that if the Jordan normal form, corresponding to some eigenvalue in the subset E of $D^{-1}A$ of (2.5), is not diagonal, and if the perturbation $Q(\omega)$ of $D^{-1}A$, for $\omega = re^{-i\psi}$ with $r > 0$ small, separates the eigenvalues associated with this Jordan block, then the optimized extrapolated Jacobi method will be asymptotically faster than the optimized successive overrelaxation method for all $r > 0$ sufficiently small.

Now, the above deduction is precisely the opposite of what one expects in the usual Stein-Rosenberg Theorem. Why? The assumptions in the usual Stein-Rosenberg Theorem require that $B := D^{-1}L + D^{-1}U$ be a nonnegative matrix, but this, by the Perron-Frobenius Theorem, insures that the Jordan blocks for B, corresponding to the subset E of $D^{-1}A = I-B$, are all 1×1 matrices and hence diagonal. Thus, the assumptions of the previous paragraph are not met!

In a later paper, we will give a more precise form of Theorem 3.1, based on perturbation theory, for optimal paths of relaxation for the matrix $\mathcal{L}_{\omega}$ in the complex plane. Included there will be results, both numerical and graphical, illustrating these comparisons of the iteration matrices J_{ω} and $\mathcal{L}_{\omega}$.

References

1. A. Berman and R. J. Plemmons, Nonnegative Matrices in the Mathematical Sciences, Academic Press, New York, 1979.

2. J. J. Buoni and R. S. Varga, "Theorems of Stein-Rosenberg Type," Numerical Mathematics (R. Ansorge, K. Glashoff, B. Werner, eds.), ISNM 49, pp. 65-75, Birkhauser Verlag, Basel, 1979.

3. A. M. Ostrowski, Solution of Equations in Euclidean and Banach Spaces, Academic Press, New York, 1973.

4. P. Stein and R. Rosenberg, "On the solution of linear simultaneous equations by iteration," J. London Math. Soc. 23 (1948), pp. 111-118.

5. R. S. Varga, Matrix Iterative Analysis, Prentice-Hall, Englewood Cliffs, New Jersey, 1962.

SPARSE VECTORIZED DIRECT SOLUTION OF ELLIPTIC PROBLEMS

D. A. Calahan*

Department of Electrical and Computer Engineering

University of Michigan

Ann Arbor, Michigan

Classifications of general vectorized sparse matrix methods are reviewed. Algorithms for the simultaneous solution of sparse systems are considered and their application to finite difference grids is presented. Timings of CRAY-1 software for solution of general simultaneous sparse systems is given.

I. INTRODUCTION

In [1] a general classification has been made of sparsity structures that are amenable to "parallel" solution, and an equation ordering for solution of PDE grids on vector machines has been studied. The implementation of the solution with this ordering in a general sparsity context is the subject of this paper.

II. SPARSE SOLUTION OF ELLIPTIC PROBLEMS

A. Vectorized Algorithm Classification

The general sparsity algorithm of Gustavson [2] can be extended to two classes of algorithms to achieve vectorization.

(a) SIN mode. If the sparse matrix is locally dense, sufficiently long vectors can be defined <u>within</u> the

*Supported by the Mathematical and Information Sciences Directorate, Air Force Office of Scientific Research, on Grant 80-0158.

ISBN 0-12-632620-7

total system (or within each subsystem) by associating vectors with dense (segments of) rows and columns. This is termed the SINgle system solution mode.

(b) SIM mode. If the sparse matrix represents systems which (1) are decoupled, and (2) have identical intraconnection patterns, then these systems can be SIMultaneously solved by defining vectors across systems [1].

B. Application to Elliptic Problems

The solution of 2-D and 3-D finite element and finite difference grids that are likely to challenge a vector processor solely by direct methods is unlikely due to the growth in computational work with (square) grid size (n). However, SIM mode direct methods have a role in such solutions.

Two-stage direct and direct/iterative solutions are becoming common [4]. The first stage involves a SIM mode direct solution, where uncoupled nodes of the grid are eliminated. The computational effort is $O(n^2)$ for a 2-D grid and $O(n^3)$ for a 3-D grid. The bandwidth of the remaining matrix does not increase due to this elimination step and the complexity of the second stage solution is at least halved.

Figure 1(a-b) illustrates, for red-black ordering of a five-point operator grid, the striped form of the resulting matrix. If the stripes are filled and lengthened with zero-values to the size of the diagonal stripe, then the vector operations involved in eliminating the red nodes can be performed as follows.

Step 1. Each (internal) red node is viewed as belonging to a cell. A template matrix is then formed that represents the coupling of each red node to the boundary nodes, without regard for the location of the cell in the grid. This process yields a full row and column template as shown in Figure 1(c) . In the template matrix, the row and column representing the red node are then reduced, creating a cell matrix representing the boundary nodes.

Step 2. The cell matrix, with each element representing a vector, is scattered and accumulated to form the familiar quintadiagonal matrix.

Step 3. The quintadiagonal matrix is solved using direct or iterative methods.

C. Application to Nested Dissection

The above procedure can be used to vectorize the initial stages of a nested dissection process [3]. If diagonal dissection is applied to a five-point operator grid, cells can be defined as shown in Figure 2(a). All similar cells of the

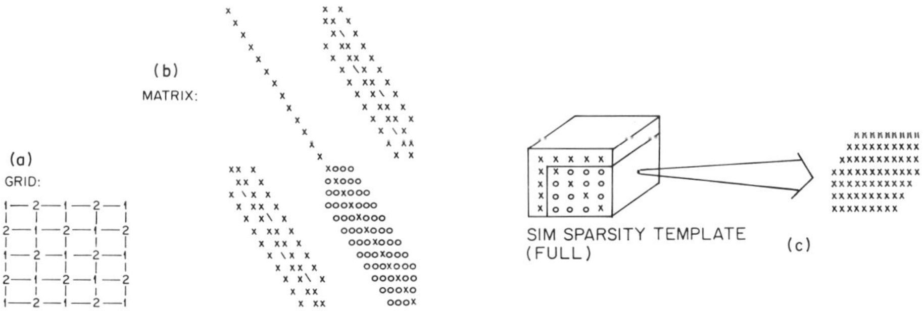

FIGURE 1. (a) 2-D grid with red nodes numbered 1; (b) related LU matrix with fills numbered 0; (c) three-step solution,producing a (compressed) band matrix.

grid can be simultaneously reduced to matrices representing only the boundary nodes (Figure 2(b)). A matrix is then created from the cell interconnections implied by the grid structure (Step 2). In the example shown, 78% of the grid nodes are eliminated in the SIM mode for large grids, versus 50% for red-black ordering.

A general simultaneous sparse equation solver implemented in assembly language on the CRAY-1 was used to reduce the rows and columns associated with the internal nodes of the cell of Figure 2(a), producing the full 16x16 cell partition in Figure 2(b). Execution rates of between 26 and 47 MFLOPS were achieved for 64 cells in steps 1 & 2 above, depending on the regularity of the cell interconnection pattern. The cell

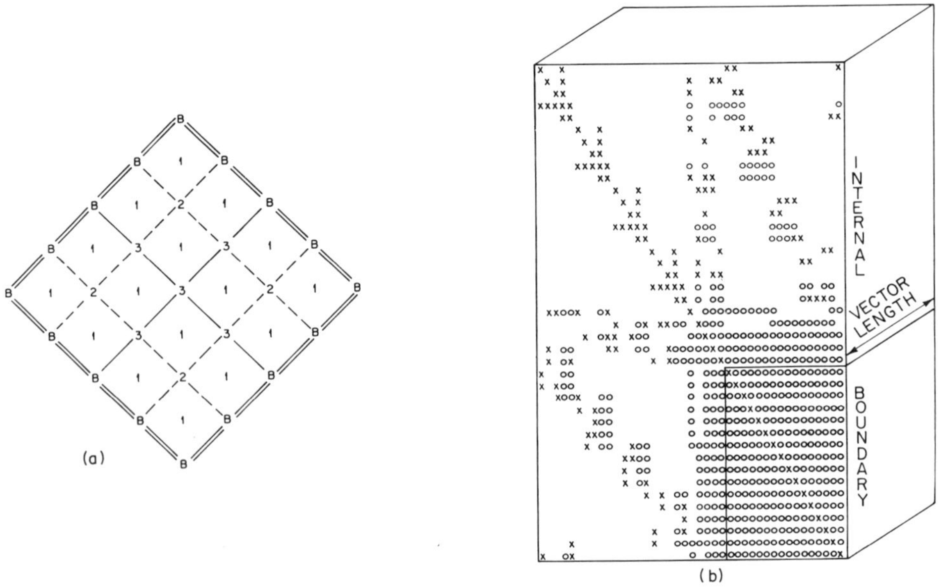

FIGURE 2. (a) Diagonally dissected cell of 5-point operators, where numbers indicate dissection stage in which internal nodes are eliminated, and B indictes boundary nodes; (b) template of simultaneous LU matrices.

sparsity was described by a bit map rather than a linked list to exploit unusual CRAY-1 instructions.

This cell-oriented procedure would appear to avoid the necessity of solving small banded matrices in the SIN mode in the initial dissection stages and incurring the resulting low execution rates, as suggested in [3].

ACKNOWLEDGMENT

Acknowledgment is given D. A. Orbits for a CRAY-1 simulator and Dan Duchamp for general programming assistance.

REFERENCES

1. Calahan, D. A., "Vectorized Sparse Elimination," Proc. Scientific Computer Information Exchange Meeting, Lawrence Livermore Laboratory, p.85, (1979).
2. Gustavson, F., "Some Basic Techniques for Solving Sparse Systems of Linear Equations" Proc. Conf. on Sparse Matrices and Their Applications, (Rose and Willoughby, Ed.), Plenum Press, New York, p. 41, (1972).
3. George, A., Poole, W., and Voigt, R., "Analysis of Dissection Algorithms for Vector Computers," Report #76-17, ICASE, NASA Langley Research Center, (1976).
4. McDonald, A. E., "Efficient Use of Mass Storage During Elimination for Sparse Sets of Simultaneous Equations," Proc. SPE-AIME Fourth Symposium on Numerical Simulation of Reservoir Performance, Los Angeles, p. 300, (1976).

MULTI-GRID AND ICCG FOR PROBLEMS WITH INTERFACES

by

J. E. Dendy, Jr. and J. M. Hyman*

In this paper we compare computation times for the multi-grid (MG) algorithm [1], the incomplete Cholesky conjugate gradient (ICCG) algorithm [4,5] and the modified ICCG (MICCG) algorithm [3] to solve elliptic partial differential equations of the form

$$-\nabla\cdot(D(x,y)\nabla U(x,y)) + \sigma(x,y)U(x,y) = f(x,y) \ , \ (x,y)\epsilon\Omega$$
$$\nu(x,y)\cdot D(x,y)\nabla U(x,y) + \gamma(x,y)U(x,y) = 0 \ , \ (x,y)\epsilon\partial\Omega \ . \qquad (1)$$

Here Ω is a bounded region in $\mathbb{R}^2$ with boundary $\partial\Omega$, ν is the outward normal to $\partial\Omega$, D is positive and σ and γ are non-negative. Moreover, the functions D, σ, and f are allowed to be discontinuous across internal boundaries Γ of Ω. We make the assumption that U and $\mu\cdot(D\nabla U)$ are continuous at almost every $(x,y)\epsilon\Gamma$, and μ is a smooth vector normal to Γ.

The application of MG to (1) is discussed in [1]. The main conclusion in that paper was that the interpolation J^K_{K-1}, from grid G^{K-1} to G^K should preserve the flux $\mu\cdot(D\nabla U)$ across the interfaces Γ, and that the coarse grid operator L^{K-1} should then be taken to be $(J^K_{K-1})^* \, L^K \, J^K_{K-1}$. This implementation of MG is more robust than the proposals in [2].

To solve Ax=b by ICCG, where A is a symmetric positive definite sparse matrix, one approximates A in factored form as LL^T where L is lower triangular and sparse. The sparseness structure $L+L^T$ is at least that of A. Following the notation of [3], we use ICCG(n) and MICCG(n) to denote how many additional diagonals, n, L has than the lower triangular part of A. In ICCG, the entries of LL^T appearing outside the sparseness structure of $L+L^T$ are taken to be zero; in MICCG they are summed [3] and added to the diagonal. Both methods are accelerated with conjugate

*Los Alamos Scientific Laboratory, University of California, MS-233, Los Alamos, N.M. 87545.

Work funded by DOE Contract W-7405-ENG-36.

ISBN 0-12-632620-7

gradient. For smooth coefficient problems the work to solve to a given level of error is $O(N^{3/2})$ for ICCG and $O(N^{5/4})$ for MICCG, where N is the number of unknowns. For MG the work is O(N).

In general, we have found the MICCG and ICCG algorithms more robust than the MG for general positive definite systems. Also, ICCG and MICCG can be implemented as reliable black boxes that work for a wide range of problems. For MG, unfortunately, this is not yet the case. MG must often be tuned to the structure of the problem at hand. A major advantage of the MG algorithm is that the structure of the problem can be exploited to significantly reduce the solution time. Problems arising from difference equations are usually rich in structure and the Brandt MG program structure is natural not only for the solution of the difference equation but also for the local mesh refinement. An added bonus of the MG structure is that it can also be used to solve companion problems (such as the eigenvalue problem) that often arise in conjunction with Eq. 1. Indeed, as many as 90% of the multi-group diffusion problems (of which (1) is a special, one group, case) solved are eigenvalue problems.

The first example is the equation

$$\Delta u - u = f \text{ on } \Omega = (0,1)X(0,1) \ , \ u = g \text{ on } \partial\Omega \ ,$$

where f and g are chosen so that $u = \log\,[(x+10^{-10})^2 + (y+10^{-10})^2]$. The fourth order nine point box is used, and ICCG(0) and MICCG(0) are compared to MG, using $u \equiv 1$ as initial guess and solving until the discrete

Entries indicate CDC 7600 C.P.U. time in seconds	Numbers of unknowns	MG $u \equiv 1$ guess to 10^{-6}	ICCG $u \equiv 1$ guess to 10^{-6}	MICCG $u \equiv 1$ guess to 10^{-6}
Problem 1	11x11	.04	.04	.03
	23x23	.14	.21	.19
	47x47	.50	1.50	1.08
	93x93	4.29	23.12	10.04

L^2 norm of the residual is less than 10^{-6}. A MG work unit (SOR sweep) took 0.0013, 0.0045 or 0.016 seconds on an 11x11, 23x23 or 47x47 point grid, respectively. The ICCG and MICCG algorithms took 0.0033, 0.011 or 0.041 seconds per iteration on the same grids.

Examples 2-5 are all simplified versions of the complicated mosaics that appear in neutron diffusion problems. In these examples we use either a second order five point difference operator, described in [1], or a piecewise bilinear nine point finite element method. The ICCG(1) and MICCG(1) algorithms are used for the five point star and the ICCG(0) and MICCG(0) algorithms are used to solve the nine point operator.

Example 2: $\Omega = (0,24)\times(0,24)$,

$$\frac{\partial u}{\partial \nu} = \begin{cases} -u/2D \text{ on } y = 24 \text{ or } x = 24 \\ 0, \text{ otherwise} \end{cases}$$

$$D(x,y) = \begin{cases} 1, \text{ if } (x,y) \in [0,12)\times[0,12) \cup (12,20]\times(12,20] \\ 1000, \text{ otherwise} \end{cases}$$

Example 3: $\Omega = (0,16)\times(0,16)$, $\frac{\partial u}{\partial \nu}$ as in Ex. 2

$$D(x,y) = \begin{cases} 1000 \text{ if } (x,y) \in (0,5)\times[5,7) \cup (5,7)\times[0,7) \\ 1, \text{ otherwise} \end{cases}$$

Example 4: $\Omega = (0,16)\times(0,16)$, $\frac{\partial u}{\partial \nu}$ as in Ex. 2

$$D(x,y) = \begin{cases} 1000, \text{ if } (x,y) \in (1,3)\times[0,5) \cup (3,7)\times(3,5) \\ \qquad \cup (5,11)\times(5,7) \cup (9,15)\times(7,9) \\ \qquad \cup (13,15)\times(9,13) \cup (15,16]\times(11,13) \\ 1, \text{ otherwise.} \end{cases}$$

Example 5: $\Omega = (0,24)\times(0,24)$ $\quad \frac{\partial u}{\partial \nu} = -\frac{1}{2D} u$ on all sides

$$D(x,y) = \begin{cases} 1000, \text{ if } (x,y) \in (11,13)\times(11,13) \\ 1, \text{ otherwise.} \end{cases}$$

In all these examples f=0 when D=1, and f=1 when D=1000. Comparisons are presented for two cases, $\sigma \equiv 0$ and $\sigma = 1/3D$, on the next page.

The MG algorithm employed differs from that described in [1] in that it uses one full MG cycle [2] to first solve to truncation error. The

Entries indicate CDC 7600 C.P.U. time in seconds	Number of unknowns	MG T.E.	ICCG T.E.	MICCG T.E.	MG T.E. to 10^{-6}	ICCG T.E. to 10^{-6}	MICCG T.E. to 10^{-6}	MG Total	ICCG 0 guess to 10^{-6}	MICCG 0 guess to 10^{-6}
Problem 2	13x13	.01	.02	.02	.03	.04	.04	.04	.04	.05
finite difference	25x25	.04	.12	.11	.11	.25	.25	.15	.27	.28
$\sigma \equiv 0$	49x49	.12	.77	.60	.48	1.75	1.23	.60	1.89	1.61
Problem 2	13x13	.01	.02	.02	.04	.04	.04	.05	.05	.05
finite element	25x25	.04	.17	.14	.15	.27	.23	.19	.29	.30
$\sigma \equiv 0$	49x49	.14	.88	.71	.69	1.95	1.29	.83	2.31	1.87
Problem 2	13x13	.01	.02	.02	.03	.04	.04	.04	.04	.04
finite difference	25x25	.03	.09	.09	.10	.22	.19	.13	.23	.25
$\sigma = \frac{1}{3D}$	49x49	.12	.54	.58	.38	1.52	1.06	.50	1.55	1.42
Problem 2	13x13	.01	.02	.02	.03	.04	.04	.04	.04	.04
finite element	25x25	.04	.10	.11	.14	.24	.21	.18	.26	.27
$\sigma = \frac{1}{3D}$	49x49	.13	.60	.68	.65	1.65	1.24	.78	1.74	1.58
Problem 3, $\sigma \equiv 0$	17x17	.02	.04	.03	.05	.08	.06	.07	.08	.08
finite difference	33x33	.05	.29	.16	.17	.53	.37	.22	.55	.47
Problem 3, $\sigma \equiv 0$	17x17	.02	.05	.03	.06	.08	.07	.08	.08	.09
finite element	33x33	.06	.36	.23	.19	.60	.38	.25	.60	.55

Problem 3, $\sigma = 1/3D$	17x17	.02	.03	.02	.05	.06	.05	.07	.05	.05
finite difference	33x33	.06	.18	.12	.16	.35	.26	.22	.36	.31
Problem 3, $\sigma = 1/3D$	17x17	.02	.03	.03	.06	.06	.06	.08	.06	.06
finite element	33x33	.06	.21	.17	.19	.40	.28	.25	.39	.36
Problem 4, $\sigma \equiv 0$	17x17	.02	.06	.03	.06	.08	.09	.08	.09	.09
finite difference	33x33	.05	.42	.16	.19	.62	.37	.24	.64	.49
Problem 4, $\sigma \equiv 0$	17x17	.02	.22	.21	.06	.10	.08	.08	.10	.09
finite element	33x33	.06	.18	.18	.21	.63	.44	.28	.64	.57
Problem 4, $\sigma = 1/3D$	17x17	.02	.05	.02	.05	.07	.05	.07	.07	.06
finite difference	33x33	.05	.31	.13	.14	.46	.28	.19	.47	.34
Problem 4, $\sigma = 1/3D$	17x17	.02	.06	.03	.05	.08	.06	.07	.08	.06
finite element	33x33	.06	.34	.17	.16	.50	.31	.22	.52	.39
Problem 5, $\sigma \equiv 0$	25x25	.03	.07	.03	.08	.17	.11	.11	.16	.14
finite difference	49x49	.12	.54	.26	.27	1.10	.61	.39	1.17	.82
Problem 5, $\sigma \equiv 0$	25x25	.04	.08	.07	.09	.16	.10	.13	.17	.15
finite element	49x49	.14	.57	.32	.28	1.28	.64	.42	1.24	.89
Problem 5, $\sigma = 1/3D$	25x25	.03	.06	.04	.06	.07	.07	.09	.09	.09
finite difference	49x49	.12	.29	.22	.19	.60	.36	.31	.60	.43
Problem 5, $\sigma = 1/3D$	25x25	.04	.06	.05	.05	.08	.07	.09	.09	.08
finite element	49x49	.13	.32	.22	.21	.67	.36	.34	.68	.46

first comparison takes this MG solution at the truncation error level and uses it as in initial guess to continue solving to a discrete L^2 residual norm of 10^{-6}; this comparison mimics having a good initial guess, as one would have in each step of a time dependent problem. The second uses $u \equiv 0$ as an initial guess. Note that MICCG does much better with a good initial guess.

The clearest conclusion from these examples is that MICCG seems preferable to ICCG, especially since if one is already using ICCG, then one can change just one line of code to obtain MICCG. Another conclusion is that MG is most impressive when solving to truncation error. We emphasize that MG has a mechanism for doing this which ICCG and MICCG do not have. When solving to smaller tolerances, the gains for MG are less impressive, especially in the $\sigma = 1/3D$ case, although for the harder problems 2 and 4, they are almost even for this case.

For problems with little structure and for one shot calculations we recommend ICCG over MG and MICCG over ICCG. For problems that are done many times, it is worth investing the effort to study methods like MG.

Acknowledgments

We thank Dr. James Sanderson for converting D. Kershaw's ICCG code [4] to standard Fortran for us.

References

[1] R. E. Alcouffee, Achi Brandt, J. E. Dendy, Jr., and J. W. Painter, The Multi-Grid Method for the Diffusion Equation with Strongly Discontinuous Coefficients, to appear.

[2] A. Brandt, Math. Comp: 31, 333-390, 1977.

[3] I. Gustafsson, BIT 18, 142-156, 1978.

[4] D. S. Kershaw, J. Comp. Physics 26, 43-65, 1978.

[5] J. A. Meijerink and H. A. Van der Vorst, Math. Comp. 31, 148-162, 1977.

ADDENDUM

It has come to the attention of the authors that following the Santa Fe elliptic conference Prof. Stanley Eisenstat at Yale University devised an improved implementation of ICCG and MICCG. Using Eisenstat's ideas the cost per iteration is reduced to 17/24 of the cost per iteration in the code we used. The computer times cited in the tables for ICCG and and MICCG should be reduced accordingly. Our basic conclusions in the comparison remain unchanged.

We have also learned that similar problems are being studied by J. A. Meijerink. He has found significant improvements can be made by accelerating the MG algorithm with conjugate gradient after each full MG cycle.

AN AD HOC SOR METHOD[1]

Louis W. Ehrlich

Applied Physics Laboratory
The Johns Hopkins University
Laurel, Maryland

Undeniably, the successive overrelaxation (SOR) method is the second simplest iterative method to implement--second only to the Gauss Seidel method, and not by much. The problem with the SOR method is the determination of the relaxation factor, ω. We present here an ad hoc method that determines a different ω for each separate unknown in a five-point approximation. Thus if, at a point,

$$a_1 x_{i+1,j} + a_2 x_{i,j+1} + a_3 x_{i-1,j} + a_4 x_{i,j-1} - a_0 x_{ij} = b_{ij}, \tag{1}$$

then we assume the difference equation is uniform throughout. The resulting Jacobi method matrix can then be separated and its eigenvalues determined via

$$\begin{aligned} &(a)\quad a_3 v_{i-1} + a_1 v_{i+1} = \mu_1 a_0 v_i \quad , \quad v_0 = v_{N+1} = 0 \\ &(b)\quad a_4 w_{j-1} + a_2 w_{j+1} = \mu_2 a_0 w_i \quad , \quad w_0 = w_{M+1} = 0 \end{aligned} \tag{2}$$

(assuming given boundary values) which leads to

$$\mu_{p,q} = \mu_1 + \mu_2 = \frac{2}{a_0}\left[\sqrt{a_1 a_3}\cos\frac{\pi p}{N+1} + \sqrt{a_2 a_4}\cos\frac{\pi q}{M+1}\right]. \tag{3}$$

It is known that for the five-point stencil, using the "natural ordering," the eigenvalue of the SOR method, λ, and the Jacobi method, μ, are related by

$$(\lambda + \omega - 1) = \omega\mu\lambda^{1/2}. \tag{4}$$

If the μ are real or pure imaginary, then the optimal ω is related to $\bar{\mu}$, the eigenvalue of largest magnitude, by

[1]This work was supported by the Department of the Navy, Naval Sea Command under Contract No. N00024-78-C-5384.

ISBN 0-12-632620-7

$$\omega_b = \frac{2}{1 + \sqrt{1 - \bar{\mu}^2}}. \tag{5}$$

However, when the μ are complex, we resort to a scheme suggested by Rigal [3], i.e.,

$$\left\{\begin{aligned} &A = \mu_r^2 + \mu_i^2;\ B = \mu_r^2 - \mu_i^2;\ a = A^2 - B^2;\ b = A^2 - B \\ &\bar{\omega} = \frac{1}{A^4 - A^2 B}\Big\{\left[3b + (a + b^2)^{1/2}\right]a^{1/3}\left[(a + b^2)^{1/2} - b\right]^{1/3} \\ &\qquad - \left[3b - (a + b^2)^{1/2}\right]a^{1/3}\left[(a + b^2)^{1/2} + b\right]^{1/3} \\ &\qquad + A^2 + 3B^2 - 4A^2 B\Big\} \\ &\omega_{ij} = -\left[\bar{\omega} \pm (\bar{\omega}^2 + 4\bar{\omega})^{1/2}\right]\Big/2 \quad \begin{cases} + \text{ if } A^2 < B \\ - \text{ if } A^2 > B. \end{cases} \end{aligned}\right. \tag{6}$$

If normal derivative conditions exist, one of the boundary conditions in (2) is replaced by $v_0 = 0$, $v_{N-1} = v_{N+1}$, and we have

$$\left\{\begin{aligned} &\mu_1 = \frac{2}{a_0}\sqrt{a_1 a_3}\cos\frac{\pi\alpha}{2} \\ &\text{where } a_3 \sin\frac{\pi\alpha(N+1)}{2} = a_1 \sin\frac{\pi\alpha(N-1)}{2} \end{aligned}\right. \tag{7}$$

for smallest positive α if $\frac{a_1}{a_3} \le \frac{(N+1)}{(N-1)}$. Otherwise

$$\left\{\begin{aligned} &\mu_1 = \frac{2}{a_0}\sqrt{a_1 a_3}\cosh\frac{\pi\alpha}{2} \\ &\text{where } a_3 \sinh\frac{\pi\alpha(N+1)}{2} = a_1 \sinh\frac{\pi\alpha(N-1)}{2}. \end{aligned}\right. \tag{8}$$

Then

$$\bar{\mu} = \mu_1 + \frac{2}{a_0}\sqrt{a_2 a_4}\cos\frac{\pi}{M+1}. \tag{9}$$

For periodic conditions, we have

$$\mu_k = \left[(a_1 + a_3)\cos\frac{2\pi k}{N} + i(a_1 - a_3)\sin\frac{2\pi k}{N}\right]/a_0 \tag{10}$$

for the eigenvalue contribution in the periodic direction.

Finally, we note that the choice of N and M is the number of unknown points on a horizontal and vertical line, thru the point in question to the nearest boundaries. (See also Brazier [1].)

The above technique was applied to six different problems on various shaped regions. The details are in Ehrlich [2]. We summarize some of these results, directly from [2].

For the Laplace equation in an L-shaped region, or a quarter ellipse, the above method converged as rapidly as the SOR at optimal ω.

For a difference equation of the form

$$(1 - Ra_{ij})x_{i+1,j} + (1 + Ra_{ij})x_{i-1,j} + (1 - Rb_{ij})x_{i,j+1} + (1 + Rb_{ij})x_{i,j-1} - 4x_{ij} = 0 \qquad (11)$$

for constant R. some representative results are in Table I. As R increased, there was a marked improvement in convergence.

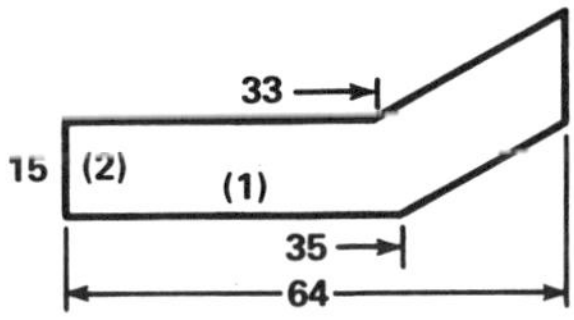

a) DIRCHLET PROB.

Re	AD-HOC SOR	SOR (ω_b)
0	84(1.77-1.07)	85(1.71)
5	55(1.75-1.07)	56(1.65)
50	34(1.61-0.65)	103(0.73)
200	90(1.03-0.22)	>200(0.097)

b) $u_n = 0$ on (1)

0	107(1.86-1.07)	125(1.81)
5	62(1.84-1.07)	64(1.72)
10	44(1.83-1.07)	49(1.57)
50	37(1.65-0.65)	103(0.73)

Table I.

IVa) 15 (rectangle) 89

α	AD-HOC SOR	SOR (ω_b)
1	43(1.72-1.40)	43(1.70)
50	41(0.93-0.084)	103(0.19)

IVb) $u_n = 0$

1	74(1.82-1.39)	68(1.78)
25	26(1.3-0.16)	93(0.39)
50	41(0.93-0.082)	

IVd) periodic (rectangle) periodic

1	43(1.72-1.38)	43(1.70)
50	43(0.93-0.084)	106(0.19)

Table II.

Table II exhibits results for a stencil of the form

$$(1 - \alpha f_{ij})x_{i+1,j} + (1 + \alpha f_{ij})x_{i-1,j} + (C - \alpha g_{ij})x_{i,j+1} + (C + \alpha g_{ij})x_{i,j-1} - 2(C + 1)x_{ij} = 0 \qquad (12)$$

for constant C. Again, there was a marked improvement with increasing α.

The ad hoc SOR matches or betters the convergence rate of other techniques for finding a different ω for each point [1,4,5].

Finally, a side effect was noted. For certain problems, the theoretical optimal ω may not produce the smallest number of iterations. Table III shows the results of a one-dimensional Laplace equation with logarithmetric mesh variation. The theoretical ω_b was determined by finding the eigenvalues of the Jacobi matrix using EISPAK and then using (5). It is a function of the given problem only. The computational ω_b was determined by choosing that ω with minimum

	N	AD-HOC SOR	SOR (ω_b–theor.)	SOR (ω_b–comp.)
(a)				
	7	18 (1.44-1.44)	18 (1.449)	16 (1.46)
	15	34 (1.67-1.67)	33 (1.675)	31 (1.70)
	31	64 (1.82-1.82)	63 (1.822)	56 (1.83)
	63	119 (1.90-1.90)	118 (1.907)	94 (1.914)
(b)			$u_n = 0$	
	7	32 (1.72-1.70)	35 (1.672)	31 (1.71)
	15	59 (1.85-1.84)	70 (1.833)	59 (1.84)
	31	129 (1.92-1.91)	135 (1.916)	117 (1.920)
	63	261 (1.96-1.95)	252 (1.958)	212 (1.960)
(c)		$u_n = 0$		
	7	27 (1.62-1.64)	28 (1.603)	26 (1.62)
	15	48 (1.79-1.80)	57 (1.789)	52 (1.80)
	31	107 (1.89-1.89)	111 (1.891)	81 (1.898)
	63	212 (1.94-1.94)	207 (1.945)	165 (1.948)

Table III.

iteration from a sequence of ω's. It is a function not only of the problem but also of the computer, the arithmetic rounding procedure, etc. It could be misleading to compare an *a priori* method with the computational ω_b results.

REFERENCE

[1] Brazier, P. H., *Comp. Meth. in App. Mech. and Engr.* 3, 335-347 (1974).

[2] Ehrlich, L. W., An ad hoc SOR method, to appear.

[3] Rigal, A., *J. of Comp. Phys.* 32, 10-23 (1979).

[4] Russel, D. B., Ministry of Aviation, Aeronautical Research Council, Reports and Memoranda No. 3331, May 1962, H.M.S.O., 1963.

[5] Strikwerda, J. C., *SIAM J. Scient. and Stat. Comp.* 1, 119-130 (1980).

ON PRECONDITIONED ITERATIVE METHODS FOR ELLIPTIC PARTIAL DIFFERENTIAL EQUATIONS

David J. Evans

Department of Computer Studies
Loughborough University of Technology
Loughborough, Leicestershire, UK

In this paper sparse and compact preconditioned iterative methods are defined for the numerical solution of the linear system Au=b derived from the discretisation on a suitable grid of a self-adjoint elliptic partial differential equation. The difference between the methods is shown to depend on whether the conditioning matrix C is derived from a splitting or approximate factorisation of A. The versatility of the preconditioning concept is shown by the application of semi-iterative and conjugate gradient techniques to accelerate the rates of convergence.

1. INTRODUCTION

This paper generalises the preconditioning concept introduced by Evans [1] and is concerned with the use of iterative methods for the solution of

$$Au = b \tag{1}$$

where the matrix A is large and its diagonal elements $a_{i,i} > 0$. Next, we premultiply (1) by a non-singular matrix C^{-1}, where C^{-1} is an approximate inverse of A. Thus, (1) is transformed into the preconditioned form,

$$C^{-1}Au = C^{-1}b \tag{2}$$

where C is a conditioning matrix such that its inverse is easily computed.

Then, we can define the general iterative scheme

ISBN 0-12-632620-7

$$u^{(n+1)} = u^{(n)} + \tau C^{-1}(b-Au^{(n)}) \quad (3)$$

where τ is a real parameter which is consistent with (2) **iff** C is non-singular and $\tau \neq 0$.

Let us now consider the possible forms for the matrix C which can possess the above qualities. These are when C is closely associated with the two alternatives:

a) Splitting of A

If we assume that A can be written (without loss of generality) as $A=I-L-U$ where L,U are strictly lower and upper triangular matrices respectively, then a suitable form of C worthy of further investigation is

$$C = (I-\omega L)(I-\omega U) . \quad (4)$$

Evidently, for different values of the parameters τ and ω we can easily construct Table 1.

From the analogy of the Jacobi and SD iterative methods we see that for $\omega=0$ the optimum value $\tau_0=2/(\bar{a}+\bar{b})$ where $\bar{a},\bar{b}$ are the minimum and maximum eigenvalues of A, an improvement in the convergence rate is assured. Thus, it follows immediately that if we can obtain the optimum parameters ω_1 and τ_1 for C given by (4) then an improvement in the convergence rate for the P.S.D. method over the S.S.O.R. method may ensue. These strategies will be referred to as Sparse Preconditioning since matrix sparseness is retained throughout.

TABLE 1

ω	τ	C	Iterative Method
0	1	I	Jacobi (J) [8]
0	τ_0	I	Simultaneous Displacement (SD) [8]
ω	$\omega(2-\omega)$	$(I-\omega L)(I-\omega U)$	S.S.O.R. [8]
ω_1	τ_1	$(I-\omega L)(I-\omega U)$	Sparse P.S.D. [2]

b) Approximate Factorisation of A

Alternatively, we can consider the factorisation of A into easily invertible lower and upper triangular matrix factors. Standard methods for such factorisations are well known, i.e. $\mathcal{L}\mathcal{U}, \mathcal{L}\mathcal{L}^T$ or $\mathcal{L}\mathcal{D}\mathcal{U}$. However, in the triangular factorisation of a sparse matrix, a large number of the zero entries of the matrix are "filled in" by non-zero values which not only are a source of rounding errors but greatly increase the work and storage requirements of the method.

Thus, in recent years, much research has been directed towards a number of indirect or iterative methods in which the strategy is to reduce the fill-in terms of the triangular factors $\mathcal{L}, \mathcal{U}$ etc. to a minimum whilst still retaining the property $C=\mathcal{L}_r\mathcal{U}_r \simeq A$ where $\mathcal{L}_r, \mathcal{U}_r$ denote the corresponding sparse triangular matrix factors in which r non-zero off-diagonal 'fill-in' vectors have been retained. Since the fill-in terms (r>1) are confined to a compact form, these strategies are referred to as Compact Preconditioning. Here a similar strategy for a normalised factorisation procedure is presented.

Finally, in addition to these basic Preconditioning strategies, the iterative process (3) can be further improved by acceleration techniques i.e. semi-iterative and conjugate gradient which will be discussed later.

2. SPARSE PRECONDITIONING METHODS

The iteration method (3) we shall consider is given alternatively by

$$u^{(n+1)} = B_{\tau,\omega}u^{(n)} + b_{\tau,\omega} \tag{5}$$

where $B_{\tau,\omega}=I-\tau A_{\omega}=I-\tau(I-\omega U)^{-1}(I-\omega L)^{-1}A$ and $b_{\tau,\omega}= (I-\omega U)^{-1}(I-\omega L)^{-1}b$. For $\omega=0$, we see that (5) reverts to the

Simultaneous Displacement method $u^{(n+1)}=(I-\tau A)u^{(n)}+\tau b$.

It is well known [7], that $S(B_{\tau,\omega})$ the spectral radius of $B_{\tau,\omega}$ is less than unity iff,

$$0<\tau<2/(\Lambda(A_{\omega}) \tag{6}$$

and the value of τ for which $S(B_{\tau,\omega})$ attains its minimum value is when

$$\tau = \tau_1 = 2/\{\Lambda(A_{\omega})+\lambda(A_{\omega})\} \quad , \tag{7}$$

where $\Lambda(A_{\omega}),\lambda(A_{\omega})$, denotes the maximum and minimum eigenvalue of A_{ω} respectively for which the rate of convergence $R(B_{\tau,\omega})$ behaves asymptotically as $-\log\{S(B_{\tau,\omega})\}$.

The problem of determining good estimates τ_1,ω_1 of τ and ω respectively such that $S(B_{\tau,\omega})$ attains its minimum value is of paramount importance. In [2], it was shown a good choice of ω_1 was given by minimising $P(A_{\omega})$, the P condition number given by

$$P(A_{\omega}) = \Lambda(A_{\omega})/\lambda(A_{\omega}) \ . \tag{8}$$

For A, a real, symmetric, positive definite matrix, the theoretical analysis determining the optimum choice of ω is given in [3] and is summarised briefly as follows.

If λ is an eigenvalue of A_{ω} and $v\neq 0$ an associated eigen-vector, then $A_{\omega}v=\lambda v$, which by taking inner products of both sides yields the following expression for λ, i.e.

$$\lambda = (v.Av)/\{v.(I-\omega L)(I-\omega U)v\} \tag{9}$$

from which we can derive the following results.

In the range $0<\omega<2$, an upper bound on $\lambda(A_{\omega})$ is $\Lambda(A_{\omega})\leq 1/\omega(2-\omega)$ whilst if M and m be numbers such that

$$-2\sqrt{\bar\beta}\leq m\leq m(B), \quad M(B)\leq M\leq \min(1,2\sqrt{\bar\beta}) \quad \text{and} \quad S(LU)\leq\bar\beta \ ,$$

where $M(B),m(B)$ denote the maximum and minimum eigenvalues of B, i.e. (L+U) respectively, then a lower bound on $\lambda(A_{\omega})$ is

$$\lambda(A_\omega) \geq \begin{cases} (1-M)/(1-\omega M+\omega^2\bar{\beta}), & \text{if } \bar{\beta}\geq 1/4 \text{ or if } \bar{\beta}<1/4 \text{ and } \omega\leq\omega^* \\ (1-m)/(1-\omega m+\omega^2\bar{\beta}), & \text{if } \bar{\beta}<1/4 \text{ and } \omega>\omega^* \end{cases} \quad (10)$$

where for $\bar{\beta}<1/4$ we define $\omega^*=2/\{1+(1-4\bar{\beta})^{\frac{1}{2}}\}$.

Moreover, the bound on $P(A_\omega)$ is minimised if we let $\omega=\omega_1$ where

$$\omega_1 = \begin{cases} 2/\{1+(1-2M+4\bar{\beta})^{\frac{1}{2}}\}, & \text{if } M\leq 4\bar{\beta} \\ 2/\{(1+(1-4\bar{\beta})^{\frac{1}{2}}\}, & \text{if } M>4\bar{\beta}. \end{cases} \quad (11)$$

Finally, further improvement in the rate of convergence is obtained from the P.S.D. semi-iterative method, at the expense of an extra vector storage, as

$$u^{(n+1)} = u^{(n-1)}+\rho_{n+1}(u^{(n)}-u^{(n-1)})+\rho_{n+1}\tau_1(I-\omega_1 U)^{-1}(I-\omega_1 L)^{-1}(b-Au^{(n)}), \quad (12)$$

where the parameter sequence $\rho_1,\rho_2,\ldots$ is given by

$$\rho_1 = 1,\ \rho_2 = (1-\sigma^2/2)^{-1},\ldots\rho_{n+1}=(1-\sigma^2\rho_n/4)^{-1}, n=2,3,\ldots$$

and $\sigma=\{P(A_\omega)-1\}/\{P(A_\omega)+1\}$.

Since the virtual spectral radius of the SI method [7] is $[T_n(1/\sigma)]^{-1}$ where $T_n(x)$ denotes the classical Chebyshev polynomial of degree n, we have that the asymptotic average rate of convergence for the PSD-SI method is

$$2/\sqrt{P(A_\omega)} \quad (13)$$

These results are confirmed by Table 2 which contains results for the model Dirichlet problem i.e. the Laplace equation in the unit square with zero boundary values and show a substantial improvement over S.O.R for optimal parameters.

It is obvious that the extensions of these techniques to block methods will yield further improvements which will be presented in a later paper.

TABLE 2

h^{-1}	ω_1	τ_1	PSD-SI. no. of iterations	ω_b	SOR. no. of iterations
20	1.7640	0.6993	17	1.7295	61
40	1.8750	0.4264	24	1.8547	121
60	1.9157	0.3031	30	1.9005	181

3. COMPACT PRECONDITIONING METHODS

It is well known [7] that for a symmetric, positive definite matrix A there exists an unique factorisation of the form

$$A = DT'TD \ , \tag{14}$$

where D is a positive diagonal matrix and T a real upper triangular matrix with unit diagonal elements and T' denotes the transpose of T.

For the sparse 5-term diagonal matrix of semi-band-width m with terms a_i on the diagonal, b_i on the sub-diagonal and c_i on the mth sub-diagonal respectively derived from the discretisation of a self-adjoint elliptic partial differential equation of 2nd order on a regular grid, the algorithmic solution is given in [3].

Recently, research into approximate factorisation techniques have yielded many new strategies similar to [4]. We now consider an approximate normalised factorisation of the form,

$$A \approx D_r T'_r T_r D_r \tag{15}$$

where T_r is a sparse triangular matrix in which only r outer-most off-diagonals (fill-in terms) of the matrix T are included (see Figure 1)

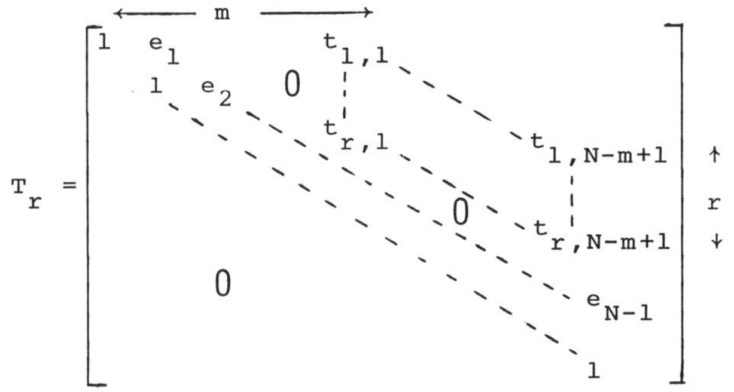

(16)

FIGURE 1

and D_r is a diagonal matrix as before. Then, with certain conditions on A the magnitude of the fill-in terms can be shown to decrease and consequently after r terms can be neglected. In [5], the analysis to determine the optimum r is presented.

Finally, the algorithmic procedure for determining the elements of the D_r and T_r component matrices can be expressed in the compact form.

First, compute

$$d_1 = \sqrt{a_1};\quad d_i = \{a_i - (b_{i-1}/d_{i-1})^2\}^{\frac{1}{2}};\quad e_{i-1} = b_{i-1}/d_{i-1}d_i$$

for $i=2,3,\ldots,m-1$.

then for $j=1,2,\ldots,N-m+1$ we have

$$w_1 = c_j/d_j;\quad v = b_{m+j-2}/d_{m+j-2}$$

and for $j \leq r-1$,

$$w_i = -e_{i+j-2}w_{i-1} \quad \text{for } i=2,3,\ldots,r+1-j. \tag{17}$$

Continuing, for $j>1$ and $r>1$,

$$w_i = -e_{i+j-2}w_{i-1} - \sum_{k=1}^{i-1} w_k t_{k-i+r+1,i+j-r-1}$$

and for either $i=(r-j+2),(r-j+3),\ldots,r$ and all $j \leq r$ or $i=2,3,\ldots,r$ for $j>r$. Then we have,

$$d_{m+j-1} = \{a_{m+j-1} - \sum_{k=1}^{r}(w_k^2 - v^2)\}^{\frac{1}{2}};\quad e_{m+j-2} = v/d_{m+j-1} \text{ and}$$

$$t_{i,j} = w_i/d_{m+j-1} \quad \text{for } i=1,2,\ldots r.$$

Finally, analogous to the preconditioned conjugate gradient method presented in [6] a normalised implicit gradient method (NICG) can be formulated incorporating compact preconditioning and can be given in algorithmic form as follows.

$$\text{Form}\quad r_0 = b - Au_0 \text{ and set } \tilde{r}_0 = D_r^{-1} r_0 .$$
$$\text{Solve}\quad (T_r'T_r)r_0^* = \tilde{r}_0 \text{ and set } \tilde{\sigma}_0 = r_0^* . \tag{18}$$

Now, for $i=0,1,2,\ldots$ calculate the vectors $u_{i+1}, \tilde{r}_{i+1}, \tilde{\sigma}_{i+1}$ and the scalar quantities $\tilde{\alpha}_i, \tilde{\beta}_{i+1}$ as follows:

$$\tilde{\alpha}_i = (\tilde{r}_1, r^*_i)/(\tilde{r}_i, A\tilde{r}_i), \quad u_{i+1} = u_i + \tilde{\alpha}_i \tilde{\sigma}_i \quad \text{and} \quad \tilde{r}_{i+1} = \tilde{r}_i - \tilde{\alpha}_i A \tilde{\sigma}_i. \tag{19}$$

Then solve

$$(T_r' T_r) r^*_{i+1} = \tilde{r}_{i+1} \tag{20}$$

and evaluate

$$\tilde{\beta}_{i+1} = (\tilde{r}_{i+1}, r^*_{i+1})/(\tilde{r}_i, r^*_i) \quad \text{and} \quad \tilde{\sigma}_{i+1} = r^*_{i+1} + \tilde{\beta}_{i+1} \tilde{\sigma}_i \tag{20}$$

to complete the computational process.

In Table 3, we present for comparison purposes the results for the model problem of the CG and NICG methods for 4 values of the fill-in parameter r which clearly shows the advantages to be gained from applying preconditioning.

TABLE 3: The Number of Iterations for the CG and NICG methods for Various Values of the Fill-in Parameter r.

h^{-1}	C.G.	N.I.C.G.			
		r=1	2	3	4
20	53	17	13	10	9
40	101	28	20	17	14
60	160	40	28	23	20

The application of compact preconditioning techniques can readily be extended to 3 dimensional and non-linear problems and these are given in [5].

REFERENCES

1. Evans, D.J., "The Use of Preconditioning in Iterative methods for solving linear equations with symmetric positive definite matrices", *J.I.M.A. 4, 295-314*, (1968).

2. Evans, D.J., and Missirlis N.M., "The Preconditioned Simultaneous Displacement (PSD) Method, *M.C.S.* in press (1980).

3. Benson, A. and Evans D.J., "A Normalised Algorithm for the Solution of Symmetric General Five-term Linear Systems of Semi-bandwidth m, *C.A.A.M.*, *5*, *299-304*, (1979).

4. Evans, D.J., "Iterative Sparse Matrix Algorithms", Conf.Proc. 'Software for Numerical Mathematics', edit. D.J. Evans, 49-84, Academic Press, (1974).

5. Evans, D.J. and Lipitakis, E.A., "A Normalised Implicit Conjugate Gradient Method for the Solution of Large Sparse Systems of Linear Equations, *C.A.M.A.M.E.*, in press (1980).

6. Evans, D.J., "The Analysis and Application of Sparse Matrix Algorithms in the Finite Element Method", Conf. Proc. 'The Mathematics of Finite Elements and Its Applications', edit. J.R. Whiteman, 427-447, Academic Press, (presented at MAFELAP conference April 1972), (1973).

7. Varga, R.S., "Matrix Iterative Analysis", Prentice Hall (1962).

8. Young, D.M., "Iterative Solution of Large Linear Systems", Academic Press, London, (1971).

BLOCK RELAXATION STRATEGIES

Vance Faber

Computer Science and Services Division
Los Alamos Scientific laboratory
Los Alamos, New Mexico

I. INTRODUCTION

In connection with finite difference approximations to the solution of the general elliptic equation $\nabla a \cdot u = -f$ on an arbitrary region in rectangular coordinates, B. Buzbee, Los Alamos Scientific Laboratory, has suggested investigating various block line relaxation strategies to determine which ones give the fastest rate of convergence.

More specifically, we wish to solve the linear system $Ax = b$, where

$$A = \begin{bmatrix} T_1 & U_1 & & & & \\ S_2 & T_2 & U_2 & & & \\ & S_3 & T_3 & U_3 & & \\ & & \ddots & \ddots & \ddots & \\ & & & S_n & T_n & U_n \end{bmatrix} \qquad (1)$$

is nxn block tridiagonal with block size m. We plan to solve this system by block line relaxation when the largest block problem with block size m we can cheaply solve directly is $\ell x \ell$.

ISBN 0-12-632620-7

II. STEPS

Each iteration of the strategies that we consider is composed of a sequence of <u>steps</u>. After the jth step, we have the approximate solution vector $x^j = (x_1^j, x_2^j, \ldots, x_n^j)$, where each x_i^j is an m-vector, and error vector $e^j = x - x^j$. We partition the matrix A as

$$A = \left[\begin{array}{c:c:c} B_1 & C_1 & 0 \\ \hdashline A_2 & B_2 & C_2 \\ \hdashline 0 & A_3 & B_3 \end{array}\right] \tag{2}$$

where B_1 is kxk block tridiagonal; B_2 is $\ell x \ell$ block tridiagonal; A_2 and A_3 have their only nonzero block, a_2 and a_3, respectively, at the upper right; and c_1 and c_2 have their only nonzero block, c_1 and c_2, respectively, at the lower left. We shall consider ℓ to be fixed, but k and m can be any nonnegative integers, depending on j, that satisfy $k + \ell + m = n$. During this step we shall relax only the vector variables $x_{k+1}, x_{k+2}, \ldots, x_{k+\ell}$. Let $z_1 = (x_1, \ldots, x_k)$; $z_2 = (x_{k+1}, \ldots, x_{k+\ell})$; $z_3 = (x_{k+\ell+1}, \ldots, x_n)$ and similarly for z_1^j, z_2^j, z_3^j. Also let $\beta_1 = (b_1, \ldots, b_k)$, $\beta_2 = (b_{k+1}, \ldots, b_{k+\ell})$, and $\beta_3 = (b_{k+\ell+1}, \ldots, b_n)$. The system we are solving in this notation is

$$\begin{aligned} B_1 z_1 + C_1 z_2 &= \beta_1 \\ A_2 z_1 + B_2 z_2 + C_2 z_3 &= \beta_2 \\ A_3 z_2 + B_3 z_3 &= \beta_3 \end{aligned} \tag{3}$$

At the jth step we relax using the equations

$$
\begin{aligned}
z_1^{j+1} &= z_1^j \\
B_2 z_2^{j+1} &= \beta_2 - A_2 z_1^j - C_2 z_3^j \\
z^{j+1} &= z^j \quad .
\end{aligned}
\tag{4}
$$

Thus e^{j+1} has vector components (define $e_0^j = e_{n+1}^j = 0$ for all j):

$$
\begin{aligned}
e_i^{j+1} &= e_i^j \quad , \; i \leq k \text{ or } i \geq \ell + 1 + k \\
e_i^{j+1} &= -(B_2^{-1})_{i-k,1}\, a_2 e_k^j - (B_2^{-1})_{i-k,\ell}\, c_2 e_{k+\ell+1}^j , \\
& \qquad k < i < \ell + 1 + k \quad .
\end{aligned}
\tag{5}
$$

A single iteration of a (periodic) strategy of length p, $(k_1, k_2, \ldots, k_p)$, is defined as a Gauss-Seidel sequence of steps, (k_j, ℓ_j, m_j), $j = 1,2,\ldots,p$, where $k_j + \ell_j + m_j = n$ and, for the purposes of our discussion, all $\ell_j = \ell$. The convergence rate per step, θ, of the strategy $(k_1, k_2, \ldots, k_p)$ is $\theta = -\frac{1}{p} \ln \rho$; where ρ is the spectral radius of the error iteration matrix of the strategy.

III. PARTICULAR STRATEGIES; TEST PROBLEMS

Certain strategies can be analytically evaluated when they are applied to the linear systems that arise from finite difference approximations to the solution of a Helmholtz-type equation. We shall call a strategy $(k_1, k_2, \ldots, k_p)$ ordered if $0 = k_1 < k_2 < \ldots < k_p = n - \ell$ (a left-to-right sweep); well ordered if it is ordered and $k_{j+1} < k_j + \ell + 1$ for every j (all variables are relaxed); and super ordered if it is well ordered and $k_j + \ell + 1 \leq k_{j+2}$ (the variables corresponding to the beginning and the end of each step are relaxed exactly once in each iteration of the strategy).

The test problems that we consider have $U_i = S_i = I$ and $T_i = T$ (for some fixed T) for all i in Eq. (1). Under these assumptions, the convergence rate per step of the strategy $(k_1, k_2, \ldots, k_p)$ depends only on the ratios

$$\alpha_j = \frac{k_{j+1} - k_j}{\ell + 1}, \quad j = 1,2,\ldots,p-1 \quad .$$

In addition, any strategy that is a permutation of a super-ordered strategy has the same convergence rate per step. If $\alpha_j = \alpha$ for all $j = 1,2,\ldots,p-1$, the strategy is called regular.

IV. RESULTS

The results deal only with the test cases. The proofs will be published elsewhere.

Result 1. Analytic expressions can be found for the convergence rates of the regular strategies in terms of the spectral radius of the matrix T.

Result 2. Asymtotically, as $n/\ell \to \infty$, the regular strategy with the fastest convergence rate per step has $\alpha = 2/3$ (the overlap from one step to the next is one-third the length of the step).

Result 3. The super-ordered strategy with the fastest rate of convergence per step is never regular when more than three steps are required per iteration.

Result 4. For the regular strategy with $\alpha = 1/2$ (the

overlap from one step to the next is one-half the length of the step) applied to the 5-point difference approximation to the solution of Poisson's equation on the unit square with uniform grid spacing ($\frac{1}{n+1} = h = \frac{1}{m+1}$), the spectral radius is $\rho \cong 1 - \frac{1}{2}\pi^2 h^2 (\ell+1)^2 + \ldots$.

V. CONCLUSIONS

The positive aspects of block relaxation schemes, namely, ease of implementation, generality, and the ability to exploit vector and parallel architecture, have traditionally been negated by the low convergence rate, high work rate, and large storage requirements of the schemes. We have shown, however, that the convergence rate of some block relaxation strategies is much faster than that of the strategies generally employed. For example, Parter and Steuerwalt (1) have shown that the spectral radius of ℓ-line relaxation applied to the 5-point difference approximation to the solution of Poisson's equation on the unit square with uniform grid spacing h is

$$\rho \cong 1 - 2\ell\, \pi^2\, h^2 + \ldots \quad .$$

So by the results given above, the regular strategy with α = 1/2 (which is far from optimal) converges faster by a factor of $\ell/4$. This improved convergence rate for these strategies warrents a deeper look into these block relaxation schemes.

REFERENCES

1. Seymour V. Parter and Michael Steuerwalt, "Another Look at Iterative Methods for Elliptic Difference Equations," Los Alamos Scientific Laboratory report LA-UR-80-202 (1980).

ON THE NUMERICAL SOLUTION
OF NONLINEAR ELLIPTIC PDEs ARISING
FROM SEMICONDUCTOR DEVICE MODELING

Wolfgang Fichtner
Donald J. Rose

Bell Laboratories
Murray Hill, New Jersey

I. INTRODUCTION AND EQUATION FORMULATION

The coupled nonlinear elliptic partial differential equations which model the intrinsic behavior of semiconductor devices provide a significant challenge for the scientific computing community. In this paper we summarize various formulations of the basic semiconductor equations and present several numerical algorithms for solving them efficiently. Some computational results are given to indicate the relative merits of different solution schemes. A more substantial discussion of this work appears in Fichtner and Rose (1).

We assume a two-dimensional model for semiconductor devices of the form:

$$-\Delta u + n - p = k(x,y) \; ; \tag{1}$$

$$\nabla \cdot \mathbf{J}_n = 0 \; ; \tag{2}$$

$$\nabla \cdot \mathbf{J}_p = 0 \; ; \tag{3}$$

where $u(x,y)$ is the electrostatic potential, $n(x,y)$ and $p(x,y)$ are the electron and hole densities, respectively, and $k(x,y)$ is the doping profile (impurity concentration). J_n and J_p are, respectively, the electron and hole current densities; they are further specified in the usual drift-diffusion form by

$$\mathbf{J}_n = -\mu_n(x,y)n \, \nabla u + D_n(x,y) \, \nabla n \; , \tag{4}$$

$$\mathbf{J}_p = -\mu_p(x,y)p \, \nabla u - D_p(x,y) \, \nabla p \; , \tag{5}$$

for given mobility coefficients μ_n, μ_p and diffusion coefficients D_n,D_p. For an MOS device [see Ref. (2)], Eqs. (1) to (3) are posed on unions of rectangular regions as in Fig. 1; Eq. (1) for $u(x,y)$ includes an oxide region (the "hat") on which $-\Delta u = 0$. Dirichlet boundary conditions are imposed on the gate (G), source (S), drain (D) and substrate (B) by the applied bias voltages; Neumann boundary conditions are assumed at the unspecified edges. The carrier densities $n(x,y)$ and $p(x,y)$ exist only in the lower rectangle and their boundary conditions are derived

ISBN 0-12-632620-7

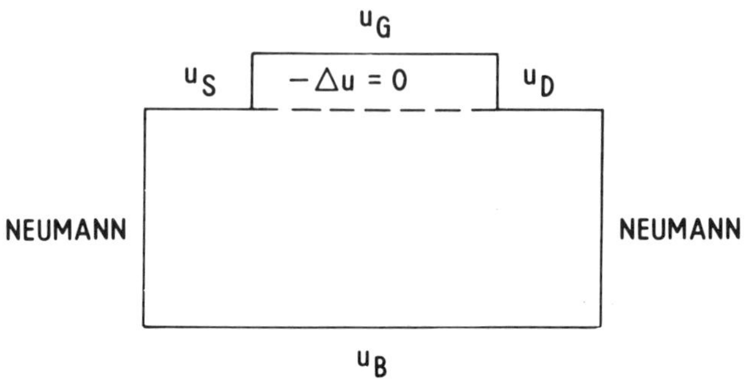

FIGURE 1. Cross-section of a planar MOS transistor.

from $k(x,y)$. The limitation to rectangular regions is not a severe restriction in our applications. We have taken the right-hand sides of the continuity Eqs. (2) and (3), to be identically zero for simplicity; our methods extend to the more general case where such right-hand sides exist and depend on the carrier densities n and p.

Assuming the validity of the Einstein relation [see Ref. (2)], Eqs. (4) to (5) can be rewritten as

$$\mathbf{J}_n = \mu_n[-n \nabla u + \nabla n] = -\mu_n e^{u-v} \nabla v \tag{6}$$

$$\mathbf{J}_p = \mu_p[-p \nabla u - \nabla p] = -\mu_p e^{w-u} \nabla w \tag{7}$$

where $n \equiv e^{u-v}$ and $p \equiv e^{w-u}$ define implicitly the quasi-Fermi levels v and w for electrons and holes, respectively. Using this change of variables in Eqs. (1) to (3) leads to

$$-\Delta u + e^{u-v} - e^{w-u} = k(x,y)\ , \tag{8}$$

$$-\nabla \cdot (\mu_n e^{u-v} \nabla v) = 0\ , \tag{9}$$

$$-\nabla \cdot (\mu_p e^{w-u} \nabla w) = 0\ . \tag{10}$$

This formulation of the semiconductor equations has proved preferable if the Einstein-relation holds mainly because the triple u,v,w are "smoother" than u,n,p, and, therefore, there is less sensitivity to the mesh (grid restrictions) and to the initial guess for the discrete solution. Note, however, that each of Eqs. (8) to (10) is nonlinear in its own variable; i.e., u,v,w respectively. In some special cases it is convenient to "reduce the nonlinearity" in Eqs. (9) and (10) by introducing a further change of variables

$$v' = e^{-v}\ ; \quad w' = e^{w}\ . \tag{11}$$

We remark that for very small devices the validity of the Einstein-relation for high electric field strengths is questionable, thus enforcing the formulation (1) to (5).

We have used a five-point finite difference discretization on a nonuniform mesh; some aspects of the discretization are physically motivated. While such a discretization is only formally first order, preliminary experiments have indicated that Richardson extrapolation can be used to improve the accuracy (and formal order) of the discrete solution relative to the solution of the PDEs.

After discretization the formulation (1) to (5) lead to a system of nonlinear algebraic equations which we write suggestively in block form

$$g_1(u,n,p) = A_1 u + n - p - k_1 = 0\,, \tag{12a}$$

$$g_2(u,n,p) = A_2(u)n - k_2 = 0\,, \tag{12b}$$

$$g_3(u,n,p) = A_3(u)p - k_3 = 0\,; \tag{12c}$$

where the g_i are vector functions; u,n,p are the vectors of the discrete unknowns at mesh points; the A_i are square matrices and the k_i represent $k(x,y)$ and the boundary conditions. We call Eq. (12) 3-block linear (3BL) since each block equation is linear in its own block variable, u,n,p, respectively. Similarly, (8) to (10) lead to the 3-block nonlinear (3BNL) system

$$g_1(u,v,w) = A_1 u + f(u,v,w) - k_1 = 0\,; \tag{13a}$$

$$g_2(u,v) = 0; \qquad g_3(u,w) = 0\,; \tag{13b}$$

while the change (11) in Eqs. (9) to (10) leads to the nonlinear Poisson 2-block linear (NLP2BL) system

$$g_1(u,v',w') = A_1 u + f(u,v',w') - k_1 = 0\,, \tag{14a}$$

$$g_2(u,v') = A_2(u)v' - k_2 = 0\,, \tag{14b}$$

$$g_3(u,w') = A_3(u)w' - k_3 = 0\,. \tag{14c}$$

Equations (12) to (14) may represent a system with $3000 \le N \le 10{,}000$ variables in a typical application.

II. DISCRETE EQUATION SOLUTION

We consider now the solution of the nonlinear system $g(z) = 0$ where g and z may be in block form

$$z = (z_1,z_2,z_3)^T; \quad g = (g_1,g_2,g_3)^T \tag{15}$$

as in Eqs. (12) to (14). We present two approaches: (i) block nonlinear iteration and (ii) full Jacobian approximate Newton.

II.A Block Nonlinear Iteration

Associate with each g_i the block variable z_i and consider the following algorithm:

Algorithm BNI

[1] guess z; set $j \leftarrow 0$

[2] for some $(i \neq j) \epsilon (1,2,3)$ do
{solve $g_i(x|z)$ approximately for i − block variable x;
set $z_i \leftarrow z_i + \omega(x - z_i)$;
set $j \leftarrow i$}

[3] if converge then return else GOTO [2].

Examples included in BNI are BNSOR,BNGS ($\omega = 1$) and BNSSOR. Note that the blocks have been chosen to reflect the PDE variables in the coupled system, and that step [2] involves the solution of systems of nonlinear equations in the 3BNL or NLP2BL formulations but not in the 3BL formulation. The advantages of the BNI approach are simplicity and small relative memory requirements; disadvantages are the asymptotic linear convergence and unreliability for conditions with high currents flowing in the device. When the nonlinear Poisson equation dominates the system, the method can be quite effective. For example, in the NLP2BL formulation, using a damped Newton-vectorized red-black SOR iteration as discussed below and the Yale Sparse Matrix Package (YSMP) for the v' and w' equations, a typical calculation requires 5-10 CPU seconds on the CRAY-1 and can be the method of choice (see example 2 in Section III).

II.B Full Jacobian Approximate Newton

The most robust and reliable methods for solving systems (12) to (14) use a Newton-like method which requires both the assembly and approximate solution of linear systems $g_k' x = -g_k$ where g_k' is the Jacobian of $g(z^k)$, g as in (15). For example, the 3BNL formulation leads to block Jacobian systems

$$\begin{bmatrix} \partial g_1/\partial u & \partial g_1/\partial v & \partial g_1/\partial w \\ \partial g_2/\partial u & \partial g_2/\partial v & 0 \\ \partial g_3/\partial u & 0 & \partial g_3/\partial w \end{bmatrix} \begin{bmatrix} x_1 \\ x_2 \\ x_3 \end{bmatrix} = - \begin{bmatrix} g_1 \\ g_2 \\ g_3 \end{bmatrix} \tag{16}$$

where the matrix and right-hand side are evaluated at the k-th iteration (u^k, v^k, w^k). The linear system, (16), can be solved by a sparse direct method (YSMP) or by iterative methods. Particularly attractive iterative methods have proved to be the Newton-Richardson iteration as described in Bank and Rose (3) [see also Sherman (4)] and the Newton-block SOR (or block SSOR) iteration. The advantage of the full Jacobian approach, in addition to reliability is the asymptotic higher order convergence. The main disadvantage, memory requirements, are somewhat mitigated by using a Newton-block iterative method.

II.C Approximate Newton Methods

We now consider the solution of a nonlinear system $h(z) = 0$ needed in both approaches A and B. We suppose $h'(z)$ is nonsingular and locally Lipschitz on R^N and that $h:R^N \leftrightarrow R^N$ is a homeomorphism [see Ortega and Rheinbolt (5) for definitions of technical terms]. The basic iteration is:

$$z^{k+1} = z^k + t_k x_k \; , \tag{17}$$

$$M_k x_k = - h_k \; , \tag{18}$$

$$t_k = (1+\kappa_k \|h_k\|)^{-1} \; . \tag{19}$$

Here $M_k = h_k'$(Newton-step) or M_k represents an iterative splitting of h_k'; $\|\cdot\|$ is any norm. Let

$$\alpha_k = \|h_k' x_k + h_k\| \, \|h_k\|^{-1} \; ; \qquad \alpha_0 < 1 \; . \tag{20}$$

The following algorithm and theorem are presented in Bank and Rose (3).

Algorithm GLOBAL

[1] input z^0, $\delta \epsilon (0,1-\alpha_0)$

[2] $k \leftarrow 0$, $\kappa \leftarrow 0$; compute h_0, $\|h_0\|$

[3] compute x_k such that $\alpha_k \leq \alpha_0(\|h_k\|/\|h_0\|)$

[4] $t_k \leftarrow (1 + \kappa\|h_k\|)^{-1}$

[5] compute z^{k+1}, h_{k+1}, $\|h_{k+1}\|$ as in (17)

[6] if $(\|h_{k+1}\|/\|h_k\| > 1 - t_k\delta)$

[7] then {if $\kappa = 0$ then $\kappa \leftarrow 1$ else {$\kappa \leftarrow 10\kappa$; GOTO [4]}}

[8] else {$\kappa \leftarrow \kappa/10$; $k \leftarrow k + 1$}

[9] if converge then return else GOTO [3].

For $h(z)$ as above convergence of algorithm GLOBAL is described in the following

Theorem: If $M_k = h_k'$ or represents a uniformly convergent splitting, then $z_k \to z^*$ with $h(z^*) = 0$ quadratically for any initial z^0.

TABLE I. Applied Boundary Conditions

Example 1	Example 2
$U_S = 0$	$U_S = 0$
$U_G = 0$	$U_G = 60$
$U_D = 120$	$U_D = 120$
$U_B = 0$	$U_B = 0$

III. EXAMPLES

In this section, we shall briefly describe the results for two typical examples of MOS devices fabricated with state-of-the-art technology (6). See Table I for the applied boundary conditions.

Example 1 is a "hard" problem in the sense that the validity of the Einstein relation is questionable. The doping profile for this so-called "buried-channel" device is shown in Fig. 2. Note the nonuniformity of $k(x,y)$ in the channel region, which results in high current densities reflecting a strong coupling of the unknowns. The electric field in Fig. 3 and the electron distribution in Fig. 4 are typical for a small size MOS device at high current flow.

Example 2 is an "easy" problem which does not have the nonuniform doping profile. For the applied boundary conditions the current flowing is very low, thus reflecting a weak coupling between the carrier densities and the electrostatic potential. Furthermore, the influence of holes [Eq. (3)] can be totally neglected. For this case, the substitution (11) can be successfully applied.

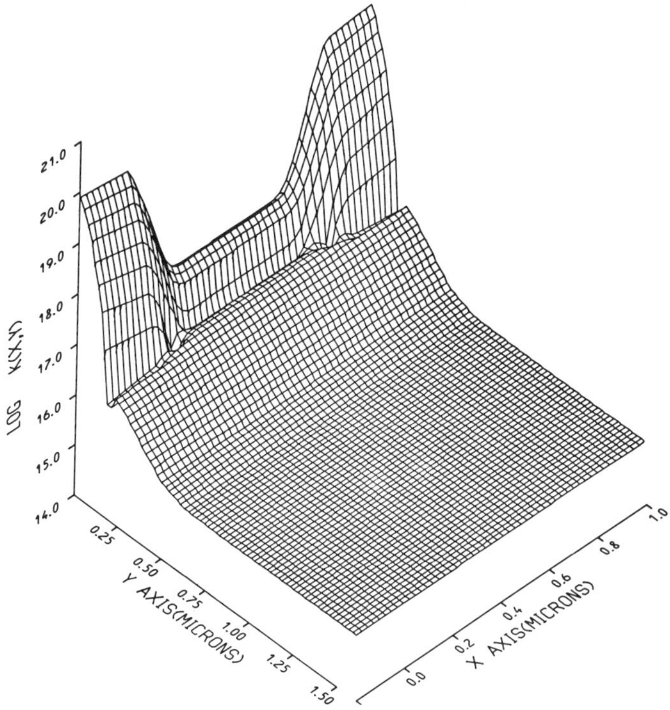

FIGURE 2. Impurity distribution for example 1.

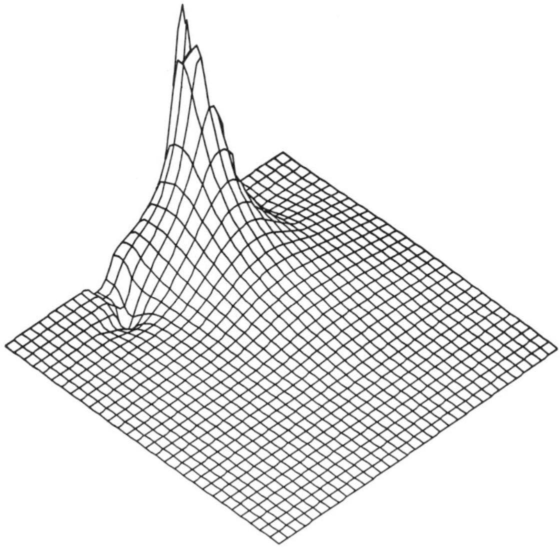

FIGURE 3. Lateral electric field for example 1.

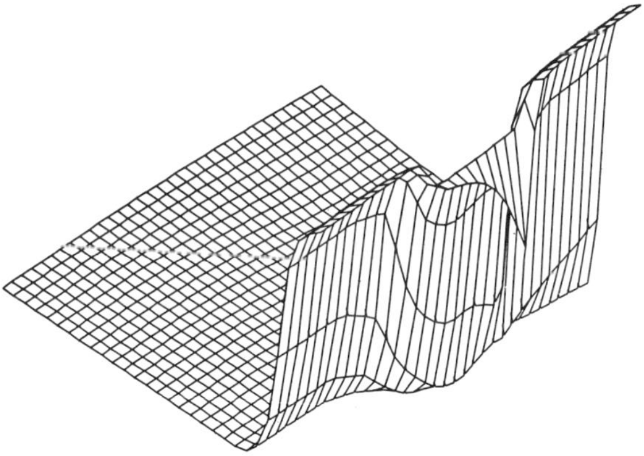

FIGURE 4. Electron distribution as seen through the oxide region.

Table II summarizes the results for CPU-time and memory requirements for the two examples. For the first problem, we considered BNGS and the full Jacobian variants NF (Newton-direct), NR (Newton-Richardson) and NB (Newton-block Gauss-Seidel); whereas, in the second problem we applied substitution (11) in the BNI case. The stopping criterion used was $\|h_k\|/N \leq 10^{-6}$.

TABLE II. Comparison of Numerical Methods

Method	Example 1 (N = 4256) u,n,p				Example 2 (N = 2954) u,v′			
	BNI	NF	NR	NB	BNI	NF	NR	NB
Symbolic Processing (sec)	0.2	4.5	4.5	0.2	0.2	2.5	2.5	0.2
Linear Equation Solution (sec)	0.2	7	0.3	0.2	0.2	4	0.21	0.2
Average Time Per Loop (sec)	0.6	8	1.6	0.7	0.4	4.6	0.9	0.6
Average Number of Equations per Loop	6	1	5	3	5	1	5	3
Total Number of Loops	100*	19	26	42	11	6	8	12
CPU-Time (sec)	63	162	43	32	8	31	11	9.5
Memory (k words)	450	900	900·	450	280	550	550	310

* Error criterion not reached.

REFERENCES

1. Fichtner, W., and Rose, D. J., "Two dimensional numerical semiconductor device modelling," to be published.
2. Sze, S. M., "Physics of Semiconductor Devices," Wiley-Interscience, New York (1969).
3. Bank, R. E., and Rose, D. J., "Global approximate Newton methods," to be published.
4. Sherman, A. H., "On Newton-iterative methods for the solution of systems of nonlinear equations." *SINUM 15,* 755 (1978).
5. Ortega, J. M., and Rheinbolt, W. C., "Iterative Solution of Nonlinear Equation in Several Variables," Academic Press, New York (1970).
6. Suciu, P. I., Fuls, E. N., and Boll, H. J., "High-speed NMOS circuits made by X-ray lithography and reactive sputter etching," *IEEE EDL-1,* 10 (1980).

NON-STANDARD MULTIGRID TECHNIQUES USING CHECKERED RELAXATION AND INTERMEDIATE GRIDS

Hartmut Foerster*
Klaus Stüben*
Ulrich Trottenberg**

*Institut für Mathematik
GMD
St. Augustin, W. Germany

**Institut für Angewandte Mathematik
and Sonderforschungsbereich 72
Bonn University
Bonn, W. Germany

1. INTRODUCTION

The multigrid (MG) techniques (cf. Brandt (1,2), Hackbusch (3,4) et al.) doubtlessly offer sensational perspectives in the numerical treatment of partial differential equations. There are, however, two essential lacks in the development of MG techniques up to now:

- The implementation of efficient and reliable software is still in the very beginning.
- The theory is not yet satisfactory: Many of the rigorous proofs give little help for practical purposes (software development), whereas the theoretical investigation of widely used algorithms mostly is only heuristic and not rigorous.

In this short survey we report on some theoretical results and on software development by which our Bonn group tries to make contributions in order to overcome this situation. Here the emphasis lies on <u>quantitative</u> theoretical

ISBN 0-12-632620-7

analysis and practical experiences in order to find "optimal" techniques for special two-dimensional elliptic boundary value problems (of the second order). One result of our investigations is that "non-standard" MG methods - as we would like to call them here - lead to considerable practical improvements and offer new possibilities for theoretical insight. Some of the developments on which we report here were motivated and influenced by discussions with Achi Brandt.

We assume the MG concept to be known in principle (see, however, Section 2). By "standard" MG methods we simply mean those algorithms which rely on pointwise lexicographic Gauß-Seidel relaxations (LEX), "usual" $h \to 2h$ restriction operators (injection; full residual weighting (5)) and linear $2h \to h$ interpolation operators.

On the other hand, the "non-standard" methods are characterized by (at least one of) the following components:

① checkered Gauß-Seidel relaxation techniques (CH),
② the use of checkered "intermediate" grids,
③ the use of the discrete differential equation in the $2h \to h$ transfer.

The components ② and ③ have been motivated by the Total Reduction (TR) method (Schröder, Trottenberg (6,7)). The disadvantages of this very fast and numerically very stable direct solver are: its mathematical complexity - due to which many numerical analysts refuse to deal with it at all - and, more important, the fact that it can be directly applied only to a rather small class of problems. It, however, can be and has been theoretically analyzed rigorously (8).

There are close connections between the MG and the TR ideas which suggest certain combinations leading to the so-called MGR methods. We give an outline in Section 2.2 stressing such versions which can be applied to "general problems". In the "optimal" MGR algorithms all components ① , ② , ③ are combined.

However, even if no checkered intermediate grids are used explicitly, considerable improvements are obtained by replacing LEX by CH relaxation techniques (component ①), see Section 2.3 and the following sections. We like to point out that - in a somewhat different way - MG algorithms with checkered relaxation techniques have been used and recommended already by Hackbusch (3); see also Brandt (9).

The quantitative results of Sections 2 and 3 refer to model problems in the unit square with Dirichlet or Neumann boundary conditions, respectively. In Section 4 we report on the program MGØ1 which solves the (non-constant) Helmholtz equation with Dirichlet boundary conditions on general 2D-domains. In Section 5 we outline further developments of our group which are related to the results presented here (zebra-line relaxation, etc.). A final remark stresses the importance of the non-standard methods in connection with parallel and vector computers.

All results of this report will be published in detail in separate papers.

2. THEORETICAL RESULTS FOR THE DIRICHLET MODEL PROBLEM

2.1 General Remarks and Assumptions

In this section we give some theoretical results on "non-standard" MG methods. With respect to both the theoretical rate of convergence and the computational effort, these methods turn out to be considerably superior to standard MG methods.

We consider convergence rates in terms of the (h,2h)-versions of the respective algorithms (two-level-analysis), symbolized by Figure 1, where • , o, ↘ , and ↗ stand for relaxing, solving exactly, fine-to-coarse and coarse-to-fine transfer, respectively. In particular, we assume that the defect equation on the 2h-grid is solved exactly in each iteration step. In practice one really uses multigrid iterations which are defined recursively by the (h,2h)-method using still coarser levels 4h, 8h, ..., the simplest one being of the form of Figure 2.

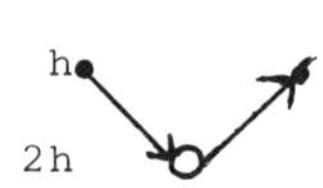

Figure 1: (h,2h)-cycle

Figure 2: multigrid cycle

All quantitative results given here refer to the discrete model problem for Poisson's equation $-\Delta_h u_h\ (P) = f(P)$ in the unit square with Dirichlet boundary conditions $u(P) = g(P)$, discretized by the usual 5-point operator $-\Delta_h$ on a square h-grid $\Omega_h (h^{-1} = N = 2^p)$. In this case the choice $h_o = 1/2$ is suitable.

We point out, however, that all non-standard methods considered are as generally applicable as the standard MG techniques themselves (apart from MGR-V1, which is reviewed mainly for motivation).

2.2 MGR: Combinations of TR and MG Techniques

The MGR methods treated in this section can be regarded from the MG and the TR point of view. Though the MGR-idea originally was motivated by TR and conceived (by the third author) in order to enlarge the range of applicability of the TR method, we here describe the MGR methods as special MG techniques. Concerning the TR point of view see (10), (11), (12).

To imbed these methods into the MG concept, we first characterize the form of a general (h,2h)-algorithm by verbally describing the components of which one iteration step $(u_h^o \to u_h^1)$ consists:

I. $u_h^o \to \tilde{u}_h^o$ by one (or more) relaxation steps on Ω_h;

II. (h→2h)-transfer of the defect $d_h = f + \Delta_h \tilde{u}_h^o$ by some operator I_h^{2h};

III. Solving the 2h defect equation exactly:
$$-\Delta_{2h} v_{2h} = I_h^{2h}\, d_h \text{ on } \Omega_{2h}$$
(with zero Dirichlet boundary conditions);

IV. (2h→h)-transfer $v_{2h} \to v_h$ of corrections;

V. $\tilde{u}_h^o + v_h \to u_h^1$ (applying eventually one or more additional relaxation steps).

A common feature of the MGR methods is the essential use of the intermediate checkered grid $\Omega_{\sqrt{2}h}$:

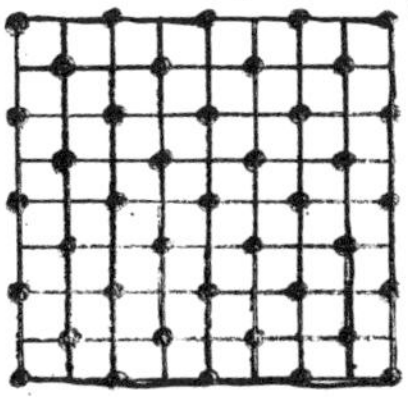

Here we consider only the following three algorithms in detail: MGR-CH-1, MGR-CH-2 and MGR-V1.

<u>MGR-CH-1:</u>

I. As above: one even-odd checkered relaxation step;

II. (h→2h)-transfer by $I_h^{2h} = I_{\sqrt{2}h}^{2h} \cdot I_h^{\sqrt{2}h}$;

III. as above;

IV. (2h→h)-transfer by two half steps using the discrete equations

$$-\Delta_{\sqrt{2}h}\, v_{\sqrt{2}h} = I_h^{\sqrt{2}h}\, d_h \text{ on } \Omega_{\sqrt{2}h} \setminus \Omega_{2h},\quad v_{\sqrt{2}h} = v_{2h} \text{ on } \Omega_{2h},$$
$$-\Delta_h\, v_h = d_h \text{ on } \Omega_h \setminus \Omega_{\sqrt{2}h},\quad v_h = v_{\sqrt{2}h} \text{ on } \Omega_{\sqrt{2}h};$$

V. $u_h^1 = \tilde{u}_h^o + v_h$ (without relaxation steps).

Here $I_h^{\sqrt{2}h} = \frac{1}{8}\begin{pmatrix} & 1 & \\ 1 & 4 & 1 \\ & 1 & \end{pmatrix}_h$, $I_{\sqrt{2}h}^{2h} = \frac{1}{8}\begin{pmatrix} 1 & & 1 \\ & 4 & \\ 1 & & 1 \end{pmatrix}_h$, $-\Delta_{\sqrt{2}h} = \frac{1}{2h^2}\begin{pmatrix} -1 & & -1 \\ & 4 & \\ -1 & & -1 \end{pmatrix}_h$.

<u>MGR-CH-2:</u> Similar to MGR-CH-1 with one additional step of even-odd checkered relaxation on the intermediate grid $\Omega_{\sqrt{2}h}$ in II for the equations

$$-\Delta_{\sqrt{2}h} v_{\sqrt{2}h} = I_h^{\sqrt{2}h}\, d_h \text{ on } \Omega_{\sqrt{2}h}$$

(starting with first approximation zero).

<u>MGR-V1:</u>

I. Cancelled (no relaxation step on Ω_h, $\tilde{u}_h^o = u_h^o$);

II. as in MGR-CH-1 with $I_{\sqrt{2}h}^{2h}$ replaced by $\hat{I}_{\sqrt{2}h}^{2h}$;

III. as above;

IV. as in MGR-CH-1 with $-\Delta_{\sqrt{2}h}$ replaced by $-\hat{\Delta}_{\sqrt{2}h}$ and $I_{\sqrt{2}h}^{2h}$ replaced by $\hat{I}_{\sqrt{2}h}^{2h}$;

V. $u_h^1 = u_h^o + v_h$ (without relaxation steps).

Here

$$\hat{I}_{\sqrt{2}h}^{2h} = \frac{1}{8}\begin{pmatrix} & & -1/2 & & \\ & 1 & 0 & 1 & \\ -1/2 & 0 & 6 & 0 & -1/2 \\ & 1 & 0 & 1 & \\ & & -1/2 & & \end{pmatrix}_h, \quad -\hat{\Delta}_{\sqrt{2}h} = \frac{1}{2h^2}\begin{pmatrix} & & -1/2 & & \\ & -1 & 0 & -1 & \\ -1/2 & 0 & 6 & 0 & -1/2 \\ & -1 & 0 & -1 & \\ & & -1/2 & & \end{pmatrix}_h .$$

MGR-V1 was the historically first method of MGR type; it can be viewed as a special "truncated" TR method (10). Using arguments similar to those in (8, Section 5.2) the convergence behaviour of MGR-V1 has been rigorously analyzed by G. Winter (12). One obtains:

Theorem: For the spectral radius $\rho = \rho_h$ of the (h,2h) iterative method MGR-V1 the following convergence statement is valid: $\rho_h \nearrow \rho^* = 1/8 \ (h \to o)$.

In the proof the fact is exploited that the spaces

$$\text{span}(\varphi_{k,l};\ \varphi_{N-k,l};\ \varphi_{k,N-l};\ \varphi_{N-k,N-l}) \quad (k,l=1,2,\ldots,N/2)$$

are invariant under the iteration process. Here the $\varphi_{k,l}$ denote the discrete eigenfunctions of Δ_h:

$$\varphi_{k,l}(x,y) = \sin(k\pi x)\cdot\sin(l\pi y) \quad (P=(x,y)\in\Omega_h; k,l=1,2,\ldots,N-1).$$

The above spaces remain invariant also under checkered relaxation processes as used in MGR-CH-1 and MGR-CH-2. Therefore these methods can be analyzed by similar arguments (see (12)). The corresponding quantitative results (convergence rates and computational effort) are contained in Table 1.

MGR-CH-1 and MGR-CH-2 differ from MGR-V1 formally in that they use simpler (approximating) operators in II and IV (note that the systems to be solved in IV are explicit!) and additionally checkered relaxation steps. This combination with CH was proposed by Achi Brandt.

MGR-CH-1 and MGR-CH-2 turn out to be the most efficient methods considered in this paper. As only 5-point operators are used in these methods, they can in principle be applied to very general problems. (The combination of checkered relaxation with MGR-V1 itself yields a method which is still more efficient in the case of the model problem. Due to the rather complicated operators $\hat{I}_{\sqrt{2}h}^{h}$ and $-\hat{\Delta}_{\sqrt{2}h}$ this method can, however, not be generalized in such a simple manner.)

2.3 MG methods

As already mentioned in the introduction, checkered relaxation techniques yield considerable improvements also if no intermediate grids are used explicitly. This is shown by the second part of Table 1 where MG methods with checkered Gauß-Seidel relaxations (CH) are compared to standard MG methods (with lexicographic Gauß-Seidel relaxations, LEX). (The theoretical investigations of this section were carried out by the second author (13).)

Method	ρ^*	Operations +/-	Operations *
MGR-CH-1	0.096	$10\frac{1}{6}$	$3\frac{1}{3}$
MGR-CH-2	0.074	$11\frac{5}{6}$	$3\frac{2}{3}$
CH(1,1)-FW	0.074	$12\frac{1}{2}$	$4\frac{1}{3}$
CH(1,1)-HW	0.125	$10\frac{5}{6}$	$3\frac{2}{3}$
LEX(1,1)-INJ	0.200 (*)	14	$4\frac{1}{3}$
LEX(1,1)-HW	0.175 (*)	$18\frac{1}{3}$	5
LEX(1,1)-FW	0.193 (*)	$20\frac{2}{3}$	$5\frac{1}{3}$

<u>Table 1:</u> Convergence rates ρ^* (for the (h,2h) process) and computational effort (per grid point of the finest grid and per multigrid cycle as explained in Figure 2).

In particular, we consider five special algorithms. Common to all of them are the following components (cf. Section 2.2): One relaxation step before and one relaxation step after the 2h-grid correction - symbolized by "(1,1)" - and linear interpolation for the (2h→h)-transfer. The algorithms are distinguished by the type of relaxation: CH or LEX, and by the choice of the (h→2h)-transfer operator I_h^{2h}. We denote by "FW" (full residual weighting), "HW" (half residual weighting), and "INJ" (injection) the following operators, respectively:

$$I_h^{2h} = \frac{1}{16}\begin{pmatrix} 1 & 2 & 1 \\ 2 & 4 & 2 \\ 1 & 2 & 1 \end{pmatrix}_h \; ; \quad I_h^{2h} = \frac{1}{8}\begin{pmatrix} & 1 & \\ 1 & 4 & 1 \\ & 1 & \end{pmatrix}_h \; ; \quad I_h^{2h} = (1).$$

FW HW INJ

The second column of Table 1 shows numerical values for ρ^*, where $\rho_h \rightarrow \rho^*$ for $h \rightarrow o$. Here ρ_h denotes the asymptotic convergence rate (= spectral radius) of the corresponding (h,2h) iterative method. In the cases CH(1,1)-HW and CH(1,1)-FW, ρ^* has been numerically evaluated from a rigorous formula, which can be derived in a similar manner as in the MGR cases. The corresponding theory (13) again relies on the fact that the $\varphi_{k,l}$-spaces characterized in Section 2.2 are invariant also under the CH multigrid process.

The LEX multigrid processes cannot be treated with this rigorous quantitative theory. The given values (*) are approximate ones derived by means of the so-called local mode analysis (1).

It can easily be proved that the convergence rates shown in Table 1 can be retained in case more than the two levels h,2h are used, if one defines a cycle in an appropriate manner (solving the coarse-grid equations with "sufficient" precision). Practical experiences, however, show that this is already approximately true for the simplest possible MG cycle as shown in Figure 2 (with $h_o = 1/2$). The computational effort for such a cycle (number of operations per point of Ω_h) is given in the last two columns of Table 1. We point out that the CH methods need less operations than the LEX methods as one can take advantage of some special features of checkered relaxation. For example, after one relaxation step the defects are zero on every second grid point, so that HW in the CH case practically means injection with the factor $\frac{1}{2}$.

Quantitatively Table 1 would look somewhat different if one would use a number of relaxation steps different from (1,1). The tendency, however, superiority of CH over LEX, is retained (see (13)).

We finally remark that the sequence of methods listed in Table 1 reflects their efficiency: it deteriorates from the first to the last method. (Here the efficiency is measured by taking into account both the (theoretical) convergence rates and the computational effort, assuming - for simplicity - that one multiplication costs about twice as much time as two sequential additions.) Nevertheless, we do not claim to know the "best" MG or MGR cycle (cf. the remark at the end of Section 2.2). However, by the combination of one of the very efficient cycles proposed here with the "Full MG" technique (see Section 4) one obtains, in any case, an extremely fast solver.

3. NUMERICAL EXPERIMENTS FOR THE NEUMANN MODEL PROBLEM (HELMHOLTZ EQUATION)

The numerical treatment of non-Dirichlet problems is well known to be more involved. Brandt pointed out already in (2) that the naive application of standard MG methods may lead to considerably slower convergence even for the Neumann model problem (see below). He suggested more sophisticated strategies for the treatment of the boundary conditions (Brandt and Dinar, Section 4.3 of (5)). On the other hand, led by our theoretical investigations, we claimed that such difficulties should not occur for non-standard MG methods.

On account of this situation some of the open questions have been investigated experimentally (by the first author in cooperation with K. Becker). In particular, some standard and non-standard MG algorithms were applied to the model problem of the Helmholtz equation $-\Delta u + cu = f$, $c \geqq o$, with Neumann boundary conditions in the unit square, discretizing $-\Delta$ as in Section 2.1 and using central 2h-differences for the normal derivative.

Although MGR techniques as described in Section 2.2 can successfully be applied to the Neumann problem (12), for brevity we report only on experiences with various MG schemes. Some strategies can be analyzed theoretically like those in Section 2.3 for the Dirichlet problem (namely the non-standard ones, for which the discrete cos-eigenfunctions of the Neumann problem form simple invariant spaces under the iteration process). But as we are mainly interested in a comparison of different procedures, we stress numerical experiments and results. For details also see (14).

All strategies in test consist of MG cycles of the form of Figure 2 being distinguished by the relaxation procedures (LEX(1,1), CH(1,1)), by the fine-to-coarse residual transfer (FW, HW, INJ) and by the treatment of the boundary conditions. The boundary conditions are either treated separately from the interior ones (SBC) or eliminated (EBC). (SBC makes essential use of the "fictitious" points beyond the boundary, whereas these points do not occur explicitly in the EBC case. For a detailed description see (14).)

We state the following results (for c>o): CH relaxations yield robust and efficient MG algorithms in all four cases: combined with HW(!) or FW, SBC or EBC.

On the other hand, LEX relaxations combined with simple INJ are not useful at all (neither in the SBC case nor in the EBC case). These deficiencies are overcome if LEX is combined with FW. Nevertheless, CH-FW remains superior to LEX-FW.

Table 2 demonstrates these statements in the case c = 10.

	Method \ ρ_h	$h=\frac{1}{32}$	$h=\frac{1}{64}$	$h=\frac{1}{128}$	+/-	*
EBC	CH(1,1)-HW	0.15	0.17	0.18	10 5/6	3 2/3
	CH(1,1)-FW	0.09	0.09	0.09	12 1/2	4 1/3
	LEX(1,1)-INJ	0.85	div.	div.	14	4 1/3
	LEX(1,1)-FW	0.15	0.15	0.15	20 2/3	5 1/3
SBC	CH(1,1)-HW	0.12	0.12	0.13	10 5/6	3 2/3
	CH(1,1)-FW	0.11	0.11	0.11	12 1/2	4 1/3
	LEX(1,1)-INJ	0.42	0.47	0.51	14	4 1/3
	LEX(1,1)-FW	0.13	0.14	0.14	20 2/3	5 1/3

Table 2: Empirical convergence rates $\rho = \rho_h$ for the MG cycle (Figure 2) and computational effort (per grid point of the finest grind and per cycle).

A LEX-strategy as recommended in (5) that combines FW along the boundary with the cheaper INJ for the interior points is also useful. In the EBC case, for example, the corresponding convergence rates are 0.17, 0.17, 0.18 for $h^{-1} = 32, 64, 128$, respectively. (Methods of this kind may be of particular interest for more general problems.)

In the case c = o the Neumann problem is solvable only if the well known compatibility condition is fulfilled (or forced to be fulfilled). A MG process can work well only if it preserves this condition on all coarser grids, at least approximately. In the case of EBC, for example, CH-FW as well as CH-HW satisfy this condition (not, however, LEX-INJ!). Indeed, practical results show that these methods are as efficient for $c = o$ as for $c > o$. (Also for $o < c << 1$ no real problems occur in using these methods.)

Finally, also staggered grids which are of interest in some applications (15), have been taken into consideration. Here the coarse grid points do not form a subset of the fine grid points, so that INJ is not applicable at all. Using a simple fine-to-coarse "averaging" instead, checkered MG strategies again turn out to work excellently.

4. THE DIRICHLET PROBLEM FOR GENERAL DOMAINS

We have already pointed out that the applicability of the non-standard methods is not restricted to the model problem. This is demonstrated here by the program MGØ1 which has been developed (by the second author (16)) in close cooperation with A. Brandt. MGØ1 solves problems of the form

$$-\Delta u + c(P)u = f(P) \qquad (P\in\Omega)$$
$$u = g(P) \qquad (P\in\partial\Omega)$$

with $c(P) \geq o$, where $-\Delta$ is discretized by the 5-point difference operator on a square h-grid using 5-point Shortley-

Weller approximations near the boundary $\partial\Omega$.

The program relies on MG cycles as described by Figure 2 which are characterized by CH relaxation, half residual weighting and linear interpolation. These cycles are applied in connection with the so-called "Full Multigrid" technique (FMG), which is assumed to be known in principle (see Brandt (5)). This technique yields a most powerful tool in the practical application of MG-(and MGR-) methods: it allows to calculate the discrete solution up to the level of discretization accuracy within a number of operations which is proportional to the number of unknowns.

In (13) a theoretical justification of the FMG technique will be given. Applied to the situation considered here it yields - under certain assumptions - an error estimation of the form

$$\| \hat{u}_h - u^* \|_2 \leq k \cdot \| u_h - u^* \|_2 \quad \text{with } 1 < k \leq 2.$$

Here u^*, u_h and $\hat{u}_h$ denote the solution of the boundary value problem, the exact solution of the difference equations, and the approximate solution calculated by MGØ1, respectively. $\| \cdot \|_2$ denotes the discrete L_2-norm. The constant k depends on the convergence rate of the underlying MG method. The above inequality specifies what is meant by "up to the level of discretization accuracy"; the corresponding assumptions can be satisfied in the case of well-behaved domains and smooth solutions.

In Table 3 we give some typical results for various domains (Poisson's equation). It shows the errors $\| \hat{u}_h - u^* \|_2$ of the solutions computed by MGØ1 as compared to the true discretization errors $\| u_h - u^* \|_2$.

In all examples the FMG algorithm uses one MG cycle on each grid combined with cubic interpolation of solutions onto each next finer grid. For the first two well-behaved domains (rectangle and circle) cycles of CH(1,1)-HW type are used. In full agreement with the above mentioned theoretical investigations we get the solution up to the level of discretization error on each grid. The total amount of work is only $16\frac{4}{9}$ additions and $6\frac{2}{9}$ multiplications per h-grid point.

	$h = \frac{1}{8}$	$h = \frac{1}{32}$	$h = \frac{1}{128}$	Domain/Solution u*
$\lVert \hat{u}_h - u^* \rVert_2$	2.88(-3)	1.62(-4)	9.92(-6)	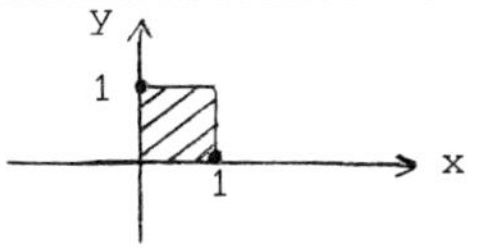
$\lVert u_h - u^* \rVert_2$	2.65(-3)	1.66(-4)	1.04(-5)	$u^*(x,y) = \sin 3(x+y)$
$\lVert \hat{u}_h - u^* \rVert_2$	1.78(-3)	1.14(-4)	7.38(-6)	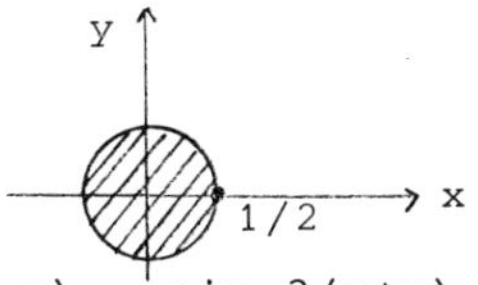
$\lVert u_h - u^* \rVert_2$	1.88(-3)	1.19(-4)	7.53(-6)	$u^*(x,y) = \sin 3(x+y)$
$\lVert \hat{u}_h - u^* \rVert_2$	3.88(-4)	1.22(-5)	5.53(-7)	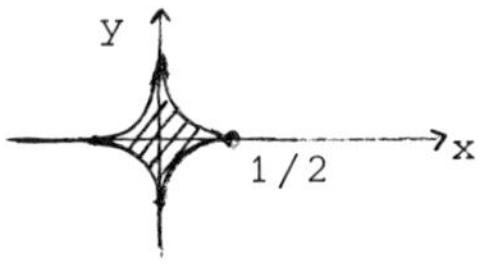
$\lVert u_h - u^* \rVert_2$	3.88(-4)	1.17(-5)	5.40(-7)	$u^*(x,y) = \sin 3(x+y)$
$\lVert \hat{u}_h - u^* \rVert_2$	6.96(-3)	1.40(-3)	2.69(-4)	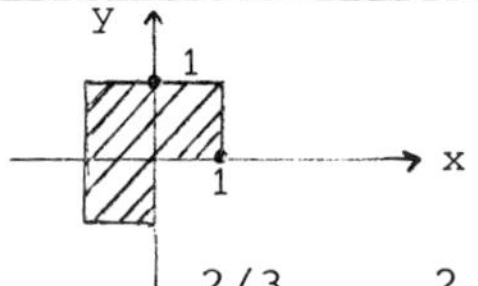
$\lVert u_h - u^* \rVert_2$	6.29(-3)	1.10(-3)	1.80(-4)	$u^*(x,y) = r^{2/3} \sin \frac{2}{3}\theta$, $x = r\cdot\cos\theta$, $y = r\cdot\sin\theta$

Table 3: MGØ1 results for various domains.

For examples 3 and 4, the assumptions for the theoretical error estimation are not satisfied (difficulties with cubic interpolation in example 3; $r^{2/3}$-singularity of the solution in example 4). In spite of this fact we obtain satisfactory results. We note, however, that we used CH(2,1)-HW cycles in these cases, which require somewhat more computational work: $21\frac{7}{9}$ additions and 8 multiplications per h-grid point.

5. ADDITIONAL REMARKS

The non-standard MG methods discussed above are all characterized by CH relaxations and checkered intermediate grids. Being as universally applicable as the MG methods with LEX pointwise relaxations they are, in particular, suitable for 2D problems as long as no direction is distinguished by the domain geometry or the differential equation.

For equations with principal part $-au_{xx} - bu_{yy}$ with $a \gg b > o$, MG methods with pointwise Gauß-Seidel relaxations are not satisfactory; line relaxation should be applied instead (1). The "non-standard" methods for this case are characterized by "zebra-like" intermediate grids and "zebra-line" relaxations.

The advantages of zebra-line against the usual line relaxation are similar to those of CH against LEX relaxation. This can be shown theoretically (13) and has been empirically verified by a variant of the MG Ø1 program for the case $a \neq b$. Furthermore, zebra-line relaxations have successfully been applied in a MG program for solving the Poisson equation in spherical shells (17).

In the same way as the use of checkered intermediate grids is related to the Total Reduction method, zebra-like intermediate grids are related to the Alternating Reduction method (18). Corresponding MGR methods which rely on the Alternating Reduction are investigated in (12).

Finally, the Cyclic (or Partial) Reduction method (Buneman's algorithm (19), also see (18)) formally corresponds to a MG technique where the mesh sizes are doubled in only one direction. The relations, however, in this case seem to be not so natural in our opinion.

Non-standard MG methods for the 3D case have not yet been investigated systematically by us. The results of Reutersberg (20) for 3D reduction methods clarify, however, the geometrical situation arising with intermediate 3D grids.

So far, we have made no assumptions about the structure of the computers used. The advantages of the considered non-standard methods still increase significantly when using computers with parallel or vector processing capabilities (9). It is characteristic for MGR and CH-MG methods that they are composed of mutually independent processes. The advantages of CH over LEX relaxation for parallel processing have already been mentioned by many authors (independently of MG applications), cf. e.g. (21).

In the case of 5-point discretizations one CH step can be viewed as a sequence of two Jacobi-like relaxation processes for half the grid points each. Using (weighted) Jacobi relaxation allows a still more extensive exploitation of parallelism. It leads, however, to significantly worse convergence rates for the MG process (1,3). (If the discretization is based on 9-point stars - e.g. in the "Mehrstellen"-method - four-colour checkered ordering (3,4) may be used to exploit parallelism.)

Beside the aspect of parallelism, the multigrid technique leads to special communication requirements for multiprocessor architectures. A concept of an "optimal" multigrid processor topology is outlined in (22).

REFERENCES

1. Brandt, A., Math. Comp. 31, 333-390 (1977).
2. Brandt, A., in "Mathematical Software III" (J.Rice, ed.), 277-318. Academic Press, New York (1977).
3. Hackbusch, W., Computing 20, 291-306 (1978).
4. Hackbusch, W., in "Special Topics of Applied Mathematics" (J. Frehse, D. Pallaschke, U. Trottenberg, eds.). North-Holland, Amsterdam (1980).
5. Brandt, A., and Dinar, N., in "Numerical Methods for Partial Differential Equations" (S. Parter, ed.). Academic Press, New York (1979).
6. Schröder, J., and Trottenberg, U., Numer. Math. 22, 37-68 (1973).
7. Schröder, J., Trottenberg, U., Witsch, K., in "Numerical Treatment of Differential Equations" (R. Bulirsch, R.D. Grigorieff, J. Schröder, eds.), 153-187. Springer, New York (1977).
8. Schröder, J., Trottenberg, U., Reutersberg, H., Numer. Math. 26, 429-459 (1976).

9. Brandt, A., in this volume.
10. Ries, M., and Trottenberg, U., Preprint no. 277 (1979). SFB 72, Bonn University, Bonn.
11. Ries, M., "Lösung elliptischer Randwertaufgaben mit approximativen Reduktionsverfahren". Dissertation to appear (1981). Bonn University, Bonn.
12. Winter, G., "Beiträge zur Konvergenztheorie iterativer schneller Löser". Dissertation to appear (1981). Bonn University, Bonn.
13. Stüben, K., Trottenberg, U., Witsch, K., "On the Convergence of Multigrid Iterations: Quantitative Results for Model Problems". To appear.
14. Becker, K., "Mehrgitterverfahren für 1D- und 2D-Modellaufgaben mit Neumannschen Randbedingungen". Diplomarbeit to appear (1980). Bonn University, Bonn.
15. Schumann, U., and Sweet, R., J. Comp. Phys. 20, 171-182 (1976).
16. Stüben, K., "MGØ1: Ein Mehrgitterprogramm zur Lösung von $-\Delta u+c(P)u=f(P)$ $(P\in\Omega)$, $u=g(P)$ $(P\in\partial\Omega)$ in allgemeinen 2D-Gebieten". Preprint to appear (1980). GMD, St.Augustin.
17. Linden, J., Trottenberg, U., Witsch, K., "Ein Mehrgitterprogramm zur Bestimmung von 2D-Lösungen für die Poisson-Gleichung im Spalt konzentrischer Kugeln, mit Dirichlet'schen oder Neumann'schen Randbedingungen". Preprint to appear (1980). SFB 72, Bonn University, Bonn.
18. Trottenberg, U., in "Computer, Fast Elliptic Solvers, and Applications" (U. Schumann, ed.), 98-111. Advance Publications, London (1978).
19. Buzbee, B.L., Golub, G.H., Nielson, C.W., Siam J. Numer. Anal. 7, 627-656 (1970).
20. Reutersberg, H., GMD-Bericht Nr. 121. Oldenburg Verlag, München, Wien (1980).
21. Heller, D., "A Survey of Parallel Algorithms in Numerical Linear Algebra", Pittsburgh, PA., Carnegie-Mellon University, Dept. of Computer Science (1976).
22. Reutersberg, H., Trottenberg, U., Witsch, K., "A Concept of a Multigrid Computer with Pyramidal Processor Topology". Preprint to appear. GMD, St. Augustin.

SOME EXPERIMENTS IN SOLVING STIFF OSCILLATORY ORDINARY DIFFERENTIAL EQUATIONS

Patrick W. Gaffney[1]

Computer Sciences Division
Oak Ridge National Laboratory[2]
Oak Ridge, Tennessee

SUMMARY

The purpose of this note is to give a brief summary of a forthcoming paper(1) that describes our numerical experience in using packaged software for solving ordinary differential equations of the form

$$\frac{dy}{dt} = F(y(t)), \qquad y(t_o) = y_o, \tag{1.1}$$

where y, y_o, and F are vectors of length n, and where the eigenvalues

$$\lambda_j = \mu_j + i\nu_j, \qquad j = 1, \ldots, n, \qquad i = \sqrt{-1}$$

of the Jacobian matrix $J = [\frac{\partial F_p}{\partial y_q}]$ p, q = 1, . . ., n, possess the following properties:

$$\mu_j < 0, \qquad j = 1, \ldots, n$$
$$\max_{1 \le j \le n} |\mu_j| \gg \min_{1 \le j \le n} |\mu_j| \tag{1.2}$$

and

$$|\mu_j| \ll |\nu_j|, \qquad 1 \le j \le n \tag{1.3}$$

---

[1]Research funded by the Fusion Energy Division of Oak Ridge National Laboratory.

[2]Operated by Union Carbide Corporation under contract W-7405-eng-26 with the U. S. Department of Energy.

Equations of this type are sometimes called stiff oscillatory ordinary differential equations.[2] They often occur when the method of lines technique is applied to a system of one dimensional partial differential equations that possess some "hyperbolic" type of behavior.[3] It is this application that we are interested in since eventually we wish to solve the reduced resistive Magnetohydrodynamic (MHD) equations[4] by the method of lines (MOL). In this case the equations are of mixed hyperbolic-parabolic type. Consequently, when the MOL technique is applied to them, we obtain a problem of the form (1.1) with properties (1.2) and (1.3).[5] It is well known that property (1.3) may cause stiffly stable/non A-stable codes to be inefficient when they are applied for the solution of (1.1).[6] Therefore, we wish to consider other types of integrators that may be suitable for solving stiff oscillatory problems. Specifically, we consider four Fortran codes, namely STRIDE,[7] BLEND,[8] STINT,[9] and DIRK,[10] that should be appropriate for this type of problem. To decide which, if any, of these codes are suitable, we have conducted some tests on a model problem.

The model problem that we use is similar to one used by Krogh,[11] Enright,[12] and Skeel et al.,[8] except that in our case we introduce a parameter α. Specifically, we solve the equation

$$\frac{dy}{dt} = Ay \qquad (1.4)$$

where A is the matrix

$$A = \begin{bmatrix} -10.0 & \alpha & 0 & 0 & 0 & 0 \\ -\alpha & -10.0 & 0 & 0 & 0 & 0 \\ 0 & 0 & -4.0 & 0 & 0 & 0 \\ 0 & 0 & 0 & -1.0 & 0 & 0 \\ 0 & 0 & 0 & 0 & -0.5 & 0 \\ 0 & 0 & 0 & 0 & 0 & -0.1 \end{bmatrix}, \quad (1.5)$$

with the initial conditions

$$y(t = 0) = \begin{bmatrix} 1 \\ 1 \\ 1 \\ 1 \\ 1 \\ 1 \end{bmatrix} \quad (1.6)$$

Because equation (1.4) is linear, the eigenvalues of the Jacobian matrix are the eigenvalues of A, and these are the quantities

$$\lambda_1 = -10 + i\alpha \qquad \lambda_2 = -10 - i\alpha$$
$$\lambda_3 = -4.0 \qquad \lambda_4 = -1.0$$
$$\lambda_5 = -0.5 \qquad \lambda_6 = -0.1$$

Thus, by varying the value of α, we can control the proximity of two of the eigenvalues to the imaginary axis. Our intention is to choose a sequence of increasing values of α and then to monitor the behavior of each code when it is applied to the resultant equations. We are particularly interested in applying "unmodified" versions of the codes. For example, we have deliberately not modified the codes so that the local error tolerance control is the same for all routines. Although this may be judged to be unfair, we wish to reproduce the type of results that a typical computer user would obtain when the unmodified codes are applied to a problem. Thus, our experiments differ from the excellent tests performed by Enright et al.,[13] since these tests serve a different purpose.

In reference 1 we present results which show the effect of using the above codes for solving the model problem on the PDP-10, CDC 7600, and CRAY-1 computers. We also compare our results with similar calculations obtained by using the new LSODE code.[14] Based on all of these results, we make some recommendations to computer users about the codes that are most suitable for solving the difficult stiff oscillatory problem. Finally, we draw some conclusions about the present state of appropriate software and indicate areas where we believe improvements can be made.

REFERENCES

1. Gaffney, P. W., "A Survey of Fortran Subroutines Suitable for Solving Stiff Oscillatory Ordinary Differential Equations," to appear.
2. Miranker, W. L., and Wahba, G., "An Averaging Method for the Stiff Highly Oscillatory Problem," Mathematics of Computation, <u>30</u>, pp. 383-399 (1976).
3. Warming, R. F., and Beam, M., "An Extension of A-stability to Alternating Direction Methods," BIT, 19 pp. 395-417 (1979).
4. Strauss, H. R., Phys. Fluids 19, p. 134 (1976).
5. Gaffney, P. W., Carreras, B., and Hicks, H. R., "The Method of Lines Solution of the Reduced Resistive MHD Equations," Poster presented at the SIAM 1979 Fall Meeting in Denver.

6. Lambert, J. D., "The Initial Value Problem for Ordinary Differential Equations," in "The State of the Art in Numerical Analysis," edited by D. A. H. Jacobs, Academic Press, pp. 451-500 (1977).
7. Butcher, J. C., Burrage, K., and Chipman, F. H., "STRIDE: Stable Runge-Kutta Integrator for Differential Equations," Report series No. 150, Department of Mathematics, University of Auckland, New Zealand.
8. Skeel, R., and Kong, A., "Blended Multistep Methods," ACM TOMS 3 No. 4, pp. 326-345 (1977).
9. Tendler, J. M., Bickart, T. A., and Picel, Z., "A Stiffly Stable Integration Process Using Cyclic Composite Methods," ACM TOMS 4 No. 4, pp. 339-368 (1978).
10. Alexander, R., "Diagonally Implicit Runge-Kutta Methods for Stiff ODE's," SIAM J. Numer. Anal. 14, No. 6, pp. 1006-1021 (1977).
11. Krogh, F. T., "On Testing a Subroutine for the Numerical Integration of Ordinary Differential Equations," Journal of the ACM, Vol. 20, No. 4, pp. 545-562 (October 1973).
12. Enright, W. H., "Using a Testing Package for the Automatic Assessment of Numerical Methods for ODE's," in "Performance Evaluation of Numerical Software," edited by Lloyd D. Fosdick, North-Holland Publishing Company, pp. 199-213 (1979).
13. Enright, W. H., Hull, T. E., and Lindberg, B., "Comparing Numerical Methods for Stiff Systems of ODE's," BIT 15, pp. 10-48 (1975).
14. Hindmarsh, A. C., "LSODE: Livermore Solver for Ordinary Differential Equations," Lawrence Livermore Laboratory report in progress, (1980).

A NUMERICAL METHOD FOR SOLVING ELLIPTIC BOUNDARY VALUE PROBLEMS IN UNBOUNDED DOMAINS*

Charles I. Goldstein

Applied Mathematics Department
Brookhaven National Laboratory
Upton, New York 11973

I. INTRODUCTION

In this paper we shall describe a numerical method for solving elliptic boundary value problems in unbounded domains. For a survey of various methods for solving such problems, see [1]. Here we shall describe a particular method obtained by first introducing an artificial boundary near infinity, say a sphere Γ_R of large radius R, and an approximate local boundary condition on Γ_R. A finite element method is then employed to discretize this approximate problem. Since R must be large in order to suitably estimate the error due to the artificial boundary, the resulting system of linear equations is usually quite large. We shall describe how to reduce the number of linear equations to the asymptotically optimal amount, while preserving optimal error estimates and the sparseness of the matrix, by grading the mesh systematically in such a way that the element mesh sizes are increased as the distance from the origin increases.

*The submitted manuscript has been authored under contract DE-AC02-76CH00016 with the U.S. Department of Energy.

ISBN 0-12-632620-7

II. THE MESH GRADING PROCEDURE

Let Ω denote a bounded subset of R^3 with smooth boundary, $\partial\Omega$. Denote the complement of $\bar{\Omega}$ by Ω^C and let $x = (x_1,x_2,x_3)$ denote an arbitary point in R^3. For simplicity, we consider the following model problem:

$$(-\Delta - K^2)\, u = f \text{ in } \Omega^C,\ \frac{\partial u}{\partial n} = 0 \text{ on } \partial\Omega$$
and
$$\frac{\partial u(x)}{\partial r} + \left(\frac{1}{r} - iK\right) u(x) = o\left(\frac{1}{r}\right) \text{ as } r = |x| \to \infty, \tag{2.1}$$

where $\frac{\partial}{\partial n}$ denotes the outward directed normal to $\partial\Omega$, f and $\partial\Omega$ are smooth, K is real and f has bounded support. This problem is well-posed. When $K \neq 0\ (K = 0)$, (2.1) is referred to as the exterior Helmholtz (Laplace) problem. Our method and results may be readily extended to other boundary value problems, including those with variable coefficients and more general boundary conditions.

We next formulate an approximate problem on a bounded domain. Define the sets S_R and Ω_R by $S_R = \{x \in R^3 : |x| \le R\}$ and $\Omega_R = S_R \cap \Omega^C$, where R is large. Now consider the following problem:

$$(-\Delta - K^2)u_R = f \text{ in } \Omega_R,\ \frac{\partial u_R}{\partial n} = 0 \text{ on } \partial\Omega$$
and
$$\frac{\partial u_R}{\partial r} + \left(\frac{1}{r} - iK\right) u_R = 0 \text{ on } \Gamma_R = \partial S_R. \tag{2.2}$$

This problem is also well-posed. Note that there are various other approximate boundary conditions that may be imposed on Γ_R (see, e.g., [1]) for which our method applies.

Set $U(x) = e^{-iKr}u(x)$, where u satisfies (2.1). It may be proved that

$$|D^{\alpha}U(x)| \le \frac{C}{r^{|\alpha|+1}} \text{ for r sufficiently large,} \tag{2.3}$$

where $D^{\alpha}U$ denotes an arbitrary derivative of U of order $|\alpha|$. In view of (2.3), we shall approximately solve for U and $U_R = e^{-iKr}u_R(x)$, where u_R satisfies (2.2). Let $H_R^E = H^1(\Omega_R)$ denote the space of complex-valued functions whose derivatives of order less than two are square-integrable over Ω_R. It follows from integration by parts that U_R satisfies

$$A_R(U_R,V) = (F,V) = \int_{\Omega_R} F(x)\overline{V(x)}dx \text{ for each } V \text{ in } H_R^E, \quad (2.4)$$

where

$$A_R(U_R,V) = a_R(u_R,v) = \int_{\Omega_R} (\nabla u_R \cdot \nabla\bar{v} - K^2 u_R\bar{v})\, dx$$

$$+ \oint_{\Gamma_R} (\tfrac{1}{R} - iK) u_R\bar{v}ds,$$

$$F(x) = e^{iKr}f(x) \text{ and } v(x) = e^{-iKr}V(x).$$

We shall approximately solve (2.4) using the finite element method. For a detailed discussion of the finite element method, see [2]. For simplicity we assume here that our one-parameter family of finite dimensional spaces, $S^h \subset H_R^E$, are defined for $h\in(0,\infty)$ as follows. Suppose that we have a partition of Ω_R into simple subsets (elements), denoted by t^h, with diameter of order $O(h)$. S^h then consists of all continuous complex-valued functions v^h such that $v^h|_{t^h} \in P_\ell$, where P_ℓ denotes the set of polynomials of degree less than ℓ for some integer ℓ greater than one. It may be readily seen that $\dim(S^h) = O(R^3h^{-3})$. Since R is large, the number of linear equations will be large.

We next construct a new family of spaces $\tilde{S}^h$, defined for $h\in(0,1]$, satisfying

$$\dim(\tilde{S}^h) \le Ch^{-3}, \quad (2.5)$$

where C is independent of R (as well as h). Set $S_j = \{x: 2^{j-1} < |x| \le 2^j\}$ and $\Omega_j^R = S_j \cap \Omega_R$, $j = 1,\ldots,J_R$, with

$J_R \approx \log_2 R$. Set $\Omega_0^R = \Omega_R - \bigcup_{j=1}^{J_R} \Omega_j^R$ and assume for simplicity that Ω and supp(f) are contained in the unit sphere and (2.3) holds outside of this sphere. We now construct a new partition of Ω_R into elements, t^{h_j}, with diameter of order $O(h_j)$ in each annular region Ω_j^R. We define the increasing sequence of positive numbers, h_j, by

$$h_0 = h \text{ and } h_j = 2^{\ell' j} h \text{ with } 1 < \ell' < \frac{\ell - \frac{1}{2}}{\ell - 1},\ j = 1, \ldots, J_R. \qquad (2.6)$$

We now define $\tilde{S}^h$ to consist of those continuous functions v^h such that $v^h|_{t^{h_j}} \in P_\ell$. The motivation for this construction and the proof of (2.5) are based on approximation theory and (2.3) and are given in [3].

III. ERROR ESTIMATES

Our main theoretical results establish the existence and uniqueness of the finite element approximation, U_R^h, satisfying

$$A_R(U_R^h, V^h) = (F, V^h) \text{ for each } V^h \text{ in } \tilde{S}^h, \qquad (3.1)$$

as well as error estimates for $U - U_R^h$. These results are easier to prove when $K = 0$ since the bilinear form, $A_R(\ ,\ )$, is non-negative definite. A detailed treatment of this case is given in [3]. When $K \neq 0$, the proofs are considerably more difficult and are given in [4] (where the case of variable coefficients and other extensions of Theorem 3.1 below are treated). Here we merely state some of our main results.

<u>Theorem 3.1</u>: Suppose that B is a fixed subset of Ω^C, $h \in (0,1]$ and R is sufficiently large. Then the following results hold.

(a) There exist positive constants, γ_1 and γ_2, such that if $KR \leq \gamma_1$ and $h_{J_R} \leq \gamma_2 R$, there exists a unique solution, U_R^h, of (3.1). Furthermore, the following estimates hold:

$$(\int_{\Omega_R} |\nabla(U-U_R^h)|^2 dx)^{\frac{1}{2}} \leq C(R^{-2} + h^{\ell-1}) \tag{3.2}$$

and

$$(\int_{\Omega_R} |U-U_R^h|^2 dx)^{\frac{1}{2}} \leq C(R^{-2} + h^{\ell} + (R^{-\frac{3}{2}} + h^{\ell-1})^2), \tag{3.3}$$

where C is independent of h and R.

(b) There are constants $C_{K,R}$ and Λ such that there exists a unique solution, U_R^h, of (3.1) provided

$$C_{K,R} K h_{J_R} \leq \Lambda. \tag{3.4}$$

Furthermore, estimates (3.2) and (3.3) hold in this case.

Observe that Theorem 3.1(b) implies that the largest mesh size h_{J_R} must satisfy the additional stability constraint, (3.4), when KR is not sufficiently small. The constant $C_{K,R}$ becomes large as $KR \to \infty$. Hence the number of equations must increase as the frequency K increases.

IV. NUMERICAL RESULTS

In collaboration with Alvin Bayliss, the author has begun a numerical investigation of the mesh grading procedure. We shall demonstrate typical results with respect to an axially symmetric problem analogous to (2.1) with $\Omega = \{x: |x| \leq .5\}$, $f = 0$ and the boundary condition on $\partial\Omega$ given by $\frac{\partial u}{\partial n} = g$. The data g corresponds to the solution, $u(x) = \frac{e^{iK|x-x_s|}}{4\pi|x-x_s|}$, where x_s is a fixed axial point such that $|x_s| = .4$. Introducing the artificial boundary Γ_R as before and employing spherical polar coordinates and axial symmetry, we consider the computational domain, $\Omega_R = \{(r,\theta): .5 \leq r \leq R, 0 \leq \theta \leq \pi\}$, for problem (2.2). We denote the last boundary condition in (2.2) by $B_1 u_R = 0$ on Γ_R and also consider the boundary condition

$$B_2 u_R = \frac{\partial^2 u_R}{\partial r^2} + (\frac{4}{r} - 2iK)\frac{\partial u_R}{\partial r} + (\frac{3}{r} - 4ik)\frac{u_R}{r} - K^2 u_R = 0 \text{ on } \Gamma_R.$$

The boundary operators, B_1 and B_2, correspond to the first two in a hierarchy of approximate boundary operators developed in [5].

We employ continuous piecewise linear finite elements on a triangular partition in (r,θ) coordinates to obtain an approximate solution using both a uniform and a graded mesh. In the following computations, we keep the grid evenly spaced in the θ-direction with 41 grid points and use 6 grid points in the r-direction with $R = 1.125$. When employing a uniform mesh we use the grid points $r_j = r_0 + .125j$, $j = 0,\ldots,5$, in the r-direction with $r_0 = .5$. When mesh grading is employed, we factor out e^{iKr} as described in Section 2 and grade the mesh in the r-direction in accordance with $r_1 = r_0 + .047$ and $r_j = r_{j-1} + 5r_{j-1}^{3/2}$, $j = 2,\ldots,5$. Only a few changes were necessary to modify the computer program so as to incorporate the mesh grading. We measure the error E_j, defined by $E_j = \frac{\|u-u_R^h\|_{L^2(\partial\Omega)}}{\|u\|_{L^2(\partial\Omega)}}$, corresponding to the boundary operator B_j, $j=1,2$, where $u_R^h = e^{iKr}U_R^h$ is computed as described in Sections 2 and 3. In the following table, we compare the errors, E_j, obtained using the graded mesh with those obtained using a uniform mesh for various values of K.

Table I.

K	E_1(%)		E_2(%)	
	Graded	Uniform	Graded	Uniform
0	2.39	3.74	1.17	3.5
3	4.26	6.08	1.33	4.03
6	4.62	7.37	2.4	5.68
9	5.9	8.1	8.17	8.8
12	8.4	21.4	9.4	17.2

Observe the substantial improvements due to mesh grading using both B_1 and B_2. These results are typical of our numerical experiments. Note that as the frequency K increases, the error increases. As we indicated in Section 3, this is to be expected since we are keeping the number of equations fixed. We also observe that the main limitation on the number of equations in our computations is the storage requirements of the banded Gauss solver. We are attempting to circumvent this difficulty by using iterative methods. A more complete discussion of our numerical results will appear elsewhere, [6].

REFERENCES

[1] Goldstein, C. Numerical Methods for Helmholtz Type Equations in Unbounded Regions. Mathematical Methods and Applications of Scattering Theory. Ed. J. A. DeSanto, A. W. Saenz, and W. W. Zachary. Lecture Notes in Physics. Springer-Verlag, Berlin-Heidelberg-New York (1980).

[2] Ciarlet, P. G. The Finite Element Method for Elliptic Problems. North-Holland, Amsterdam (1978).

[3] Goldstein, C. I. The Finite Element Method with Non-Uniform Mesh Sizes for Unbounded Domains. (To appear), BNL Report Number 27307, AMD Report Number 839.

[4] Goldstein C. I. The Finite Element Method with Non-Uniform Mesh Sizes Applied to the Exterior Helmholtz Problem. (To appear), AMD Report Number 861.

[5] Bayliss, A., M. Gunzberger, and E. Turkel. Boundary Conditions for the Numerical Solution of Elliptic Equations in Exterior Regions. ICASE Report 80-1 (1979).

[6] Bayliss, A. and C. I. Goldstein. (To appear).

APPLICATIONS OF TRANSFINITE ("BLENDING-FUNCTION") INTERPOLATION TO THE APPROXIMATE SOLUTION OF ELLIPTIC PROBLEMS

William J. Gordon[1]
Sandra J. Kelly

Department of Mathematical Sciences
Drexel University
Philadelphia, PA 19104

Dedicated to Garrett Birkhoff on the occasion of his 70th birthday.

1. INTRODUCTION

Our main objective in this paper is to show how the so-called "*blending function methods*" [2,5,6] of bivariate and multivariate interpolation may be usefully employed in the approximate numerical solution of elliptic boundary value problems. The advantage which these interpolation techniques have over standard schemes such as tensor product methods is that they are *not* finite dimensional schemes. Rather than interpolating at a finite number of discrete points, the transfinite "blending function" methods actually match bivariate or multivariate functions *exactly* along certain lines or on certain planes. As we will see, this means that interpolants which exactly satisfy rather general boundary conditions for an elliptic partial differential equation can be very easily constructed. Once this has been done, the interpolating

[1]This author's work was supported in part by the Office of Naval Research and the Air Force Office of Scientific Research.

ISBN 0-12-632620-7

function can be symbolically subtracted from the true solution so as to obtain a problem with *homogeneous boundary conditions*. As a practical matter, the reduction to homogeneous boundary conditions enables one to reduce the total amount of computational work required to solve a problem numerically, and we are presently developing preprocessing software to do this.

As is well-known (cf. [9], p. 34), the number of iterations required to solve a discrete system of equations arising from an elliptic problem depends upon, among other factors, the initial estimate of the solution. In lieu of anything better, it is common practice ([8] - [11]) to take all unknowns interior to the problem domain equal to the same value (i.e., u_{ij} = const.) and to set the boundary values of u_{ij} equal to those given by the boundary conditions. As an alternative, we propose first to construct a transfinite interpolant which exactly matches the boundary conditions and to use this as the initial approximation. Numerical experiments carried out with the ITPACK collection of programs ([9, 10]) show that dramatic reductions in the number of iterations can be achieved at modest computational expense by using the techniques described in Sections 2, 4, 5 and 6 to obtain a good initial approximation.

In Section 4 we demonstrate how, in addition to the boundary conditions, other available information concerning the behavior of the solution may be incorporated into transfinite interpolants. This notion was first proposed by Gordon and Hall in [8].

Section 5 deals with the combination of exact matching of boundary conditions and *collocation* of the differential equation. The results there show that, by making only minor modifications, existing collocation codes can be easily adapted to incorporate the exact satisfaction of boundary conditions. Although there is reason to expect that the computational savings to be gained by *iteratively* solving the reformulated set of collocation equations will be even greater than the savings associated with our finite difference experiments using ITPACK, we have not conducted numerical experiments to substantiate this conjecture.

The basic ideas developed herein for elliptic problems are also applicable to time-dependent (parabolic) problems. In Sec. 6, we briefly indicate how one may employ transfinite interpolation to advance a numerical solution from time t to t + Δt. (The solution can be computed to higher accuracy using standard iterative methods.) This method seems closely related to the ideas discussed in [3].

The formulas presented in Sec. 6 are also of value in the commonly occurring situation where one is solving a series of elliptic problems which differ only slightly from one another. In this case, one can use the previously computed solution to a similar problem and *automatically adjust* it so as to exactly satisfy the new boundary conditions, cf. (6.1).

2. EXACT MATCHING OF BOUNDARY CONDITIONS

To begin, we consider the problem of constructing an interpolant U which exactly matches given Dirichlet boundary conditions on the perimeter of the unit square $S = [0,1] \times [0,1]$. As is well known [2,5,6,8], the function U given by the following *bilinearly blended* expression interpolates to any continuous bivariate function F on ∂S:

$$\begin{aligned} U(x,y) = {} & (1 - x)F(0,y) + xF(1,y) \\ & + (1 - y)F(x,0) + yF(x,1) \\ & - (1 - x)(1 - y)F(0,0) - (1 - x)yF(0,1) \\ & - x(1 - y)F(1,0) - xyF(1,1). \end{aligned} \tag{2.1}$$

Since, unlike finite dimensional interpolation formulas, the function U of (2.1) incorporates a nondenumerable number of *scalar* samples of F (i.e., it incorporates the *four univariate boundary functions*), U is termed a *transfinite* interpolant.

An introduction to the algebraic theory of transfinite interpolation is given in [5] and [6]. In terms of the projector algebra developed therein, we see that U is simply the function which results by operating on F with the *Boolean sum* of two elementary projectors P_x and P_y. That is:

$$\begin{aligned} U &= (P_x \oplus P_y)[F] \\ &\equiv P_x[F] + P_y[F] - P_xP_y[F] \end{aligned} \tag{2.2}$$

where

$$\begin{aligned} P_x[F] &= (1 - x)F(0,y) + xF(1,y) \\ P_y[F] &= (1 - y)F(x,0) + yF(x,1). \end{aligned} \tag{2.3}$$

In [5] it is shown that a collection of commutative projectors $P_1, P_2, \ldots, P_N$ generate a *distributive lattice* whose algebraically maximal element is $P_1 \oplus P_2 \oplus \ldots \oplus P_N$, and whose algebraically minimal element is $P_1P_2\ldots P_N$. Among all the elements of the lattice, the projector $P_1 \oplus P_2 \oplus \ldots P_N$ extracts more information from the function F being approximated than does any other element; and, the product operator $P_1P_2\ldots P_N$ uses the least amount. Thus, the maximal projector provides, in this sense, the "best approximation". Clearly, the bilinearly blended interpolant $(P_x \oplus P_y)[F]$ above uses more information extracted from F than do any of the other three elements in the lattice: P_x, P_y and $P_xP_y = P_yP_x$. Note that $P_xP_y[F]$ is, in fact, a finite dimensional interpolant; namely, it represents four-parameter bilinear interpolation to F at the corners of S.

The so-called "blending functions" in P_x and P_y need not be linear in order to achieve interpolation to F on ∂S. For, consider any P_x and P_y

of the form:

$$P_x[F] = \phi_0(x)F(0,y) + \phi_1(x)F(1,y)$$
$$P_y[F] = \psi_0(y)F(x,0) + \psi_1(y)F(x,1) \tag{2.4}$$

where ϕ_0, ϕ_1, ψ_0 and ψ_1 are any four functions which satisfy the *cardinality conditions*:

$$\begin{aligned} \phi_0(0) &= 1 & \phi_1(0) &= 0 \\ \phi_0(1) &= 0 & \phi_1(1) &= 1 \\ \psi_0(0) &= 1 & \psi_1(0) &= 0 \\ \psi_0(1) &= 0 & \psi_1(1) &= 1\ . \end{aligned} \tag{2.5}$$

In certain applications, instead of taking $\phi_0 = 1 - x$, $\phi_1 = x$ as in (2.1), it may be advisable to employ, for example, "exponential blending":

$$\phi_0(x) = \frac{1 - e^{1-x}}{1 - e}\ , \qquad \phi_1(x) = \frac{e^{1-x} - e}{1 - e}\ ; \tag{2.6}$$

or, "trigonometric blending":

$$\phi_0(x) = \cos \frac{\pi x}{2}\ , \qquad \phi_1(x) = \sin \frac{\pi x}{2}\ . \tag{2.7}$$

Note too that it may be advantageous to take the ϕ_i to be from one class of functions and the ψ_i to be from another. The main ideas of the paper are amply illustrated by considering only polynomial blending functions. However, any set of functions which satisfy the appropriate cardinality conditions (e.g., (2.5), (2.13), (4.14), etc.) may be used instead.

We now consider more general boundary conditions than the Dirichlet conditions to which the function (2.1) interpolates, and we show that one can actually construct functions U which exactly match any *mixed linear boundary conditions* of the rather general form:

$$L_0[F] = \alpha_0 F(0,y) + \beta_0 F_x(0,y) = g_0(y) \qquad \text{along } x = 0$$

$$L_1[F] = \alpha_1 F(1,y) + \beta_1 F_x(1,y) = g_1(y) \qquad \text{along } x = 1$$

$$\begin{aligned} M_0[F] &= \tilde{\alpha}_0 F(x,0) + \tilde{\beta}_0 F_y(x,0) = h_0(x) \quad \text{along } y = 0 \\ M_1[F] &= \tilde{\alpha}_1 F(x,1) + \tilde{\beta}_1 F_y(x,1) = h_1(x) \quad \text{along } y = 1 \end{aligned} \tag{2.8}$$

provided that the α_i, $\tilde{\alpha}_i$, β_i and $\tilde{\beta}_i$ are constants and that the boundary conditions are *consistent*, i.e.:

$$L_i M_j[F] = M_j L_i[F] \qquad (i,j = 0,1). \tag{2.9}$$

In the above case of Dirichlet conditions, the linear operators L_i and M_j are just:

$$\begin{aligned} L_0[F] &= F(0,y) = g_0(y), \qquad L_1[F] = F(1,y) = g_1(y) \\ M_0[F] &= F(x,0) = h_0(x), \qquad M_1[F] = F(x,1) = h_1(x) \end{aligned} \tag{2.10}$$

and the consistency requirement simply means that the boundary conditions are *continuous* at the four corners:

$$M_j[g_i(y)] = L_i[h_j(x)] \qquad (i,j = 0,1). \tag{2.11}$$

Theorem: Let the L_i and M_j be as in (2.8) and (2.9) and define two projectors P_x and P_y as follows:

$$\begin{aligned} P_x[F] &= \phi_0(x) L_0[F] + \phi_1(x) L_1[F] \\ P_y[F] &= \psi_0(y) M_0[F] + \psi_1(y) M_1[F], \end{aligned} \tag{2.12}$$

where the functions ϕ_i and ψ_j satisfy the conditions that:

$$\begin{aligned} L_i[\phi_k] &= \delta_{ik} \qquad \text{for } i,k = 0,1 \\ M_j[\psi_\ell] &= \delta_{j\ell} \qquad \text{for } j,\ell = 0,1. \end{aligned} \qquad \text{(Kronecker Delta)} \tag{2.13}$$

Then, the function U obtained from the Boolean sum of P_x and P_y exactly satisfies all of the specified boundary conditions.

Proof: The function U is given by

$$U = (P_x \oplus P_y)[F]$$

$$= \phi_0(x)L_0[F] + \phi_1(x)L_1[F] + \psi_0(y)M_0[F] + \psi_1(y)M_1[F]$$
$$- \phi_0(x)\psi_0(y)L_0M_0[F] - \phi_0(x)\psi_1(y)L_0M_1[F] \tag{2.14}$$
$$- \phi_1(x)\psi_0(y)L_1M_0[F] - \phi_1(x)\psi_1(y)L_1M_1[F].$$

The proof consists of a straighforward verification of the facts that $L_i[U] = g_i(y)$ and $M_j[U] = h_j(x)$. We have, for example:

$$\begin{aligned} L_0[U] &= L_0[F] + \psi_0(y)L_0M_0[F] + \psi_1(y)L_0M_1[F] \\ &\quad - \psi_0(y)L_0M_0[F] - \psi_1(y)L_0M_1[F] \\ &= L_0[F] \\ &= g_0(y) \end{aligned} \tag{2.15}$$

in which we have used the *cardinality conditions* (2.13). To show that $M_j[U] = M_j[F] = h_j(x)$, we need also to use the consistency hypotheses (2.9). Q.E.D.

As an example, consider the following set of mixed boundary conditions on the perimeter of S:

$$\begin{aligned} &F_x(0,y) = 0 \\ &F_x(1,y) - 2\alpha F(1,y) = 0 \\ &F_y(x,0) = 0 \\ &F(x,1) = \exp(\alpha(x^2 - 1)) \end{aligned} \tag{2.16}$$

In the notation of the linear operators (2.8), we have:

$$\begin{aligned} L_0[F] &= F_x(0,y) = g_0(y) = 0 \\ L_1[F] &= F_x(1,y) - 2\alpha F(1,y) = g_1(y) = 0 \\ M_0[F] &= F_y(x,0) = h_0(x) = 0 \\ M_1[F] &= F(x,1) = h_1(x) = \exp(\alpha(x^2 - 1)). \end{aligned} \tag{2.17}$$

The functions ϕ_i and ψ_j must satisfy:

$$\begin{aligned} L_0[\phi_0] &= \phi_0'(0) = 1 & L_0[\phi_1] &= \phi_1'(0) = 0 \\ L_1[\phi_0] &= \phi_0'(1) - 2\alpha\phi_0(1) = 0 & L_1[\phi_1] &= \phi_1'(1) - 2\alpha\phi_1(1) = 1 \\ M_0[\psi_0] &= \psi_0'(0) = 1 & M_0[\psi_1] &= \psi_1'(0) = 0 \\ M_1[\psi_0] &= \psi_0(1) = 0 & M_1[\psi_1] &= \psi_1(1) = 1. \end{aligned} \tag{2.18}$$

The simplest functions which come to mind as candidates for the ϕ_i and ψ_j are first degree polynomials. Thus, we take $\phi_i(x) = a_i x + b_i$ $(i = 0,1)$ and from (2.18) we find that we must have:

$$\phi_0(x) = x + \frac{1}{2\alpha} - 1 , \qquad \phi_1(x) = -\frac{1}{2\alpha} , \tag{2.19}$$

and for the ψ_j $(j = 0,1)$ we obtain:

$$\psi_0(y) = y - 1 , \qquad \psi_1(y) = 1. \tag{2.20}$$

The reader may easily verify that the commutativity conditions $L_i M_j[F] = M_j L_i[F]$ $(i,j = 0,1)$ are satisfied. Thus, the desired function U which satisfies the given boundary conditions is, from (2.14), given by:

$$U(x,y) = \exp(\alpha(x^2 - 1)). \tag{2.21}$$

This example is particularly simple since three of the four functions $g_i(y)$ and $h_j(x)$ in (2.8) vanish identically. Note, however, that the blending functions $\phi_i(x)$ and $\psi_j(y)$ depend *only* upon the linear operators L_i and M_j, and *not* upon the right hand sides of (2.8).

As a second example, suppose we wish to satisfy the boundary conditions

$$\begin{aligned} L_0[F] &= F(0,y) = 0 \\ L_1[F] &= F_x(1,y) = 2ye^{2y} \\ M_0[F] &= F(x,0) = 0 \\ M_1[F] &= F_y(x,1) = 2xe^{2x}. \end{aligned} \tag{2.22}$$

We find that the following blending functions satisfy the necessary cardinality conditions (2.13):

$$\phi_0(x) = 1\ , \qquad \phi_1(x) = x$$
$$\psi_0(y) = 1\ , \qquad \psi_1(y) = y \tag{2.23}$$

so that the function U of (2.14) is given by:

$$U(x,y) = 2xy(e^{2x} + e^{2y}) - 6xye^2. \tag{2.24}$$

This function may be readily seen to satisfy the requisite boundary conditions.

With the following boundary conditions:

$$\begin{aligned} L_0[F] &= F(0,y) = 0 \\ L_1[F] &= F_x(1,y) = 2ye^{2y} \\ M_0[F] &= F(x,0) = 0 \\ M_1[F] &= F_y(x,1) - 2F(x,1) = 2 - 2(1 - x)e^{2x}, \end{aligned} \tag{2.25}$$

the appropriate blending functions are found to be:

$$\phi_0(x) = 1\ , \qquad \phi_1(x) = x$$
$$\psi_0(y) = 1 - 2y\ , \qquad \psi_1(y) = -y \tag{2.26}$$

and the function U which satisfies these boundary conditions is:

$$U(x,y) = 2xy(e^{2y} - e^{2x} + e^2) + 2y(e^{2x} - 1). \tag{2.27}$$

In practice, the consistency constraints on the boundary conditions (2.11) may not be satisfied. As an expedient, one may replace the four terms $L_iM_j[F]$ in (2.14) by their averages: $\frac{1}{2}[L_iM_j[F] + M_jL_i[F]]$. The bivariate function U(x,y) so obtained *will not* satisfy the boundary conditions, but may nevertheless be useful as a first approximation to the solution of a boundary value problem which is to be discretized and solved by an iterative scheme.

For the model fourth order elliptic operator, namely the *biharmonic operator* ∇^4, the same ideas apply. Suppose, for example, that the function values and normal derivatives of the solution F are given on the perimeter of S. If we consider the projectors P_x and P_y given by:

$$\begin{aligned} P_x[F] &= \phi_0(x)F(0,y) + \phi_1(x)F_x(0,y) + \phi_2(x)F(1,y) + \phi_3(x)F_x(1,y) \\ P_y[F] &= \psi_0(y)F(x,0) + \psi_1(y)F_y(x,0) + \psi_2(y)F(x,1) + \psi_3(y)F_y(x,1) \end{aligned} \tag{2.28}$$

where $\phi_i(x)$ and $\psi_j(y)$ are the cardinal basis functions for cubic Hermite interpolation, then the "*bicubically blended function*" obtained from $U = (P_x \oplus P_y)[F]$ will exactly match the given boundary conditions provided that they are consistent, e.g., $\frac{\partial}{\partial x}F_y(x,0) = \frac{\partial}{\partial y}F_x(0,y)$ at (0,0), etc. The requisite functions $\phi_i(x)$ are:

$$\begin{aligned} \phi_0(x) &= (x-1)^2(2x+1) \qquad & \phi_1(x) &= x(x-1)^2 \\ \phi_2(x) &= x^2(3-2x) \qquad & \phi_3(x) &= x^2(x-1), \end{aligned} \tag{2.29}$$

and analogously for the $\psi_j(y)$.

The techniques of transfinite interpolation described above for bivariate functions extend easily to trivariate. We give here the details only for the case of Dirichlet boundary conditions given on the six faces of the unit cube $[0,1] \times [0,1] \times [0,1]$. The relevant projectors are

$$\begin{aligned} P_x[F] &= (1-x)F(0,y,z) + xF(1,y,z) \\ P_y[F] &= (1-y)F(x,0,z) + yF(x,1,z) \\ P_z[F] &= (1-z)F(x,y,0) + zF(x,y,1). \end{aligned} \tag{2.30}$$

Clearly, the function $P_x[F]$ interpolates to the given values of F on the two faces x=0 and x=1, and $P_y[F]$ and $P_z[F]$ have analogous properties.

The *tri*linearly blended function which interpolates F on all six faces of the cube is given by:

$$\begin{aligned} U(x,y,z) &= (P_x \oplus P_y \oplus P_z)[F] \\ &= P_x[F] + P_y[F] + P_z[F] - P_xP_y[F] \\ &\quad - P_yP_z[F] - P_zP_x[F] + P_xP_yP_z[F]. \end{aligned} \tag{2.31}$$

Computation of the explicit 26 term expression for U will be left to the interested reader. We note that the interpolant U actually reproduces the function F exactly for a rather wide class of functions. Specifically, $U \equiv F$ if F is of the form:

$$\begin{aligned} F(x,y,z) &= f_0(y,z) + xf_1(y,z) + g_0(x,z) + yg_1(x,z) \\ &\quad + h_0(x,y) + zh_1(x,y), \end{aligned} \tag{2.32}$$

where the *bivariate* functions f_i, g_j, and h_k are arbitrary.

3. HOMOGENIZATION OF BOUNDARY CONDITIONS

The transfinite interpolation techniques outlined in the previous section provide a means for the *a priori* exact satisfaction of the rather general type linear boundary conditions expressed by (2.8). To be specific, consider any second order linear elliptic differential operator D and the partial differential equation

$$D[U] = \rho(x,y) \tag{3.1}$$

defined over the unit square S with consistent boundary conditions of the form (2.8) specified on ∂S. As we proved, the function U_1 given by the right-hand side of (2.14) will exactly match the boundary conditions (2.8). Therefore, with a slight change in the notation from the previous section, we may write the exact solution U to equation (3.1) as:

$$U = U_1 + U_r \tag{3.2}$$

where U_r is the "residual" function. That is, U_r is the difference between the exact solution U and the approximation U_1 which was designed to satisfy the boundary conditions.

From the differential equation we have:

$$D[U_1 + U_r] = \rho \tag{3.3}$$

which yields the following equation for U_r:

$$D[U_r] = \rho - D[U_1] \tag{3.4}$$

in which $D[U_1]$ is *explicitly computable*. Since U_1 satisfies the original boundary conditions, U_r must satisfy the *homogeneous* boundary conditions:

$$L_0[U_r] = L_1[U_r] = M_0[U_r] = M_1[U_r] \equiv 0. \tag{3.5}$$

For example, consider the Poisson Equation

$$\nabla^2 U = 4(x^2 + y^2)e^{2xy} \tag{3.6}$$

subject to the boundary conditions (2.25). (The exact solution to this problem is $U = e^{2xy} - 1$.) Instead of solving this problem with inhomogeneous boundary conditions, we first construct the function U_1 given by expression (2.27) and reduce the original problem to the following for U_r:

$$D[U_r] = 4(x^2 + y^2)e^{2xy} - \nabla^2 U_1 \tag{3.7}$$

where

$$U_1 = 2xy(e^{2y} - e^{2x} + e^2) + 2y(e^{2x} - 1) \tag{3.8}$$

and the residual function U_r satisfies the homogeneous boundary conditions:

$$\begin{aligned} U_r(0,y) &= 0 , \qquad & \frac{\partial}{\partial x}U_r(x,y)\Big|_{x=1} &= 0 \\ U_r(x,0) &= 0 , \qquad & \frac{\partial}{\partial y}U_r(x,y)\Big|_{y=1} - 2U_r(x,1) &= 0. \end{aligned} \tag{3.9}$$

In this example, we can compute the analytic expression for $\nabla^2 U_1$:

$$\nabla^2 U_1 = 8xy(e^{2y} - e^{2x}) + 8xe^{2y} \tag{3.10}$$

so that the differential equation for U_r is seen to be:

$$D[U_r] = 4(x^2 + y^2)e^{2xy} + 8xy(e^{2x} - e^{2y}) - 8xe^{2y}. \tag{3.11}$$

In the important special case of the Poisson Equation with Dirichlet boundary conditions, the function U_1 which we construct is the bilinearly blended function of (2.1) and the original problem

$$\nabla^2 U = \rho(x,y) \quad \text{with } U = F \text{ on } \partial S \tag{3.12}$$

reduces to

$$\nabla^2 U_r = \rho(x,y) - \nabla^2 U_1 \quad \text{with } U_r \equiv 0 \text{ on } \partial S, \tag{3.13}$$

and with

$$\begin{aligned} \nabla^2 U_1 = {} & (1 - x)F_{yy}(0,y) + xF_{yy}(1,y) \\ & + (1 - y)F_{xx}(x,0) + yF_{xx}(x,1). \end{aligned} \tag{3.14}$$

As a practical matter, unless the boundary functions $F(0,y)$, etc. are particularly simple, the second derivatives in (3.14) should be computed numerically. It can be arranged to do these computations automatically to an accuracy comparable to the discretization error of the numerical approximation to the partial differential equation.

If the exact solution U is of the form

$$U(x,y) = g_0(y) + xg_1(y) + h_0(x) + yh_1(x) , \tag{3.15}$$

in which the univariate functions $g_i(y)$ and $h_j(x)$ are arbitrary, then we will have completely solved the boundary value problem simply by constructing the interpolant U_1 since $U_1 \equiv U$ for functions of the form

(3.15). In general, however, we cannot expect U_1 to be a "good" approximation to U since the construction of U_1 takes no cognizance at all of the differential equation, just the boundary conditions. Even so, our numerical experiments indicate that taking U_1 as an initial approximation in an iterative solution procedure is almost invariably better, in the sense of requiring fewer iterations, than the common practice of taking the null vector $u_{ij} = 0$ (or any constant value) for all interior points (x_i, y_j).

4. INCORPORATION OF AUXILIARY CONSTRAINTS

The goal of this section is to extend the transfinite interpolation techniques used to obtain exact satisfaction of boundary conditions to allow also for interpolation to point-value data in the interior of the region. Although such interpolatory methods are new and may be of general interest and applicability, our main purpose in this paper is to demonstrate how they can be used to obtain good initial approximations to be used in iterative methods for the solution of linear and nonlinear boundary value problems. In addition, as we show in the following section, these new interpolatory techniques lead to a promising new class of collocation methods.

In [8], Gordon and Hall contend that the applied analyst confronted with a boundary value problem almost invariably knows something about the solution beyond that mathematical information required for a well-posed problem. It may be known, for example, that the solution satisfies a Maximum Principle, as with Laplace's Equation; or, physical considerations may dictate that the solution is everywhere non-negative. The analyst may sometimes know the value of the solution to within 10 or 20% at some interior point of the region. Such information may be heuristic or qualitative, based upon empirical knowledge and experience. To the extent that a good initial approximation reduces the amount of work required to obtain an accurate solution via an iterative scheme, it is desirable to quantify *all* of the available information and to incorporate it into an initial approximation. The techniques developed here are aimed at providing the tools necessary to do this. Numerical experiments demonstrate that starting with a good initial approximation does indeed lead to a substantial decrease in the number of iterations required to achieve a specified accuracy.

Our immediate objective is to extend the results of the previous section to allow for both exact transfinite interpolation of boundary conditions *and* interpolation to a finite number of function values interior to the region S. The interpolation methods developed here will be used in the following section to obtain a new class of collocation methods for the approximate solution of elliptic partial differential equations. *For the time being*, however, we adopt the viewpoint that some interior values of the solution are somehow known. Rather than obscure the main ideas by treating the very general boundary conditions of (2.8), we will illustrate the methods for the case of Dirichlet boundary conditions on ∂S.

To begin, we consider again the projectors P_x and P_y of (2.3), but with slightly changed notation:

$$\begin{aligned} P_x^1[F] &= \phi_0^1(x)F(0,y) + \phi_1^1(x)F(1,y) \\ P_y^1[F] &= \psi_0^1(y)F(x,0) + \psi_1^1(y)F(x,1) \end{aligned} \tag{4.1}$$

where the ϕ_i^1 and ψ_j^1 satisfy the cardinality conditions of (2.5). (Most commonly we would take $\phi_0(x) = 1 - x$, $\phi_1(x) = x$, $\psi_0(y) = 1 - y$ and $\psi_1(y) = y$, unless special considerations indicate that other blending functions are more suitable.) Thus, our transfinite interpolant to the given boundary values of F is:

$$U_1 = (P_x^1 \oplus P_y^1)[F]. \tag{4.2}$$

Now, suppose for the moment that we know the value of the exact solution U at the center of the unit square. That is, suppose we are given the numerical value $F(\frac{1}{2}, \frac{1}{2})$. We seek to construct a function $V = U_1 + U_2$ which both matches the boundary conditions *and* interpolates to the (assumed known) value $F(\frac{1}{2}, \frac{1}{2})$. Such a function is given by:

$$V(x,y) = U_1(x,y) + \phi_1^2(x)\psi_1^2(y)[F(\tfrac{1}{2}, \tfrac{1}{2}) - U_1(\tfrac{1}{2}, \tfrac{1}{2})] \tag{4.3}$$

where $\phi_1^2(x)$ and $\psi_1^2(y)$ are such that:

$$\begin{aligned} \phi_1^2(0) &= 0 & \psi_1^2(0) &= 0 \\ \phi_1^2(\tfrac{1}{2}) &= 1 & \psi_1^2(\tfrac{1}{2}) &= 1 \\ \phi_1^2(1) &= 0 & \psi_1^2(1) &= 0. \end{aligned} \tag{4.4}$$

The simplest such functions are $\phi_1^2(x) = 4x(1 - x)$ and $\psi_1^2(y) = 4y(1 - y)$. The reader may verify that V does, in fact, satisfy:

$$V \equiv F \text{ on } \partial S \quad \text{and} \quad V(\tfrac{1}{2}, \tfrac{1}{2}) = F(\tfrac{1}{2}, \tfrac{1}{2}). \tag{4.5}$$

We wish to rewrite (4.3) in a form more suggestive of the general case. To do so, we consider the projectors P_x^2 and P_y^2 defined by:

$$\begin{aligned} P_x^2[F] &= \phi_0^2(x)F(0,y) + \phi_1^2(x)F(\tfrac{1}{2},y) + \phi_2^2(x)F(1,y) \\ P_y^2[F] &= \psi_0^2(y)F(x,0) + \psi_1^2(y)F(x,\tfrac{1}{2}) + \psi_2^2(y)F(x,1) \end{aligned} \tag{4.6}$$

in which the ϕ_i^2 and ψ_j^2 are the cardinal quadratic polynomials for Lagrange interpolation at 0, $\frac{1}{2}$ and 1. Clearly, the product $P_x^2P_y^2[F]$ provides an interpolant to the values of F at the 9 Cartesian product points $(0,0)$, $(0,\frac{1}{2})$, ..., $(1,1)$. Since $V \equiv F$ on ∂S, note the important fact that:

$$P_x^2P_y^2[F - U_1] \equiv \phi_1^2(x)\ \psi_1^2(y)[F(\tfrac{1}{2}, \tfrac{1}{2}) - U_1(\tfrac{1}{2}, \tfrac{1}{2})]. \tag{4.7}$$

Or, in operation notation:

$$P_x^2 P_y^2 (I - (P_x^1 \oplus P_y^1))[F] \equiv \phi_1^2(x)\psi_1^2(y)[F(\tfrac{1}{2}, \tfrac{1}{2}) - U_1(\tfrac{1}{2}, \tfrac{1}{2})]. \quad (4.8)$$

If we substitute the left-hand side of this last equation into (4.3), we find that:

$$\begin{aligned} V &= U_1 + P_x^2 P_y^2 (I - (P_x^1 \oplus P_y^1))[F] \\ &= ((P_x^1 \oplus P_y^1) + P_x^2 P_y^2 - P_x^2 P_y^2 (P_x^1 \oplus P_y^1))[F] \\ &= ((P_x^1 \oplus P_y^1) \oplus P_x^2 P_y^2)[F]. \end{aligned} \quad (4.9)$$

This shows that the function V, which *combines* the interpolatory properties of exact matching of boundary conditions *and* matching of F at the point $(\frac{1}{2}, \frac{1}{2})$, is just the Boolean sum of the transfinite interpolant $(P_x^1 \oplus P_y^1)[F]$ and the tensor product interpolant $P_x^2 P_y^2[F]$.

This result generalizes easily. Let Q be any projector which interpolates F at an arbitrary number of distinct points (x_k, y_k) interior to the unit square S and such that

$$Q(P_x^1 \oplus P_y^1)[F] = Q[F] \quad \text{on } \partial S. \quad (4.10)$$

Then, the function V(x,y) given by

$$V = (Q \oplus (P_x^1 \oplus P_y^1))[F] \quad (4.11)$$

satisfies the interpolation conditions:

$$\begin{aligned} &V \equiv F \quad \text{on } \partial S \\ &V(x_k, y_k) = F(x_k, y_k) \quad k = 1,2,\ldots,K. \end{aligned} \quad (4.12)$$

Note that the projector Q and the projector $P_x^1 \oplus P_y^1$ need not commute; all that is required is that (4.10) hold. However, (4.10) is a rather strong condition since it implies that the values which Q[F] assumes on ∂S depend only upon the boundary values of F and not upon any interior values.

In the sequel, we will consider projectors of the general form:

$$P_x^m[F] = \sum_{i=0}^{m} \phi_i(x)F(x_i,y)$$
$$P_y^n[F] = \sum_{j=0}^{n} \psi_j(y)F(x,y_j) \qquad (4.13)$$

where the ϕ_i and ψ_j are cardinal polynomials (*or polynomial splines*) for Lagrange interpolation at the (m + 1) *equally spaced* nodes $x_i = i/m$ and the (n + 1) equally spaced nodes $y_j = j/n$, respectively:

$$\phi_i(x_k) = \delta_{ik} \quad (i,k = 0,1,\ldots,m)$$
$$\psi_j(y_\ell) = \delta_{j\ell} \quad (j,\ell = 0,1,\ldots,n). \qquad (4.14)$$

As simple, specific cases of the types of projectors Q which satisfy (4.10), we take:

$$Q = P_x^m P_y^n \qquad (m,n > 1) \qquad (4.15)$$

and

$$Q = P_x^m P_y^{2m} \oplus P_x^{2m} P_y^m \qquad (m > 1). \qquad (4.16)$$

Clearly, for such projectors Q, we do have commutativity with $P_x^1 \oplus P_y^1$, and thus the results that:

$$((P_x^1 \oplus P_y^1) \oplus P_x^m P_y^n)[F] = F \quad \text{on } \partial S$$
$$((P_x^1 \oplus P_y^1) \oplus (P_x^m P_y^{2m} \oplus P_x^{2m} P_y^m))[F] = F \quad \text{on } \partial S. \qquad (4.17)$$

Moreover, we find that

$$((P_x^1 \oplus P_y^1) \oplus P_x^m P_y^n)[F] = F \quad \text{on } \pi_x^m \times \pi_y^n$$
$$((P_x^1 \oplus P_y^1) \oplus (P_x^m P_y^{2m} \oplus P_x^{2m} P_y^m))[F] = F \quad \text{on } (\pi_x^m \times \pi_y^{2m}) \cup (\pi_x^{2m} \times \pi_y^m) \qquad (4.18)$$

where $\pi_x^m = \{x_0,x_1,x_2,\ldots,x_m\}$ and $\pi_y^n = \{y_0,y_1,y_2,\ldots,y_n\}$.

These results provide transfinite interpolation formulas which exactly match Dirichlet boundary conditions on the edges of the unit square and match a finite number of discrete values of F in the interior. In the next section, we derive collocation techniques based on these formulas.

5. COLLOCATION USING TRANSFINITE INTERPOLATION FORMULAS

The model problem is the Poisson Equation with Dirichlet boundary conditions, equation (3.12). As an approximation to the exact solution U we take

$$U \simeq V = U_1 + U_2 \tag{5.1}$$

where U_1 is the bilinearly blended function of (2.1):

$$U_1 = (P_x^1 \oplus P_y^1)[F] \tag{5.2}$$

and U_2 is the function

$$\begin{aligned} U_2 &= P_x^m P_y^n (I - (P_x^1 \oplus P_y^1))[F] \\ &= P_x^m P_y^n [F - U_1] \\ &= \sum_{i=0}^{m} \sum_{j=0}^{n} \phi_i(x)\psi_j(y)[F(x_i,y_j) - U_1(x_i,y_j)]. \end{aligned} \tag{5.3}$$

Since $U_1 = F$ for all grid points (x_i,y_j) which are on the perimeter of S, all of the terms in (5.3) pertaining to boundary points vanish and we have:

$$U_2 = \sum_{i=1}^{m-1} \sum_{j=1}^{n-1} \phi_i(x)\psi_j(y)[F(x_i,y_j) - U_1(x_i,y_j)]. \tag{5.4}$$

The interior values $F(x_i,y_j)$ in (5.4) are *unknown* and we wish to determine them in some way. One simple method is that of *collocation* at the points (x_i,y_j) interior to S. From the original differential equation we obtain the approximate relation:

$$\nabla^2(U_1 + U_2) = \rho \tag{5.5}$$

which implies:

$$\nabla^2 U_2 = \rho - \nabla^2 U_1. \tag{5.6}$$

By using (2.1) and (5.4) in this last equation and setting $\nabla^2 U_2$ equal to $\rho - \nabla^2 U_1$ at the (m-1)(n-1) points (x_k, y_ℓ) ($k = 1,2,\ldots,m-1$; $\ell = 1,2,\ldots, n-1$), we obtain the following *collocation equations*:

$$\begin{aligned}\sum_{i=1}^{m-1} \phi_i''(x_k)\alpha_{i\ell} + \sum_{j=1}^{n-1} \psi_j''(y_\ell)\alpha_{kj} = \rho(x_k, y_\ell) &- (1 - x_k)F_{yy}(0,y_\ell) \\ &- x_k F_{yy}(1,y_\ell) - (1 - y_\ell)F_{xx}(x_k,0) \\ &- y_\ell F_{xx}(x_k,1)\end{aligned} \tag{5.7}$$

in which the (m-1)(n-1) algebraic unknowns α_{ij} represent the *difference* between the unknown function F and the bilinearly blended function U_1:

$$\alpha_{ij} = F(x_i,y_j) - U_1(x_i,y_j). \tag{5.8}$$

Although the values $U_1(x_i,y_j)$ are known, it is more efficient to solve (5.7) for the α_{ij} than for the $F(x_i,y_j)$. To compute the second derivatives of the boundary values which appear on the right-hand side of (5.7) we suggest using the standard 3-point finite difference approximation with an h chosen so as to make the truncation error smaller than the error in the subsequent discretization of the continuum problem.

It is interesting to note that the coefficient matrix associated with the collocation equations (5.7) is *identical* to that for ordinary collocation using the tensor product function $P_x^m P_y^n[F]$. To see this, let

$$\begin{aligned}U &\simeq P_x^m P_y^n[F] \\ &= \sum_{i=0}^{m} \sum_{j=0}^{n} \phi_i(x)\psi_j(y)F(x_i,y_j)\end{aligned} \tag{5.9}$$

and put this function into the differential equation (3.12). After some manipulation, we obtain the following (m-1)(n-1) collocation equations:

$$\sum_{i=1}^{m-1} \phi_i''(x_k)F(x_i,y_\ell) + \sum_{j=1}^{n-1} \psi_j''(y_\ell)F(x_k,y_j) = \rho(x_k,y_\ell)$$

$$- \phi_0''(x_k)F(0,y_\ell) - \phi_m''(x_k)F(1,y_\ell) \tag{5.10}$$

$$- \psi_0''(y_\ell)F(x_k,0) - \psi_n''(y_\ell)F(x_k,1).$$

In these equations, unlike (5.7), one solves directly for the (m-1)(n-1) unknowns $F(x_i,y_j)$. *The important practical point is that, since the coefficient matrix for our new method is identical to that for standard collocation techniques, any existing collocation code can be very easily modified to solve the system of equations (5.7).* All that needs to be done is to input the right-hand side of (5.7) in place of the right-hand side of (5.10) and properly interpret the solution vector $\underline{\alpha} = (\alpha_{ij})$.

6. TIME-DEPENDENT AND RELATED PROBLEMS

Each time-step in the numerical solution of the model parabolic problem $\partial U/\partial t + \nabla^2 U = \rho$ amounts to solving an elliptic boundary value problem. Let F(x,y,t) be the *exact solution to the discretized problem* and suppose that F has been computed at all mesh points for all time-steps up to and including time t_0. To abbreviate, let F_0 represent the (already known) solution at t_0 and F_1 be the unknown solution at $t_1 = t_0 + \Delta t$. We assume for simplicity that the problem domain, which is independent of t, is the unit square S and that the boundary conditions are Dirichlet. We seek a good first approximation for F_1 which can be used to compute a higher accuracy solution by iteration.

Specifically, we wish to find a fuction U which satisfies the boundary conditions on F_1 *and* incorporates the essential "shape" of the solution F_0. A function with these properties is given by:

$$U = F_0 + (P_x \oplus P_y)[F_1 - F_0], \tag{6.1}$$

with P_x and P_y given by expressions (2.3). Note that the term $(P_x \oplus P_y)$ $[F_1 - F_0]$ uses only information extracted from the boundary conditions on F_0 and F_1. It is easy to see that $U \equiv F_1$ on ∂S. U is really nothing more than the function F_0 *adjusted* so that the new boundary conditions are satisfied. In this sense, U represents the minimal perturbation of F_0 needed to force satisfaction of the boundary conditions. One suspects that the use of this technique for parabolic problems may allow for larger time steps Δt. For closely related considerations, see [3].

Another context in which this technique promises to be of value is the frequently occurring situation where an analyst wishes to compute the solutions to a series of very similar problems. If each of the problems is to be solved iteratively, it is often arranged to use the output file from one problem as input to the next. If F_0 denotes the previously computed solution, this standard method amounts to setting $U(x_i,y_j) = F_0(x_i,y_j)$ for mesh points (x_i,y_j) in the *interior* of S and $U(x_i,y_j) = F_1(x_i,y_j)$ for the *boundary* mesh points. Loosely speaking, this provides an initial approximation which has sharp discontinuities at the boundary. As an alternative, we propose using (6.1) to initialize the solution at *all* mesh points. The starting function U given by (6.1) obviously makes full use of both the known output file for F_0 and the boundary conditions for F_1. Again speaking loosely, equation (6.1) provides a "smooth" function which is continuous at the boundary.

ACKNOWLEDGMENTS

In the course of this research the authors have benefitted from correspondence and conversations with Professors David M. Young and David Kincaid of the University of Texas at Austin, Professor Richard S. Varga of Kent State University, and Dr. Louis Hageman of Westinghouse Electric Corp.

REFERENCES

1. Bank, R.E., "Multi-Level Iterative Methods - Some Recent Results," These *Proceedings*.
2. Birkhoff, G. and Gordon, W.J., "The Draftsman's and Related Equations," *J. Approx. Theory*, *1* (1968), pp. 199-208.
3. Berger, M., Oliger, J., and Rodrigue, G., "Predictor-Corrector Methods for the Solution of Time-Dependent Parabolic Problems on Parallel Processors," These *Proceedings*.
4. Brandt, A., "Multigrid Solvers on Vector and Parallel Computers," These *Proceedings*.
5. Gordon, W.J., "Distributive Lattices and the Approximation of Multivariate Functions," *Proc. of the Symposium on Approximation with Special Emphasis on Spline Functions* held at Madison, Wisconsin, May 5-7, 1969. Academic Press, 1969, pp. 223-277.
6. Gordon, W.J., "Blending-Function Methods of Bivariate and Multivariate Interpolation and Approximation," SIAM *J. Num. Anal.*, *8*, No. 1, March 1971, pp. 158-177.
7. Gordon, W.J., and Hall, C.A., "Construction of Curvilinear Coordinate Systems and Application to Mesh Generation," *J. Num. Methods in Eng.*, *7*, 1973, pp. 461-477.
8. Gordon, W.J., and Hall, C.A., "Exact Matching of Boundary Conditions and Incorporation of Semi-Quantitative Solution Characteristics in Initial Approximations to Boundary Value Problems," *J. Comp. Physics*, *25*, No. 2, Oct. 1977, pp. 151-162.
9. Grimes, R.G., Kincaid, D.R., MacGregor, W.I., and Young, D.M., "ITPACK Report: Adaptive Iterative Algorithms Using Symmetric Sparse Storage," CNA-139, Univ. of Texas at Austin, Aug. 1978.
10. Grimes, R.G., Kincaid, D.R., and Young, D.M., "ITPACK 2.0 User's Guide," CNA-150, Univ. of Texas at Austin, Aug. 1979.
11. Kincaid, D.R., and Young, D.M., "Survey of Iterative Methods," CNA-135, Univ. of Texas at Austin, April, 1978.

APPLICATION OF A PARALLEL PROCESSOR TO THE SOLUTION OF FINITE DIFFERENCE PROBLEMS

David J. Hunt
Stephen J. Webb
Alan Wilson

International Computers Limited
London

1. INTRODUCTION TO DAP

The search for ever greater processing power for solution of such problems as field equations has led to the development of new computer architectures. One such machine is the ICL Distributed Array Processor (DAP) which is truly an array processor in that it comprises many processing units operating concurrently on different data streams but executing a common instruction stream (1). The first production model has 4096 processing elements (PEs) arranged in a 64 x 64 array. Each PE has a random access memory of 4096 bits and the totality of those memories forms a memory module of a standard computer system which acts as a host.

An essential feature is "activity control" effectively enabling each PE to be switched off temporarily according to some geometrical or data dependent condition. Hence the conditional operations common in many physical processes may be handled naturally and efficiently. Individual PEs are very simple, having just three one-bit registers and a

ISBN 0-12-632620-7

one-bit arithmetic-logic unit. Data paths connect each PE to its four neighbours and to highways running across the array in two orthogonal orientations.

Programming is usually done in a Fortran extension known as DAP-Fortran (2). This has matrices and vectors as well as scalars as basic elements.

II. APPLICATION TO FIELD PROBLEMS

The two-dimensional structure of DAP is ideal for field problems; the PE array can be mapped directly onto a two-dimensional problem or used to process plane-by-plane appropriate sections of a three-dimensional one. Access to neighbouring data values is expressed using shift functions in DAP-Fortran, for example SHEP meaning SHift East Planar. Thus the most elementary relaxation method for laplaces equation can be written as:

```
X(INTERIOR) = 0.25 * (SHEP(X) + SHWP(X) + SHNP(X) + SHSP(X))
```

Here X is the result matrix and INTERIOR is a logical matrix. The effect of this masked assignment is to leave the boundary values unchanged.

Tests for convergence and other global operations are implemented rapidly using functions that apply logical operations across the array. For example if E is a scalar criterion:

```
IF( ALL (ABS (X - OLDX ) .LT. E )) GOTO 9
```

It is well known that simultaneous over relaxation on an entire grid is divergent, but that parallel operations can be applied alternately to half the data points selected according to the colours of a chessboard: the red-black method. Less obvious is that implicit methods are amenable to Parallel processing, but it will be seen that DAP is effective at such work.

III. TECHNIQUES FOR IMPLICIT METHODS

A major part of implicit techniques is the solution of sets of tridiagonal equations. On a conventional serial computer the usual technique is to factorise the tridiagonal matrix representing a set of equations into upper and lower bidiagonal matrices each being solved in turn as a recurrence relation. A parallel processor conveniently holds one result value, and hence one equation, in each processor and for this case there is a method very suitable for DAP implementation known as "odd-and-even reduction" (3).

The method involves $\log_2 M$ steps, M being the number of equations. At the first step all the off-diagonal elements in the equations are zeroised sumultaneously by adding to each equation appropriate multiples of its two neighbours. Naturally this introduces new non-zero elements but the resulting system is still tridiagonal with off-diagonal terms at a distance of two points from the diagonal rather

```
      REAL LL( , ) , DD( , ) , UU( , ) , RR( , )
      INTEGER K
      K = 1
  100 CONTINUE
      LL = LL / DD
      UU = UU / DD
      RR = RR / DD
      DD = 1.0  -  LL * SHEP (UU,K) - UU * SHWP(LL,K)
      RR =  RR  -  LL * SHEP (RR,K) - UU * SHWP(RR,K)
      IF (K, EQ. 32) GOTO 200
      LL = LL * SHEP(LL,K)
      UU = UU * SHWP(UU,K)
      K = K + K
      GOTO 100
  200 RR = RR / DD
```

FIGURE 1. Code for Solution of Sets of Tridiagonal Systems

than one. This process is repeated, with distances doubling at each step. The off-diagonals have progressively fewer elements and finally disappear. Solution of the resulting diagonal matrix is trivial.

A complete DAP-Fortran code for solving 64 independent sets of 64 equations is given in FIG.1. Matrices LL, DD and UU are the current lower, diagonal and upper elements respectively, each PE holding the set of coefficients for one equation. Line 9 performs operations on the right-hand-side RR, the remaining statements in the loop being operations on the matrix coefficients.

Clearly the $\log_2 M$ steps mean that more work is being done than on a sequential machine. However, this effect is quite tolerable in the context of an overall calculation.

IV. THE EXAMPLE PROBLEM

Programs were written in DAP-Fortran to illustrate the SOR, LSOR and ADI methods on grid sizes of 64 x 64, 128 x 128 or 256 x 256. The demonstration problem, defined by the equation below, was one timestep of a simple diffusion problem (4).

$$\nabla . (\underline{T} . \nabla P) - q = \emptyset . C. \frac{\partial P}{\partial t}$$

In general the transmissibility $\underline{T}$ is a function of position and direction, and the implementation assumes the general case. As with other machines, times per iteration could be dramatically improved by taking advantage of special cases like Poissons equation.

Conventional finite difference approximations and solution techniques were used, the only changes for parallel processing being in the use of the parallel tridiagonal solver and of red-black ordering for SOR and LSOR.

V. RESULTS

With "easy" coefficients the largest problem took 24 iterations of LSOR or 12 of ADI to achieve an accuracy of about 1 in 10^5, solution times being about five seconds. Of more interest are the times per iteration given in Table I and times per gridpoint per iteration in Table II.

The relatively low performance of SOR and LSOR on a 64^2 problem are due to the use of red-black ordering. With the larger grids this performance loss is avoided; also for the implicit methods a suitable choice of mapping meant that relatively less time was spend on the $\log_2 M$ steps in tridiagonal solution and more on fully parallel arithmetic.

Comparisons with other machines are difficult. The nearest relevant figures given in (4) are around one micro-second per grid point per iteration for LSOR problems having a few thousand grid points run on CDC STAR 100 or CYBER 203. However, this latter figure seems to assume that the various multiplier factors arising during the tridiagonal solution are computed only once, whereas in the DAP implementation

TABLE I. Times Per Iteration (In Seconds)

Method	Problem Size 64^2	128^2	256^2
SOR	0.00854		0.053
LSOR	0.0589		0.231
ADI	0.0497	0.142	0.420

TABLE II. Times Per Grid Point Per Iteration (In Microseconds)

Method	Problem Size 64^2	128^2	256^2
SOR	2.1		0.81
LSOR	14.4		3.5
ADI	12.1	8.6	6.4

they are re-computed at each iteration. Another way of expressing performance is that the DAP is achieving about 15 million floating point results per second.

VI. CONCLUSIONS

The odd-and-even reduction algorithm for solving tridiagonal systems is very suitable for DAP implementation, particularly for problem grids larger than the DAP array. This permits DAP to be applied to solution of field equations by LSOR and ADI as well as SOR. DAP-Fortran implementations give high performance using concise and readable code.

REFERENCES

1. Flanders P.M., Hunt D.J., Parkinson D., Reddaway S.F. Symposium on High Speed Computer and Algorithm Organisation, University of Illinois. Academic Press (1977).
2. ICL 2900 series publication TP6755.
3. Jordan T.L. Los Alamos Scientific Laboratory report LA-5803 (1975).
4. Nolen J.S., Kuba D.W., Kasac K.J. (Jr.) AIME Fifth Symposium on Reservior Simulation, Denver Colorado (1979).

VECTOR ALGORITHMS FOR ELLIPTIC PARTIAL DIFFERENTIAL EQUATIONS BASED ON THE JACOBI METHOD

Olin G. Johnson
George Paul

Systems Technology Department
IBM Thomas J. Watson Research Center
Yorktown Heights, New York 10598

I. INTRODUCTION

The Jacobi method is often used to introduce students to the study of iterative solutions of equations; then is quickly replaced by the faster converging but less parallel Gauss-Seidel family. More sophisticated variations of Jacobi, however, exhibit higher rates of convergence while retaining the parallel nature of Jacobi.

If the system of equations to be solved is:

$$Az = b$$

where, $A = D^{1/2}(L + I + U)D^{1/2}$ is an n by n matrix, then the Jacobi method is given by:

$$x^{(k+1)} = Gx^{(k)} + c$$

where, $G = -(L + U)$, $c = D^{-1/2}b$ and the solution is $x = D^{1/2}z$.

The spectral radius, μ, for the Jacobi method applied to the model problem [8] approaches one as the grid size increases.

By comparison, the spectral radii of Gauss-Seidel and point SOR are respectively μ^2 and

$$(\mu/(1 + (1 - \mu^2)^{1/2}))^2.$$

ISBN 0-12-632620-7

The simultaneous relaxation Jacobi method, JOR, given by the iterative equation:

$$x^{(k+1)} = \omega G x^{(k)} + (1-\omega) x^{(k)} + (1+\omega) c$$

does not improve the rate of convergence for the model problem over that of Jacobi. The two-step (second-degree) Jacobi method, J2, given by the iterative equation:

$$x^{(k+1)} = \omega G x^{(k)} + (1-\omega) x^{(k-1)} + (1+\omega) c$$

however, achieves a respectable spectral radius of:

$$\mu / (1 + (1 - \mu^2)^{1/2}) .$$

Hence, the two-step Jacobi method is to point SOR as the Jacobi method is to Gauss-Seidel. It is also an attractive method for any vector processor in which the vector performance is at least twice its scalar performance. The method is also essentially a long-vector algorithm.

Stationary multi-step Jacobi algorithms beyond the second degree do not improve convergence and the optimal n-step Jacobi method is one which reduces to the optimal two-step method.

Semi-iterative Jacobi methods are enveloped by the Chebyshev semi-iterative Jacobi method in the usual way. However, convergence is not significantly better than the two-step Jacobi method.

The conjugate gradient Jacobi method envelopes the set of non-stationary single-step Jacobi relaxation schemes. Its rate of convergenge is better than two-step Jacobi and compares favorably with that of point SOR.

Most of what has been stated up to this point is well known and can be found in [6], [8] and elsewhere.

In the next section, we limit our discussion to the previous remarks on congugate gradient Jacobi methods and introduce a new class of methods called incomplete inverse conjugate gradient

Jacobi (IICGJ) methods. We present numerical results in the final section.

II. CONJUGATE GRADIENT JACOBI METHODS

One can apply a variant of the conjugate gradient method to any iterative scheme as suggested in [6]. In the case of the Jacobi method this amounts to the standard conjugate gradient algorithm with the original matrix scaled to have ones on the diagonal. Scaling in the conjugate gradient method is known to be advantageous and is discussed briefly in [7]. ITPACK [4] includes conjugate gradient Jacobi as one of its seven iterative techniques.

The resulting method envelopes the set of non-stationary JOR methods [6]. The literature points out that rounding error is a problem only in cases where the original matrix A is not sparse [3] and [7].

The numerical experiments presented below indicate that the intermediate values of the solution at each step compare quite well with those of SOR for the model problem. In grids larger than eight by eight, for instance, the results were uniformly better than SOR. More research is required to understand the intermediate convergence of this technique for matrices which arise in elliptic problems.

The class of methods ICCG(n) introduced in [5] has proved to be highly successful. The incomplete factorization of A, however, does not lend itself to parallel computations. In [2], a form of the incomplete inverse of A was used in place of the incomplete factorization. We now suggest a new form of the incomplete inverse which appears to be better. After scaling, we must solve:

$$(I - G)x = c$$

but

$$(I - G)^{-1} = I + G + G^2 + \dots$$

Consider the splitting

$$I - G = (I + G + \dots + G^n)^{-1} - R$$

and the method

$$(I + G + \dots + G^n)^{-1} x^{(k+1)} = R x^{(k)} + c$$

The iterative matrix $(I + G + \dots + G^n)R$ has as its spectral radius the $(n+1)^{st}$ power of the spectral radius of G. As an iterative method, this splitting is useless since it amounts to precisely $(n + 1)$ applications of Jacobi. However, as a preconditioning for the conjugate gradient algorithm it is quite effective. Numerical results indicate that, per unit work, it is competitive with the ICCG family. We note that the splitting is not regular for $n>0$, and that it is a scaled version of the method suggested in [2]. The effectiveness of these methods is due to the fact that the inverse of $(I - G)$, though full, exhibits the same <u>fuzzy</u> structure as $(I - G)$. That is, the large elements of the inverse are on the diagonals of $(I - G)$. The early terms in the power series reproduce most of the significance of the inverse. We can do better still by other forms of the incomplete inverse such as:

$$I + G + G^2 + \overline{G}^3 + \overline{G}^4 + \dots$$

where the bars represent incomplete matrix multiplies or

$$uI + vG + wG^2$$

where the parameters u, v and w are chosen to <u>complete</u> the inverse as closely as possible on the non-zero diagonals. Numerically, the parameterized version of the Neumann series appears to be the most efficient and the fastest. The parameters:

$$u = 1 - \tfrac{1}{4}\omega$$

$$v = 1 + \omega$$

$$w = 1 + \tfrac{1}{2}\omega$$

where ω is the optimal SOR relaxation parameter as above, work well for the model problem. We call these methods IICGJ(u,v,w), incomplete inverse conjugate gradient Jacobi with parameters u, v and w in the inner iteration. We can of course use any parallel algorithm for the inner computation.

III. NUMERIC RESULTS

In the figures below, the natural logarithm of the L_1 error for the model problem on a 24 by 24 grid is given as a function of equivalent iterations of the ICCG(3) method. Each figure compares IICGJ with the above parameters to other methods. Figure 1 compares IICGJ with the scalar methods IICG(3) and point SOR. Figure 2 compares IICGJ with the vector methods CGJ, CSIJ and J2. The final graph shows convergence rates of IICGJ for three different sets of parameters: the ones above, (1,1,1) and (1,1.5,1). As can be seen, IICGJ with the above parameters converges as fast as ICCG(3) and faster than any of the other methods. Similar results were obtained for other grid sizes.

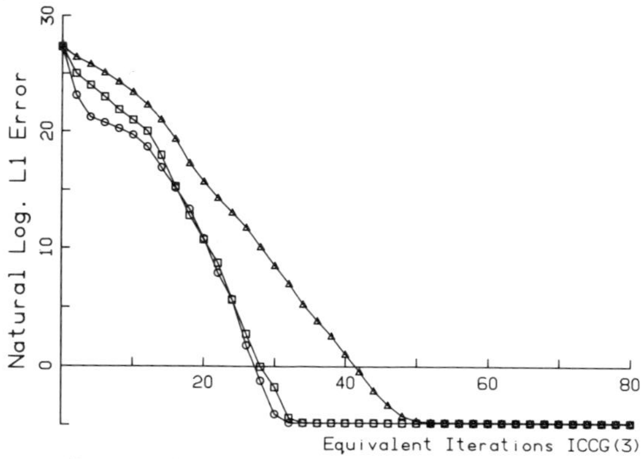

Figure 1: Comparison with Scalar Methods
24x24 Grid

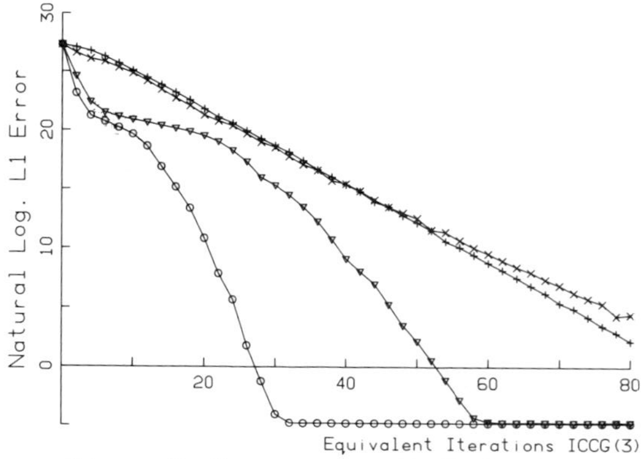

Figure 2: Comparison with Vector Methods
24x24 Grid

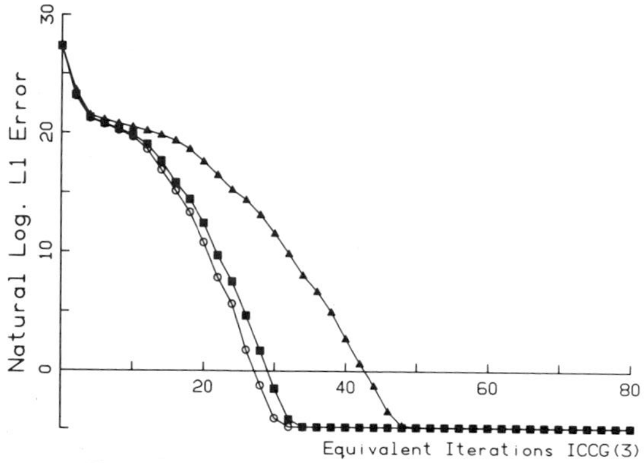

Figure 3: Comparison of IICGJ Methods
24x24 Grid

LEGEND

○ IICGJ - incomplete inverse conjugate gradient Jacobi with the parameters (u,v,w) given in Section II above

□ ICCG(3) - incomplete Cholesky conjugate gradient

△ SOR - point successive over-relaxation with the optimal relaxation factor

▽ CGJ - conjugate gradient Jacobi

+ CSIJ - Chebyshev semi-iterative Jacobi

× J2 - second degree Jacobi

■ IICGJ - incomplete inverse conjugate gradient Jacobi with the parameters (1,1.5,1)

▲ IICGJ - incomplete inverse conjugate gradient Jacobi with the parameters (1,1,1)

REFERENCES

1. F.S. Beckman, "The Solution of Linear Equations by the Conjugate Gradient Method," in Mathematical Methods for Digital Computers, ed. by A. Ralston and H.S. Wilf, John Wiley and Sons, Inc., New York, (1960).

2. P.F. Dubois, A. Greenbaum and G.H. Rodrigue, "Approximating the Inverse of a Matrix for Use in Iterative Algorithms on Vector Processors," Computing, Vol. 22 (1979), pp. 257-268.

3. M. Hestenes and E. Stiefel, "Method of Conjugate Gradients for Solving Linear Systems," Journal of Research of the National Bureau of Standards, Vol. 49 (1952), pp. 409-436.

4. D.R. Kincaid and R.G. Grimes, "ITPACK: Numerical Studies of Several Adaptive Iterative Algorithms," University of Texas Center for Numerical Analysis, ITR-CNA-126 (1977).

5. J.A. Meijerink and H.A. van der Vorst, "An Iterative Solution Method for Linear Systems of Which the Coefficient Matrix is a Symmetric M-Matrix," Mathematics of Computation, Vol. 31 (1977), pp. 148-162.

6. W.L. Miranker, "Enveloping an Iteration Scheme," IBM Journal of Research and Development, Vol. 16, No. 4 (July 1972), pp. 389-392.

7. J.K. Reid, On the Method of Conjugate Gradients for the Solution of Large Sparse Systems of Linear Equations," Proc. Conference on Large Sparse Sets of Linear Equations, Academic Press, New York (1971).

8. D.M. Young, Iterative Solution of Large Linear Systems, Academic Press, New York (1971).

ADAPTING ITERATIVE ALGORITHMS DEVELOPED FOR SYMMETRIC SYSTEMS TO NONSYMMETRIC SYSTEMS[1]

David R. Kincaid
David M. Young

Center for Numerical Analysis
University of Texas
Austin, Texas

I. INTRODUCTION

The iterative algorithms utilized in the ITPACK computer package for solving large sparse linear systems of equations assume that the coefficient matrix is symmetric positive definite. (See [5] for additional information on this package.) As with many packages, the need to handle a slightly larger class of problems, namely, nearly symmetric systems, soon became evident. This required adapting the routines involving storage of the linear system so as to allow both a symmetric and a nonsymmetric storage mode.

Numerical solutions of boundary-value problems with self-adjoint partial differential equations may result in symmetric systems when low-order discretizations are used; however, the use of higher-order discretizations may result in nonsymmetric systems which are nearly symmetric. Also, non-self-adjoint equations or mixed boundary conditions may result in nonsymmetric systems.

An attempt was made to apply the existing routines of ITPACK to nonsymmetric systems. This attempt was not successful at first for the programs which involve conjugate gradient (CG) acceleration. The Chebyshev

[1]Work was supported in part by National Science Foundation Grant MCS76-03141 at The University of Texas at Austin.

ISBN 0-12-632620-7

acceleration and the SOR algorithms worked successfully when applied to nearly symmetric systems. A modification of the CG algorithms was developed which did not significantly alter the behavior for the symmetric case and which, though not guaranteed to converge in all cases, did work successfully for a large number of mildly nonsymmetric problems.

Since the idea behind these modifications to ITPACK was to expand the scope of the package without having to do a great deal of new coding, a simple change to the conjugate gradient procedure was sought to handle nearly symmetric systems. Such a modified conjugate gradient method is discussed in section II. Various discretizations applied to a model problem and solved by iterative methods are presented in sections III-V.

II. MOTIVATION OF MODIFIED FORMULAS FOR CONJUGATE GRADIENT ACCELERATION

In this section we describe the formulas for conjugate gradient acceleration which are used in ITPACK 2.0 and the modified formulas which are used in ITPACK 2A. We consider the acceleration of the basic iterative method

$$u^{(n+1)} = Gu^{(n)} + k$$

where, for some matrix W, the matrix $W(I-G)W^{-1}$ is positive real. (In the <u>symmetrizable</u> case $W(I-G)W^{-1}$ is symmetric positive definite (SPD).) The conjugate gradient acceleration procedure used in ITPACK 2.0 is given by

$$u^{(n+1)} = \rho_{n+1}\{\gamma_{n+1}\delta^{(n)} + u^{(n)}\} + (1-\rho_{n+1})u^{(n-1)} \tag{1}$$

where $\delta^{(n)}$ is the pseudo-residual vector $\delta^{(n)} = Gu^{(n)} + k - u^{(n)}$ and where

$$\gamma_{n+1} = \left[1 - \frac{(W\delta^{(n)}, WG\delta^{(n)})}{(W\delta^{(n)}, W\delta^{(n)})}\right]^{-1}$$

$$\rho_{n+1} = \left[1 - \frac{\gamma_{n+1}(W\delta^{(n)}, W\delta^{(n)})}{\gamma_n\rho_n(W\delta^{(n-1)}, W\delta^{(n-1)})}\right]^{-1}, \quad (\rho_1 = 1) \tag{2}$$

In the symmetrizable case the conditions on the $\{\rho_{n+1}\}$ and $\{\gamma_{n+1}\}$ are necessary and sufficient that for the method (1) the $\{\delta^{(n)}\}$ should be mutually W-orthogonal, i.e., that $(W\delta^{(i)}, W\delta^{(j)}) = 0$ for $i \neq j$. However, the orthogonality does not hold in general for the nonsymmetrizable case. In fact, while $(W\delta^{(n+1)}, W\delta^{(n)}) = 0$, it is not true in general that $(W\delta^{(n+1)}, W\delta^{(n-1)}) = 0$.

We can make $\delta^{(n+1)}$ be W-orthogonal to both $\delta^{(n)}$ and $\delta^{(n-1)}$ if we replace (2) by

$$\rho_{n+1} = \left[1 - \gamma_{n+1} \frac{(WG\delta^{(n)}, W\delta^{(n-1)})}{(W\delta^{(n-1)}, W\delta^{(n-1)})}\right]^{-1} \tag{3}$$

Both schemes are equivalent for the symmetrizable case, but not in general for the nonsymmetrizable case. The scheme involving (3) has been found to be much more effective for many nonsymmetric problems. We remark that the modified scheme is one of several procedures developed in [4].

III. MODEL PROBLEM AND CENTRAL DIFFERENCE DISCRETIZATION

We consider as a model problem the Convection-Diffusion Problem of the form

$$L[u] \equiv u_{xx} + u_{yy} + \beta u_x = f, \qquad (x,y) \in \Omega$$

$$u = xy(1-x)(1-y), \qquad (x,y) \in \partial\Omega$$

in the unit square Ω with boundary $\partial\Omega$. Here β is constant and $f = 2x(x-1) + y(y-1)[2-\beta(1-2x)]$ so that the values on the boundary satisfy the partial differential equation. It is interesting to note that this equation is not self-adjoint but is essentially self-adjoint. Using standard central difference approximations of $\mathcal{O}(h^2)$ accuracy at each grid point (i,j), we have

$$L_h[u] \equiv (u_{i+1,j} + u_{i-1,j} - 2u_{ij})h^{-2} + (u_{i,j+1} + u_{i,j-1} - 2u_{ij})h^{-2}$$

$$+ \beta(u_{i+1,j} - u_{i-1,j})(2h)^{-1} = f$$

Here (i,j) corresponds to (x_i,y_j) for uniform spacing h. Hence, the typical equation at each grid point is of the form

$$-u_{i,j-1} - \left(1-\frac{h\beta}{2}\right)u_{i-1,j} + 4u_{ij} - \left(1+\frac{h\beta}{2}\right)u_{i+1,j} - u_{i,j+1} = -h^2 f_{ij}$$

The stencil corresponding to this equation is Figure 1(a).

Notice when $\beta = 0$ the coefficient matrix is symmetric, and when $\beta \neq 0$ it is nonsymmetric. Hence, we use this simple problem as a test of the implementation of the nonsymmetric storage mode in ITPACK 2A. The numerical results in Table I are the numbers of iterations for convergence to 10^{-6} accuracy with $h = 1/5$. (To conserve space, similar results with smaller grids and execution times are omitted in this paper.) In the line with $\beta = 0$, we see that the results are independent of the particular storage mode, as they should be. Recall that CG procedures failed for $\beta \neq 0$ without the modification in the computation of ρ_{n+1} from (2) to (3).

As is known [1,2], the use of central differences in this problem is not recommended. The reason for this is that the strict diagonal dominance of the coefficient matrix is lost when $\beta \geq 10$. This fact led to either the failure or nonconvergence within 100 iterations of each of the iterative procedures before β reached thirty. This problem can be corrected by using a different five-point stencil as outlined in the next two sections.

TABLE I. Iterations for Convergence: Sparse Nonsymmetric (Symmetric) Storage

SOR	JSI	JCG	SSORSI	SSORCG	RSSI	RSCG	
15(15)	25(25)	3(3)	11(11)	6(6)	12(12)	3(3)	$\beta = 0$
15	24	14	11	7	11	8	$\beta = 1$
16	37	19	9	7	23	10	$\beta = 9$

IV. MODIFIED UPWIND DIFFERENCE REPRESENTATION

For the convection-diffusion operator $L[u]$, we use the standard $\mathcal{O}(h^2)$ central differences for u_{xx} and u_{yy} but only the $\mathcal{O}(h)$ right-side or upwind difference approximation for u_x. To obtain higher accuracy, we insert certain factors, similar to a scheme in [2], into the result

$$L_h[u] \equiv \left(1+\frac{\beta h}{2}\right)^{-1}(u_{i+1,j}+u_{i-1,j}-2u_{ij})/h^2 + (u_{i,j-1}+u_{i,j-1}-2u_{ij})/h^2$$
$$+\ \beta(u_{i+1,j}-u_{ij})/h$$

Using Taylor series for $u_{i\pm 1,j\pm 1}$ and $\left(1+\frac{\beta h}{2}\right)^{-1} = 1 - \frac{\beta h}{2} + \frac{\beta^2 h^2}{4} - \ldots$, we have $L_h[u] = L[u] + \mathcal{O}(\beta h^2)$. Hence, the typical equation at grid point (i,j) is

$$-u_{i,j-1} - \left(1+\frac{\beta h}{2}\right)^{-1} u_{i-1,j} + \left[2+\beta h+2\left(1+\frac{\beta h}{2}\right)^{-1}\right]u_{ij} - \left[\beta h+\left(1+\frac{\beta h}{2}\right)^{-1}\right]u_{i+1,j}$$
$$-\ u_{i,j+1} = -h^2 f_{ij}$$

and the stencil is in Figure 1(b).

Repeating our numerical experiments with this modified upwind difference, the results in Table II are quite different. Only the reduced system methods either fail or have nonconvergence within 100 iterations but at much larger β values.

V. MODIFIED CENTRAL DIFFERENCE DISCRETIZATION

A modification of the central difference representation suggests itself from the analysis of the upwind representation. We replace $\left(1-\frac{\beta h}{2}\right)$ with $\left(1+\frac{\beta h}{2}\right)^{-1}$ and 4 with $2+\left(1+\frac{\beta h}{2}\right)+\left(1+\frac{\beta h}{2}\right)^{-1}$ to obtain

$$-u_{i,j-1} - \left(1+\frac{\beta h}{2}\right)^{-1} u_{i-1,j} + \left[2+\left(1+\frac{\beta h}{2}\right)+\left(1+\frac{\beta h}{2}\right)^{-1}\right]u_{ij} - \left(1+\frac{\beta h}{2}\right)u_{i+1,j}$$
$$-\ u_{i,j+1} = -h^2 f_{ij}$$

Clearly this representation is $\mathcal{O}(\beta^2 h^2)$ accurate. The stencil here is in Figure 1(c).

Again repeating the numerical experiments, we see in Table II in parentheses that larger values of β are allowable for the reduced systems. It is interesting to observe that for both modified schemes the Jacobi matrix B for the natural ordering tends to a zero matrix with nonzero constants on the superdiagonal. Hence, the spectral radius of B approaches zero as β increases to infinity. This was observed in [2] for SSOR. In the numerical experiments with $\epsilon = 10^{-6}$, the number of correct digits in the computed solution were 6 to 7 for all methods which converged except for the SOR method. For this method, they increased from 6 to 13 as β increased.

We remark that the modified upwind scheme and the modified central difference scheme are actually symmetrizable and performance could have been improved by using this fact.

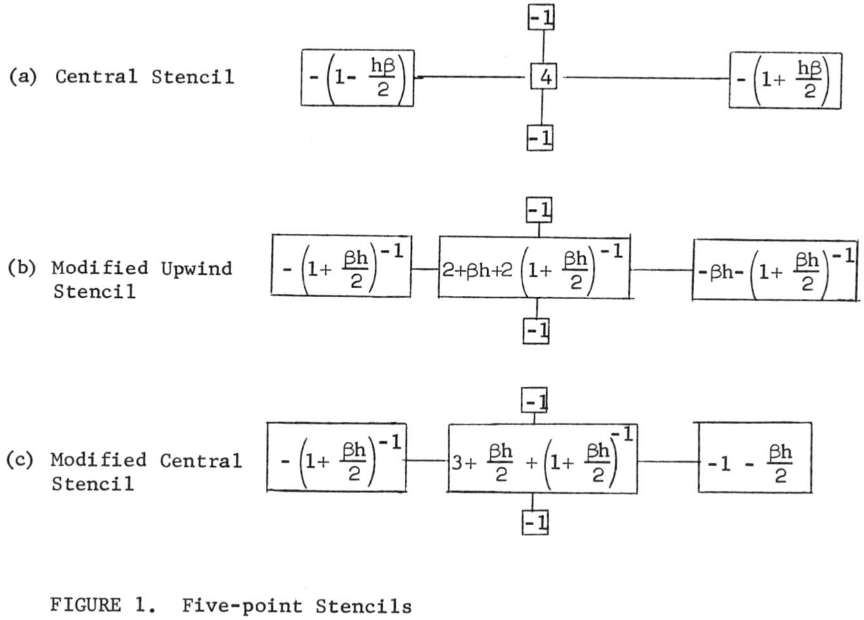

FIGURE 1. Five-point Stencils

TABLE II. Iterations for Convergence: Modified Upwind (Central) Differences

β	SOR	JSI	JCG	SSORSI	SSORCG	RSSI	RSCG
20	12(15)	31(28)	27(24)	8(9)	6(6)	20(13)	12(13)
30	11(13)	34(29)	21(26)	7(8)	6(6)	58(17)	12(13)
40	9(11)	39(31)	16(19)	6(8)	5(6)	46(21)	- (12)
50	9(10)	43(34)	19(15)	5(7)	5(6)	- (30)	- (12)
60	8(10)	47(40)	19(17)	5(6)	5(6)	- (47)	- -
70	8(10)	49(42)	19(17)	5(6)	5(6)	- (98)	- -
700	8(8)	56(52)	19(19)	3(3)	3(3)	- -	- -
1000	8(8)	56(54)	19(19)	3(3)	3(3)	- -	- -

ACKNOWLEDGEMENTS

Some of the numerical experiments for this paper were programmed by John Respess and others by Kang Jea at The University of Texas at Austin.

REFERENCES

1. Axelsson, O., and Gustafsson, I., "A modified upwind scheme for convective transport equations and the use of a conjugate gradient method for the solution of nonsymmetric systems of equations," Department of Computer Sciences, Chalmers University of Technology (1977).

2. Eisenstat, S., Elman, H., Schultz, M., and Sherman, "Solving approximations to the convective diffusion equation," Symposium on Reservoir Simulation, SPE (1979).

3. Grimes, R., Kincaid, D., and Young, D., "ITPACK 2.0 user's guide," CNA-150, Center for Numerical Analysis, University of Texas (1979).

4. Young, D., and Jea, K., "Generalized conjugate gradient acceleration of nonsymmetrizable iterative methods," to appear in Special Issue on Large Scale Matrix Problems, SIAM Journal on Linear Algebra and Its Applications (1980).

5. Young, D., and Kincaid, "The ITPACK Package for Large Sparse Linear Systems," this volume.

COMPARISON OF METHODS OF SOLUTION OF THE FINITE ELEMENT EQUATIONS FOR THE LARGE DISPLACEMENT ANALYSIS OF ARCHES

Lois Mansfield[1]

Department of Applied Mathematics and Computer Science
University of Virginia
Charlottesville, Virginia

We compare the relative efficiencies of two finite element methods for the large displacement analysis of arches. This note summarizes some of the results contained in a paper [3] in preparation. While [3] will contain results concerning the relative efficiencies of methods for the large displacement analysis of plates and shells as well as for arches, for simplicity we restrict our results in this note to arches.

The first method we compare comes from the minimum potential energy formulation where the stress-strain and strain-displacement relations are substituted into the equilibrium equations resulting in a system of equations to be solved for the displacements alone. This system represents the Euler-Lagrange equations associated with the minimization of the strain energy functional. The other method we consider, the so-called mixed method, results from using instead a modified form of the Hellinger-Reissner stationary variational principle where the stresses and bending moments appear as independent variables in addition to the displacements.

The method of solution on which our comparison is based consists of taking the load to be given by $f = \lambda f_0$ and tracing sucessive solutions for a sequence of values of λ.

[1]Research supported by NSF Grants MCS78-27482 and MCS80-16532.

ISBN 0-12-632620-7

At each load step we solve the finite element systems by Newton's method. At each Newton iteration step we solve the linear system

(1) $$J(x_n)(x_{n+1}-x_n) = -F(x_n)$$

by the multigrid method. Our comparison will be based primarily on the number of arithmetic operations required to set up the Jacobian matrix $J(x_n)$ and the system $F(x_n)$. Our conclusion is that the mixed method is considerably more efficient than the method based on the minimum potential energy formulation.

In defining the equations for the large displacement analysis of arches, we assume the presence of large displacements but small strains and that Kirchhoff's hypothesis holds. We assume that body forces f act on the arch and take as the boundary conditions

$$u(0) = w(0) = w_n(0) = u(L) = w(L) = w_n(L) = 0.$$

With these assumptions the strain energy functional is given by

(2) $$W(\underline{u}) = \frac{E\delta}{2(1-\nu^2)} \int_0^L (u' + kw + \tfrac{1}{2}(ku-w')^2)^2 dx + \frac{E\delta^3}{24(1-\nu^2)} \int_0^L (ku'-w'')^2 dx - \int_0^L \underline{f}\cdot\underline{u}\,dx,$$

where $\underline{u} = (u,w)$ are the displacements, E is Young's modulus, ν is Poisson's ratio, δ denotes the thickness of the arch, and k is the curvature.

The finite element method based on the minimum potential energy formulation consists of choosing finite dimensional subspaces $S_u^h \subset H_0^1(0,L)$, $S_w^h \subset H_0^2(0,L)$, and minimizing (2) over all functions in $S_u^h x S_w^h$. In the mixed formulation one seeks a stationary value of

$$S(\underline{\phi}) = \frac{-(1-\nu^2)}{2E\delta}\int_0^L N^2dx - \frac{6(1-\nu^2)}{E\delta^3}\int_0^L M^2dx$$

(3) $$+ \int_0^L [(u' + kw + \tfrac{1}{2}(ku-w')^2)N + (ku' - w'')M]dx$$

$$- \int_0^L \underline{f}\cdot\underline{u}dx,$$

where $\underline{\phi} = (\underline{u},N,M)$ where N is the direct stress and M is the bending moment. The mixed finite element method consists of choosing, in addition to S_u^h and S_w^h, finite dimensional subspaces $S_N^h \subset L^2(0,L)$ and $S_M^h \subset L^2(0,L)$ and seeking a stationary point of (3) in $S_u^h x S_w^h x S_N^h x S_M^h$.

Our comparison of the relative efficiencies of these two finite element methods is based primarily on the number of arithmetic operations required to set up the Jacobian matrix $J(x_n)$ and the system $F(x_n)$ in order to perform each Newton iteration step. The procedure we follow for evaluating $J(x_n)$ and $F(x_n)$ is the standard procedure of proceeding interval by interval and inserting into $J(x_n)$ and $F(x_n)$ the contributions from that interval. This procedure relies on a table of integrals for each interval which needs to be set up once and stored.

In our comparison we take S_w^h to consist of piecewise cubic Hermite polynomials and S_M^h to consist of piecewise linear polynomials with jumps at the grid points. We take S_u^h to consist of C^0 - piecewise linear polynomials, in POT1 and MXD1, or C^0 - piecewise quadratic polynomials, in POT2 and MXD2, and S_N^h to consist of piecewise constants in MXD1 or piecewise linear polynomials with jumps at the grid points in MXD2. The number of operations per interval required to set up $J(x_n)$ and $F(x_n)$ are given in Table 1, under the assumption that all necessary tables of integrals have been set up and exist in memory. POT1 and POT2 denote the finite element methods determined from the potential energy formulation using the finite element subspaces just mentioned, while MXD1 and MXD2 denote the mixed finite element methods. We have split up the number of operations required to set up $J(x_n)$ into a linear part and a nonlinear part since only the nonlinear part needs to be updated at each iteration step.

Table 1 indicates that 26.6 times more operations are required to set up $J(x_n)$ in POT1 than in MXD1 and 14.2 times more operations are required in POT2 than in MXD2. To set up $F(x_n)$ 12.97 times more operations are required in POT1 than in MXD1 and 10.9 times more are required in POT2 than in MXD2. The savings in the number of operations required to set up $J(x_n)$ and $F(x_n)$ in the mixed method as opposed to the potential energy method becomes even greater for plates and shells as will be shown in [3]. For plates and shells approximately 30 times more

Table 1

No. Arith. Op. per interval

	set up $J(x_n)$			set up
	linear	nonlinear	total	$F(x_n)$
POT1	63	2523	2586	4176
POT2	84	4374	4458	6902
MXD1	4	93	97	322
MXD2	6	308	314	644

operations are required to set up $J(x_n)$ in the potential energy method as compared to the mixed method while approximately 90 times more operations are required in the potential energy method to set up $F(x_n)$.

The reason that so many more operations are required in the potential energy method is that each nonzero entry in $J(x_n)$ contains quadratic polynomial terms in the unknowns x_n while in the mixed method each nonzero entry in $J(x_n)$ contains at most linear polynomial terms in the unknowns. In the potential energy method $F(x_n)$ contains cubic polynomial terms in the unknowns while in the mixed method $F(x_n)$ contains quadratic polynomial terms in the unknowns.

For POT1 and MXD1 we have counted the number of multiplications needed to set up the tables of integrals required. The integrals for POT1 required 1102 multiplications while those for MXD1 required 70 multiplications. The integrals were computed using a Gauss quadrature rule of sufficient accuracy to carry out all integrations exactly assuming k and f are constants. This insures that the integrals are computed with the same order of accuracy as the finite element methods themselves. See [5]. Results concerning orders of accuracy of the finite element methods described here are given in [2].

The multigrid method for the solution of linear systems consists of alternating iteration sweeps, whose purpose is to smooth the error, with corrections obtained from approximating the solution of the residual equation on a coarser grid. See, for example, [1] and [4]. For the mixed method $J(x_n)$ is indefinite so the usual type of iteration scheme, such as the Gauss-Seidel method or under-relaxed Jacobi method cannot be used directly. For mixed methods the linear systems have the form

$$\begin{pmatrix} -M & B \\ B^T & C \end{pmatrix} \begin{pmatrix} \underline{\sigma} \\ \underline{u} \end{pmatrix} = \begin{pmatrix} \underline{f} \\ \underline{g} \end{pmatrix}$$

where the vector $\underline{\sigma}$ represents those unknowns associated with the stresses and bending moments and $\underline{u}$ represents the unknowns associated with the displacements. Since elements of our subspaces S_N^h and S_M^h have support on one interval only, the matrix M is block diagonal with the size of the blocks determined by the number of basis functions with common support on a given interval.

For mixed methods we propose the iterative procedure

$$(4) \qquad \text{solve} \quad M\underline{\sigma}^i = B\underline{u}^{i-1} - \underline{f} \quad \text{for} \quad \underline{\sigma}^i,$$

$$(5) \qquad \text{set} \quad \underline{u}^i = \underline{u}^{i-1} - \alpha(B^T\underline{\sigma}^i + C\underline{u}^{i-1} - \underline{g}), \quad i=1,\ldots,n.$$

Because M is block diagonal (4) is inexpensive to solve. To determine α, note that for $\lambda = 0$, $J(x_n)$ is the same matrix as for the linear small displacement problem. Because of our choice of subspaces, for $\lambda = 0$, $B^TM^{-1}B = K$, where K is the stiffness matrix for the linear minimum potential energy method. Thus α should be chosen to be whatever is appropriate for K. If $\lambda \neq 0$, of course, $B^TM^{-1}B$ is not the same as K. In numerical experiments we have undertaken, however, we were still able to obtain rapid convergence with the original value of α used throughout.

References

1. Brandt, A., "Multi-level adaptive solutions to boundary value problems," Math. Comp., 31 (1977), 333-390.

2. Mansfield, L., "Finite element methods for nonlinear shell analysis," Numer. Math. (to appear).

3. Mansfield, L., "On the solution of the finite element equations for nonlinear shell analysis," (in preparation).

4. Nicolaides, R. A., "On the ℓ^2 convergence of an algorithm for solving finite element equations," Math. Comp., 31 (1977), 892-906.

5. Strang, G. and Fix, G. J., An Analysis of the Finite Element Method, Prentice-Hall, Englewood Cliffs, N.J., 1973.

MESH GENERATION BY CONFORMAL AND QUASICONFORMAL MAPPINGS*

C. W. Mastin and J. F. Thompson

Mississippi State University
Mississippi State, Mississippi

I. INTRODUCTION

Many recent advances in the finite-difference solution of elliptic equations have been limited to regions whose boundary contours coincide with coordinate lines of the Cartesian coordinate system. The reason for this lies in the fact that when an arbitrary curvilinear coordinate system is used, the original equation becomes much more complex. For example, Poisson's equation is transformed to an equation with variable coefficients and a mixed derivative term. There is no added complexity, or additional computer storage needed for the coefficients, if an orthogonal coordinate system is generated from a conformal mapping. An early application of conformal mappings for coordinate generation was proposed by Godunov and Prokopov [3]. An arbitrary region was transformed onto a rectangular region by solving a variational problem; however, the curvilinear coordinate lines were not the level lines of the conformal mapping and hence were not orthogonal. Later Barfield [1] developed a method for constructing orthogonal coordinates. That method involved several calculations including the solution of an integral equation, a Schwarz-Christoffel transformation, and the solution of Laplace's equation on a rectangle. It is much easier to generate a curvilinear coordinate system by constructing a harmonic transformation since the boundary correspondence may be arbitrarily assigned (see [5]). Recently this same computational procedure has been extended to construct conformal mappings. These conformal mapping schemes by Mobley and Stewart [6], Pope [7], and Yen and Lee [8] use the same numerical methods but differ in the way the boundary values and

*Research sponsored by NASA Langley Research Center under grant NSG 1577.

ISBN 0-12-632620-7

the conformal module of the region are computed. In all cases an iterative procedure determines the module of the region, the boundary correspondence, and the position of the interior coordinate lines.

Quasiconformal mappings were initially conceived to reduce any second order linear elliptic partial differential equation to canonical form (i.e., the principal part of the differential operator reduces to the Laplacian). Although they have been studied extensively by complex analysts, very little work has been done on the numerical construction of quasiconformal mappings. With the method of Belinskii et. al. [2] the mapping parameters are determined from the boundary correspondence, and the method of Mastin and Thompson [4] would be difficult to implement on arbitrary regions. Modeling the conformal mapping procedure described above, it will be shown how the same techniques can be used to construct quasiconformal mappings. This would allow one to simultaneously fit the boundary contours with coordinate lines and simplify the original equation when transforming to the new coordinate system. The equation, in canonical form, could possibly be solved more efficiently or by methods which would not be applicable to the original equation. Except for the method of Godunov and Prokopov [3], this appears to be the only conformal mapping method which can be easily adapted to handle the problem of constructing quasiconformal mappings.

II. BOUNDARY VALUE PROBLEM FOR QUASICONFORMAL MAPPINGS

Let D be a bounded simply-connected region in the xy-plane whose boundary C is a simple closed contour. Let z_1, z_2, z_3 and z_4 be distinct boundary points ordered by the orientation on C. There exists a unique quasiconformal mapping of D onto the interior of a rectangle such that the points z_i map to the vertices which are also ordered by the orientation of the rectangle. The ratio of the length to the width of the rectangle is a quasiconformal invariant called the module of D and will be denoted by m. The quasiconformal mapping of the region D onto a rectangular region can be obtained by constructing the mapping of D onto a square region S which satisfies the linear system

$$\begin{aligned} \xi_x &= m(c\eta_y + b\eta_x) \\ \xi_y &= -m(a\eta_x + b\eta_y) \end{aligned} \tag{1}$$

where $ac - b^2 = 1$.

It is easily shown that x and y, as functions of ξ and η, satisfy

$$\alpha x_{\xi\xi} - 2\beta x_{\xi\eta} + \gamma x_{\eta\eta} + \delta x_{\xi} + \varepsilon x_{\eta} = 0 \tag{2}$$

$$\alpha y_{\xi\xi} - 2\beta y_{\xi\eta} + \gamma y_{\eta\eta} + \delta y_{\xi} + \varepsilon y_{\eta} = 0$$

where

$$\alpha = a y_{\eta}^{2} - 2b x_{\eta} y_{\eta} + c x_{\eta}^{2}$$

$$\beta = a y_{\xi} y_{\eta} - b(x_{\xi} y_{\eta} + x_{\eta} y_{\xi}) + c x_{\xi} x_{\eta}$$

$$\gamma = a y_{\xi}^{2} - 2b x_{\xi} y_{\xi} + c x_{\xi}^{2}$$

$$\delta = J[(b_x + c_y)x_{\eta} - (a_x + b_y)y_{\eta}]$$

$$\varepsilon = J[(b_x + c_y)x_{\xi} - (a_x + b_y)y_{\xi}]$$

$$J = x_{\xi} y_{\eta} - x_{\eta} y_{\xi}.$$

At each boundary point of D one of the functions, ξ or η, is constant while the other satisfies an oblique derivative condition. This implies that the condition $\beta = 0$ must be satisfied by x and y at each boundary point of S (except for vertices). Note that in the solution of (1) we would have $\beta = 0$ throughout S. Thus the computed value of β can serve as a test of the accuracy of our solution. It also follows from (1) that $m = \sqrt{\alpha/\gamma}$. All computations have used finite differences with SOR iteration. (See [5] - [8]).

For conformal mappings $(a = c = 1,\ b = 0)$ the equations simplify since $\delta = \varepsilon = 0$. Furthermore, if we set $\beta = 0$ and $\alpha = m^2\gamma$, the equations in (2) can be easily solved by direct methods. The value of m and the boundary values must still be determined iteratively. In a recent communication, C. D. Mobley has reported good results in using an iterative scheme with cyclic reduction.

III. EXAMPLES

We conclude by giving some examples of meshes generated by quasiconformal mappings. When the mapping is conformal, the accuracy in computing m is also given.

Each region was normalized by setting the diameter of a boundary component equal to two. For the doubly-connected regions, the usual technique of identifying opposite sides of the square was employed in the computations. In that case the diameter of the interior boundary component was two. In all regions an elliptic equation of the form

$$au_{xx} + 2bu_{xy} + cu_{yy} = f(u,u_x,u_y)$$

would transform to an equation of the form

$$m^2 u_{\xi\xi} + u_{\eta\eta} = g(u,u_\xi,u_\eta)$$

and all computations can be done on a square computational region.

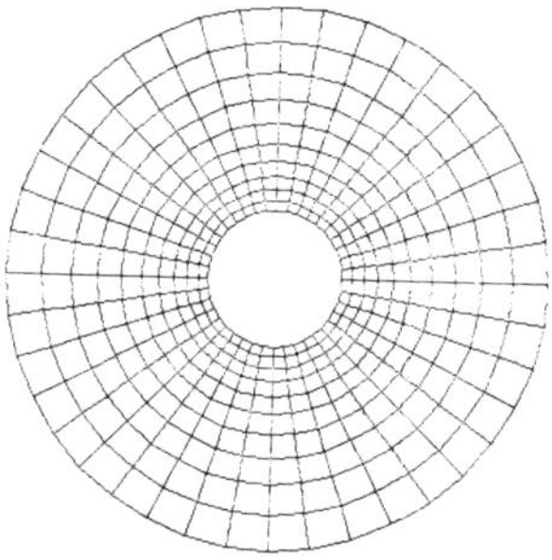

Figure 1. $a = c = 1,\ b = 0$

Exact: $m = 0.22$

Computed: $0.279 < \sqrt{\alpha/\gamma} < 0.283$

$|\beta| < 10^{-10}$

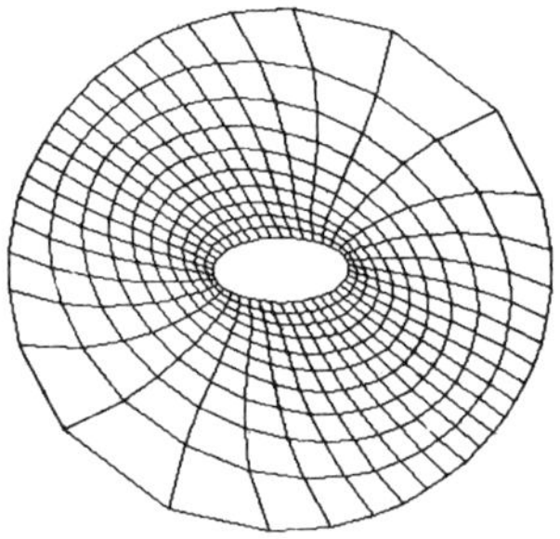

Figure 2. $a = 1,\ b = 1/2,\ c = 5/4$

Computed: $|\beta| < 2 \times 10^{-4}$

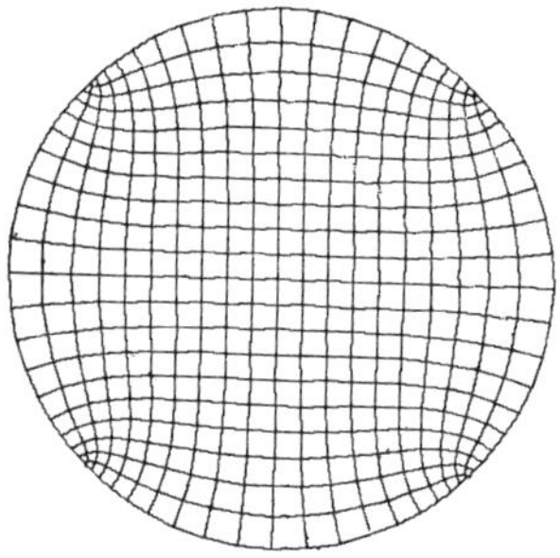

Figure 3. $a = c = 1,\ b = 0$

Exact: $m = 1$

Computed: $|1 - \sqrt{\alpha/\gamma}| < 0.01$

$|\beta| < 2 \times 10^{-7}$

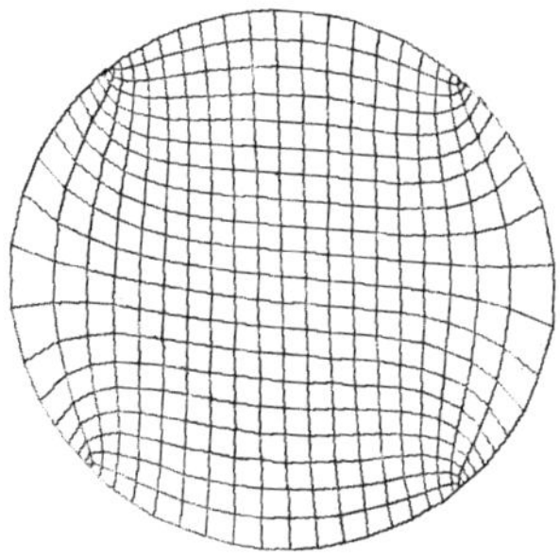

Figure 4. $a = 1,\ b = 1/2,\ c = 5/4$

Computed: $|\beta| < 3 \times 10^{-6}$

IV. CONCLUSIONS

A finite difference method developed for the construction of conformal mappings has been generalized to construct quasiconformal mappings. It is expected that the use of more sophisticated numerical algorithms could lead to improvements in both speed and accuracy. Further research in this area would certainly be justified since quasiconformal mappings have applications not only in the solution of elliptic equations but also in other areas such as orthogonal mesh generation on surfaces and the solution of certain fluid flow problems.

REFERENCES

1. Barfield, W. D., J. Comp. Phys. 5 (1970), 23-33.
2. Belinskii, P. P., Godunov, S. K., Ivanov, Y. B., and Yanenko, I. K., USSR Comp. Math. Math. Phys. 15 (1975), 133-144.
3. Godunov, S. K., and Prokopov, G. P., USSR Comp. Math. Math. Phys. 7 (1967), 89-124.
4. Mastin, C. W., and Thompson, J. F., Z. Angew. Math. Phys. 29 (1978), 1-11.
5. Mastin, C. W., and Thompson, J. F., J. Math. Anal. Appl. 61 (1978), 52-62.
6. Mobley, C. D., and Stewart, R. J., J. Comp. Phys. 34 (1980), 124-135.
7. Pope, S. B., J. Comp. Phys. 26 (1978), 197-217.
8. Yen, S. M., and Lee, K. D., in "Proc. Sixth Int. Conf. on Numer. Methods in Fluid Dynamics." (A. I. van de Vooren and P. J. Zandbergen, ed.), pp. 579-586. Springer-Verlag, New York, (1979).

BLOCK ITERATIVE METHODS

Seymour V. Parter[1]

Department of Mathematics
University of Wisconsin-Madison
Madison, Wisconsin

I. INTRODUCTION

Some 15-20 years ago there was considerable interest in iterative methods for the solution of elliptic difference equations. The interest in finite-element techniques and direct methods for the solution of large sparse systems made most of us put aside our interest in iterative methods. However, these questions have taken on a new life because of (i) the interest in 3-dimensional problems and (ii) novel computer archetures. During the summer of 1977 Bill Buzbee of the Los Alamos Scientific Laboratory convinced us that there were many important problems in the theory of iterative methods that were worthy of further serious consideration.

In this note we will summarize our activities of these past four years and some of our current research activities.

Four reports have been written. Two are concerned with the model problem

$$\Delta_h U = f, \quad \text{in} \quad 0 < x,y < 1$$

[1]Supported by ONR Contract N00014-76-C-0341 and LASL under Contract W-7405-ENG-36.

ISBN 0-12-632620-7

and give special emphasis to k-line schemes and $k \times k$ block schemes. The first [1] is by Dan Boley, Bill Buzbee and S. V. Parter. The second [2] is by S. V. Parter and Mike Steuerwalt. Another [3] by S. V. Parter and Mike Steuerwalt is concerned with the model problem

$$\Delta_h U + \frac{2}{h^2} U = f, \quad \text{in} \quad 0 < x,y < 1$$

which arises when one considers "plane" iterative methods for the 3-dimensional model problem, [3]. The fourth report [4] by Dan Boley and S. V. Parter is concerned with finite-element equations.

II. THE MODEL PROBLEM

An integer $P \geq 1$ is chosen and we set

$$h = \Delta x = \Delta y = \frac{1}{(P+1)} .$$

The equations become

$$4U_{ij} - [U_{i+1,j} + U_{i-1,j} + U_{i,j+1} + U_{i,j-1}] = h^2 f_{ij},$$

$$1 \leq i,j \leq P .$$

Let k be an integer which divides P, i.e.

$$P = kQ .$$

In the k-line iterative schemes one takes the sets of unknowns

$$U_s = \{U_{ij};\ 1 \leq i \leq P,\ k(s-1)+1 \leq j \leq ks\},\quad s = 1,2,\ldots,Q$$

as the basic blocks. In the $k \times k$ block schemes we take the sets of unknowns

$$U_{rs} = \{U_{ij};\ k(r-1)+1 \leq i \leq kr,\ k(s-1)+1 \leq j \leq ks\},$$

$$1 \leq r,s \leq Q$$

as the basic blocks. In both cases, these block structures

give rise to systems which satisfy "block property A", see [5]. Thus, from the basic theory developed by David Young [6], [7] it suffices to consider the block Jacobi schemes. The results for the associated Gauss-Seidel and SOR schemes may then be derived from those for the Jacobi schemes (see [7], [8]).

The k-line schemes were studied many years ago in [9] and [10]. This latter paper is extremely general. However, partly because of the generality of that work (variable coefficients, general domains, etc.) it is by no means a transparent discussion.

Let $\rho(kL)$ and $\rho(kB)$ denote the dominant eigenvalues of the k-line Jacobi scheme and $k \times k$ block Jacobi scheme respectively. The basic results of [1] are

$$\rho(kL) \approx 1 - k\pi^2 h^2 \tag{2.1}$$

$$\rho(kB) \approx 1 - \frac{k}{2}\pi^2 h^2 . \tag{2.2}$$

While these results were obtained in a relatively straightforward and direct way, the methods used in [1] depended on estimates which are valid only in the rectangle in the plane. Because of this limitation, the problem was attacked again in [2]. The estimates (2.1), (2.2) were reobtained by methods which extend to general 2nd order elliptic equations, in general smooth domains and in any number of dimensions.

III. A MODIFIED 3-d PROBLEM

Consider the 3-dimensional model problem. Suppose one decides to use a plane iterative scheme in which the basic blocks of unknowns are the sets

$$U_n = \{U_{i,j,n};\ 1 \le i,j \le P\}, \qquad n = 1,2,\ldots,P .$$

A straightforward tensor product argument shows that: for the Block Jacobi scheme based on these blocks one has

$$\rho \approx 1 - \frac{3}{2}\pi^2 h^2 .$$

If one used k-planes, then (see [2]) the general theory yields

$$\rho(k\pi) \approx 1 - \frac{3}{2} k\pi^2 h^2 .$$

A variety of arguments based on the limited memory, the data structure and data transfer rates (say in the CRAYI) suggest $k = 1$. Thus, we have the equation

$$6U_{ij} - [U_{i+1,j} + U_{i-1,j} + U_{i,j-1} + U_{i,j+1}] = f_{ij} \tag{3.1}$$

for the "inner problem" which must be solved in each plane. In the light of our work for the 2-dimensional problem, it seemed like a natural idea to consider k-line and $k \times k$ block schemes for this problem. We discovered that the general approach of [2] broke down for this problem. In an effort to get <u>some</u> understanding of this problem we studied the model problem in [3]. The methods used are limited to the model problem. However, as is the case with model problems which must be studied by special methods, the <u>hope</u> is: similar results hold for the general problems even if this particular mathematical analysis cannot prove such results.

The main result is that: <u>in both cases</u>, for moderate k, say $k \ge 8$,

$$\rho(k,h) \approx \rho(\infty,h) > 0$$

and $\rho(\infty,h)$, the asymptotic value, is quite insensitive to h. Computational results bear this out in a rather dramatic way. In our test cases we had

$$P = 128, \quad h = \frac{1}{129} .$$

For technical reasons we used the Gauss-Seidel method. Thus we calculated ρ^2. The results are given in the following two tables:

<u>k-line:</u>

k	ρ^2(theory)	ρ^2(computed)
4	.07184 ± .0026	.07662
8	.071797 ± .35(-9)	.07164
32	.071797	.07164
64	.071797	.07164

<u>k × k block:</u>

k	ρ^2(theory)	ρ^2(computed)
4	.1465 ± .0143	.16620
8	.145898 ± .296(-3)	.14631
16	.145898 ± .134(-6)	.14590
32	.145898	.14590
64	.145898	.14590

<u>Remark</u>: In the k-line case, the last entry in our table is the case where there are only 2 blocks each containing 8192 unknowns. In the k × k block case, the last entry in our table is the case where there are only 4 blocks each containing 4096 unknowns. These calculations were carried out by Molly Mahaffy of the Los Alamos Scientific Laboratory on the CRAY I. The code is a modification of the code developed by Dan Boley.

IV. FINITE ELEMENT PROBLEMS

In an effort to extend the general theory to finite-element equations Boley and Parter studied the following simple problem. Let $u = u(x)$ be the unique solution of

$$u''(x) = f(x), \quad 0 \leq x \leq 1, \quad u(0) = u(1) = 0 .$$

Let $h = 1/ + 1$ and let X_h be the space of Hermite cubics which satisfy the boundary conditions. We seek a function $U_h \in X_h$ which is the Ritz-Galerkin approximant to $u(x)$. Using "point" relaxation, i.e. each block of unknowns consists of the 2 values $U_h(x_k)$, $U_h'(x_k)$. From the matrix view we look at 2×2 blocks. The results are

$$\rho \approx 1 - \frac{5}{12} \pi^2 h^2 .$$

If one takes "2-points" as the basic block we obtain

$$\rho \approx 1 - \frac{5}{6} \pi^2 h^2 .$$

Suppose one considers the model problem in the unit square and takes X_h to be the space of Tensor Products of such Hermite cubics. The above analysis extends to allow one to discuss k-line iterative schemes. Each block consists of $4P$ values $U_h(x_i, y_j)$, $U_x(x_i, y_j)$, $U_y(x_i, y_j)$ and $U_{xy}(x_i, y_j)$ with j fixed and $1 \leq i \leq P$.

This analysis leads to some general questions. First of all, the methods developed here should be extendable. There are some technical questions to be resolved. But those technical mathematical questions are not the real issue. When considering finite-element methods one must remember that part of the power of such methods is the flexibility to use irregular triangulations. The questions then are: (i) Given a rather general irregular partitions, is there a good way to look at blocks of unknowns which would lead to useful block iterative schemes? (ii) Given such useful blocks, can the analysis be extended to these cases?

These are difficult questions and being given a great deal of thought.

V. FURTHER RESEARCH

The main thrust of our ongoing research on these problems is to extend the analysis of [2] to parabolic problems and non-self-adjoint problems. In the self adjoint parabolic case one is led to problems of the form

$$\left(\frac{c}{\Delta t} - \Delta_h\right)U = f \ .$$

If

$$\frac{ch^2}{\Delta t} = c_0 h^\alpha, \qquad 0 < \alpha < 2$$

then it is an easy matter to extend the methods of [2] to complete the analysis. For example, in the k-line case we have

$$\rho \approx 1 - \frac{k}{2} c_0 h^\alpha + 0(h^\alpha) \ .$$

REFERENCES

1. Boley, D. L., Buzbee, B. L., and Parter, S. V., On block relaxation techniques, University of Wisconsin MRC Technical Summary Report #1860 (1978).

2. Parter, S. V., and Steuerwalt, M., Another look at iterative methods for elliptic difference equations, University of Wisconsin Computer Sciences Department Report #358.

3. Parter, S. V., and Steuerwalt, M., On k-line and $k \times k$ block iterative schemes for a problem arising in 3-d elliptic difference equations. To Appear: SIAM J. on Numerical Analysis.

4. Boley, D. L., and Parter, S. V., Block relaxation techniques for finite-element elliptic equations: An example, Los Alamos Scientific Laboratory Report LA-7870-MS.

5. Arms, R. J., Gates, L. D., and Zondek, B., A method of block iteration, J. SIAM 4 (1956), 220-229.

6. Young, David M., Iterative Solution of Large Linear Systems, Academic Press, New York (1971).

7. Young, David M., Iterative methods for solving partial difference equations of elliptic type, Trans. Amer. Math. Soc. 76 (1954), 92-111.

8. Varga, R. S., Matrix Iterative Analysis, Prentice-Hall, Englewood Cliffs, NJ (1962).

9. Parter, S. V., On estimating the 'Rates of Convergence' of iterative methods for elliptic difference equations, Trans. Amer. Math. Soc. 114 (1965), 320-354.

10. Parter, S. V., Multi-line iterative methods for elliptic difference equations and fundamental frequencies, Numerische Math. 3 (1961), 305-319.

A MESH - PARAMETER - CONTINUATION METHOD

S.J. Polak

A. Wachters

Th. Beelen

N.V. Philips' Gloeilampenfabrieken

Eindhoven, The Netherlands

P.W. Hemker

Mathematisch Centrum

Amsterdam, The Netherlands

I. INTRODUCTION

The most important aspect in the construction of program packages for P.D.E. problems [1-3] is Robustness. Other criteria (in order of importance) are user-friendliness and speed. This paper describes our present efforts to construct Robust algorithms for the solution of a class of nonlinear P.D.E. problems.

The algorithms are only considered with respect to a special class of problems (section 2) but extensions to other problems are obvious. The predictor - corrector - continuation (PCC) method [4] is one of the most Robust algorithms known to us for the solution of nonlinear equations. Combinations of the PCC method and spatial discretizations are therefore obvious choices when constructing Robust algorithms for P.D.E. problems [5]. Two different combinations are considered in this paper. First a continuation parameter t is introduced in the continuous problem to solve the

ISBN 0-12-632620-7

problem on a single given mesh (section 4). Secondly a PCC method is used for the solution of the residual problem arising from a two mesh algorithm (section 5). In section 6 we consider the simultaneous use of both algorithms. In section 7 a practical implementation of this combination is described.

II. SEMICONDUCTOR PROBLEMS

The equations involved in the analysis of semiconductor problems are discussed in detail in [6], [7] and [3]. Here we only consider the Poisson problem for the case of negligible currents. In [3] this problem is treated for physically realistic 2-D composite regions.

In this paper we use only a 1-D simplified problem as an example. The equation has the form

$$-\Delta u = H(u), \tag{1}$$

where

$$H(u) = a(\exp b(c_1-u) - \exp b(u-c_2)) + D(x),$$

with

$x \in [20\mu, 100\mu], \mu = 10**-6, u(20\mu) = 0, u(100\mu) = 700,$

$a = 10**13,\ b = 40,\ c_1 = 0,\ c_2 = 700$ and

$D(x) = -a$ if $x < 50\mu, D(x) = a$ if $x > 50\mu$.

The problem will also be denoted by $L(u) = 0$. There exists a unique bounded solution for this problem. This is also the case for the more general problem treated in [3]. To understand this, a theorem from [8] is used. The basic property there is monotony. Our operator is monotonic because $-\Delta$ is coercive and $H_u(u)$ is negative definite.

The upper and lower bound of the solution of the problem are $c_1 - \exp(b(c_1-c_2))/b$ and $c_2 + \exp(b(c_1-c_2))/b$ respectively, as can be easily shown by techniques similar to those used for the proof of a maximum principle [10]. These bounds imply that, within the range of floating point representable numbers, the solution of (1) is a monotonous function.

In the general case, a priori lower and upper bounds for the solution are known on physical grounds. Therefore, a, c_1 and c_2 are always such that, within rounding errors, $c_1-u, u-c_2 \le 0$; hence, in (1) the function exp(x) may be replaced by 1+x for $x \ge 0$. This does not change the solution of the problem and it eliminates the problems caused by the large exponential factor b.

III. THE PCC METHOD

The PCC method is described in [3] and [4]. Its use for semiconductor potential problems is discussed in [3]. The algorithm is summarized here and a new step strategy is presented.
Suppose we want to solve the equation

$$F(u) = 0. \tag{2}$$

Then the continuation parameter t is introduced giving

$$\tilde{F}(u(t),t) = 0, \ 0 \le t \le 1, \tag{3}$$

with $\tilde{F}(u(1),1) = F(u(1)) = 0$ and the initial solution u(0) is known.
Differentiation with respect to t gives

$$\tilde{F}_t + \tilde{F}_u u_t = 0. \tag{4}$$

The solution of (3) is approximated at $0 = t_0 < \dots < t_n < \dots < t_n = 1$, the steplength is denoted by $\tau_n = t_n - t_{n-1}$.
With equation (4) forward Euler is used to compute the prediction $\bar{u}(t_n)$ and Newton's method is applied to (3) to find the solution $u(t_n)$.

Because we were not satisfied with the existing step strategies we devised a new one, which proved to be better both in the 1-D model and in the realistic 2-D problems.

Set $e_{i,n} = \| u_i(t_n) - u_{i-1}(t_n) \|$, $i = 1,\dots,J+1$, $u_0(t_n) = \bar{u}(t_n)$, define $\alpha_{i,n} = e_{i+1,n}/e_{i,n}^2$. Then the mean quadratic convergence factor is defined by $\bar{\alpha}_n = \sqrt[m]{P}$ with $P = \Pi_{i=1}^{J} \alpha_{i,n}^{s_i}$, $s_i = 2^{J-i}$ and $m = 2^{J-1}-1$ where $u_{J+1}(t_n)$ is the final value of the Newton iterations.
Suppose we want to accept $u_I(t_{n+1})$ if $e_{I,n} < \delta$ for some chosen tolerance δ and number of iterations I. Suppose $u_0(t_{n+1}) = u(t_{n+1})$ is calculated with an integration method, e.g. forward Euler. Then

$$e_{0,n+1} \le \|u_0(t_{n+1})-u_I(t_{n+1})\|+\alpha\|u_0(t_{n+1})-u_I(t_{n+1})\|^2,$$

where α is the convergence factor of Newton at t . Now we use $\bar{\alpha}_n$ as approximation for α. Thus

$$e_{0,n+1} \le \varepsilon(\tau) + \bar{\alpha}_n\, \varepsilon(\tau)^2,$$

where $\varepsilon(\tau)$ is the truncation error of the integration method and from

$$e_{I-1,n+1} \le \bar{\alpha}_n^{\,2^I-1}\, e_{0,n+1}^{\,2^I} \le \bar{\alpha}_n^{\,2^I-1}\, (\varepsilon(\tau) + \bar{\alpha}_n(\varepsilon(\tau)^2)^{2^I}) \le \delta,$$

τ can be calculated. The step changes are limited by $\tau_{n-1}/2 \le \tau_n \le 3\tau_{n-1}$. We start with trying $\tau_1 = 1$. If $\tau_1 = 1$ fails then $\tau_1 = \tau_1/5$ etc. until τ_1 is accepted. A t-step is rejected if

$$F(u_{i+1})/F(u_i) > .9 \text{ and } F(u_{i+2})/F(u_{i+1}) > .9$$

for some $i \ge 1$. Then the step is divided by 2.

IV. PARAMETER - CONTINUATION (PC)

The parameter t is introduced in (1) by replacing H(u) by t×H(u). Thus at t = 0 the problem has a smooth solution (linear) and at t = 1 the resulting singular perturbation problem has a solution with an interior boundary layer. To approximate the solution, the problem is discretised a FEM on a uniform mesh. The resulting equations are solved with the PCC method. For 3 different meshes the number of t-steps, rejected steps and Newton iterations are shown in table I.

TABLE I.

number of mesh intervals	8	32	128
Newton iterations	9	125	418
t-steps accepted (rejected)	1(0)	27(2)	92(5)

V. MESH - CONTINUATION

Let G_i and G_{i+1} be two meshes, G_i being a submesh of G_{i+1}. The operator I^i_{i+1} interpolates a function defined on G_i, yielding a function defined on G_{i+1}. The discretization of $L(u)$ on G_j, is indicated by $L_j(u)$, $j = i, i+1$. Suppose $L_i(u_i) = 0$, then we calculate u_{i+1}, defined by $L_{i+1}(u_{i+1}) = 0$, from

$$(1-t)L_{i+1}(I^i_{i+1}\, u_i) - L_{i+1}(\tilde{u}_{i+1}(t)) = 0, \qquad (5)$$

with $\tilde{u}_{i+1}(0) = I^i_{i+1}\, u_i$ and $u_{i+1} = \tilde{u}_{i+1}(1)$. Again we use the PCC algorithm to solve this problem.

Table II shows the number of Newton iterations needed to proceed from the 8 to the 32 interval mesh, from the 32 to the 128 interval mesh and from the 8 to the 128 interval mesh.

The interpolation operator is defined by locally solving the problem with coarse grid solutions as boundary values and, again, a PCC method.

TABLE II. Number of Newton iterations

number of mesh intervals	8	32	128
parameter continuation (PC)	9	125	418
PC at coarse mesh + mesh continuation (MPC)	9 →	+ 3 →	+ 6
	9 →	→	+ 2

If the solutions on the different meshes are sufficiently close, no continuation method is required to solve $L_{i+1}(u_{i+1}) = 0$, i.e. the mesh continuation is simply Newton with $I^i_{i+1}\, u_i$ as the initial estimate. However, if $I^i_{i+1}\, u_i$ is not sufficiently close (and for most realistic problems we do not have this a priori knowledge), then the mesh-continuation improves the robustness essentially.

VI. MESH - PARAMETER - CONTINUATION (MPC)

The algorithms from sections 5 and 6 may be combined to give a Robust and fast algorithm. The basic idea, of course, is to solve small sets of equations first, using parameter continuation on the coarse mesh. Then the solution is calculated on a sufficiently fine mesh using mesh-continuation. In the following section we present a practical implementation of this combination.

If we take the coarse mesh too coarse, the solution may be a very bad approximation of fine mesh solution. Then the work for the MPC algorithm may be more than the work necessary for the simple PC algorithm on the fine mesh. Therefore, in future we shall try to construct algorithms which use a sequence of meshes and optimize the path followed in the two parameter embedding provided by the combination of mesh and parameter continuation. Of course, in this context [5] and [9] should be mentioned. However, in those papers, Robustness is not the primary concern.

VII. 2-D MESHES

The 2-D meshes used e.g. in the program packages MAGGY2 [1] and SEMMY2 [3] are very suited for the algorithm as described in section 6. The meshes used in these packages are discussed in [1]. The meshes consist of quadrilaterals and triangles. They may be considered as a distortion of a rectangular mesh. Coinciding points yield triangles. These meshes are used for FE discretizations and they allow an easy description in the problem oriented language as can be seen in Figure 1.

We see that a coarse and a fine mesh are given at the same time by the user of the package [1], the fine mesh being constructed from the coarse mesh by linear interpolation on the coordinate space. Realistic problems with complicated geometries, solved with such meshes, can be found in [1] and [2].

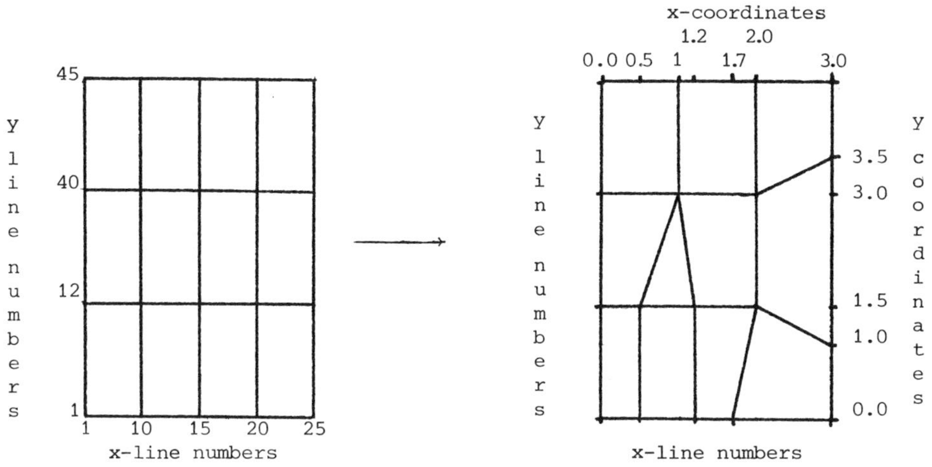

FIGURE 1. The mesh is specified in the SEMMY2 input language by:
x(1) = 0,x(10) = 2*.5,2*1,x(15) = 2*1.2,2*1,x(20) = 1.7,3*2,
x(25) = 3,
y(1) = 0,y(12) = 4*1.5,1,y(40) = 4*3.,3.5,y(45) = 4.5.

REFERENCES

1. Polak, S.J., de Beer, A., Wachters, A. and van Welij, J.S. (1980). *MAGCY2, A programpackage for two dimensional magnetostatic problems.* Int. J. for Num. Meth. in Eng. Vol. 15, 113-127.
2. Polak, S.J., Schrooten, J. and Barneveld Binkhuysen, C. (1978). *TEDDY2, A programpackage for parabolic composite region problems.* ACM TOMS vol. 4, no. 3, 209-227.
3. Polak, S.J., Wachters, A., Vaes, H.J.M., de Beer, A. and den Heyer, C. (1979). *A continuation method for the calculation of electrostatic potentials in semiconducturs.* In: Proc. of the NASECODE I Conference, Trinity College, Boole Press, Dublin.
4. Rheinboldt, W.C. (1976). In: "*Mathematics of FEM and applications*" J. Whiteman, ed.. Academic Press, London, pp. 465-482.
5. Babuska, I. and Rheinboldt, W.C., *Reliable error estimation and mesh adaption for the FEM.* Tech. note BM910, Univ. of Maryland, april 1979.
6. Slotboom, J.W. (1977). *Analysis of bipolar transistors,* Thesis, Technische Hogeschool, Eindhoven.

7. Mock, M.S. (1972). *On equations describing steady-state carrier distributions in a semiconductor device*. Comm. Pure and Appl. Math. Vol.25, 781-792.
8. Ciarlet, P.C., Schultz, M.H. and Varga, R.S. (1969). Numer. Math. 13, 51-77.
9. Zave, P. and Rheinboldt W.C. (1979). *Design of an adaptive parallel finite element system*. ACM-TOMS, Vol. 5, no. 1, 1-17.
10. Protter, H.H. and Weinberger, H.F. (1967). *Maximum principles in differential equations*. Prentice Hall.

CAPACITANCE MATRIX METHODS - A BRIEF SURVEY

Wlodzimierz Proskurowski

Department of Mathematics
University of Southern California
Los Angeles, California

Since the late sixties fast algorithms to solve the Helmholtz equation

$$(1.1) \qquad - \Delta u + cu = f$$

on rectangular regions are widely available. We shall call such algorithms Fast Helmholtz Solvers (FHS). These solvers, based on the Fast Fourier Transforms (FFT) or cyclic reduction, or both (FACR), produce a solution at a cost proportional to NlogN airthmetic operations and using N storage locations, where N is the number of mesh points inside the rectangle. For a recent survey of FHS see Temperton (1979).

The Capacitance Matrix Methods (CMM) allow for the extension of the above mentioned techniques to arbitrary bounded regions. The solution here is obtained as a low rank modification of the solution on a rectangle in which the arbitrary region is imbedded. The iterative variant of the CMM requires modest storage requirements and the total computational expenses are of the order of 10 calls of a FHS. A first brief description of CMM can be found in Hockney (1968), who credits Buneman for the idea. Although the Hockney 1968 paper described CMM in terms of potentials and electrodes, most popular in the numerical community is the

Supported by NSF Grant MCS-8003382

ISBN 0-12-632620-7

algebraic presentation of Buzbee-Dorr-George-Golub (1971). Here the low rank modification is expressed in terms of the Woodbury formula and its generalizations. If the resulting linear system of equations of order $P=\mathcal{O}(N^{1/2})$ is solved by a direct method of the Gaussian elimination type, then the algebraic framework is satisfactory in most cases, and produces stable algorithms. Unfortunately, in many cases of practical interest a rather large number of grid points is required, either to describe the boundary adequately or to obtain satisfactory accuracy in the solution. The order of the capacitance system, P, grows then up to a point when the fast core memory is saturated (on computers that have no virtual memory). A typical such limit for P is below 200. The use of larger values of P (finer meshes) becomes then prohibitively expensive. George (1970) was the first to try an iterative procedure to solve the capacitance system without explicitly generating a capacitance matrix. The question of a proper approach that guarantees the fast convergence of the iterations was solved when the problem was analyzed in potential theoretical framework, see Proskurowski-Widlund (1980).

Since then several authors have made important contributions in the development of the CMM ; for the history up to 1976 see Section 8 of Proskurowski-Widlund (1976) and also Widlund (1978). We shall now report briefly on the new work in the area. Shieh (1978, 1979) analyzed the efficiency of the CMM and proved that the number of the conjugate gradient iterations to solve the capacitance matrix system grows as $\mathcal{O}(\log N)$. Banegas (1978) developed a Sparse Helmholtz Solver (SHS). Proskurowski (1979) implemented an iterative version of the CMM for 2-D regions using SHS and compared the efficienty of different CMM solvers. O'Leary-Widlund

(1979) made extensive studies of the iterative CMM focused on the 3-D case. Proskurowski-Widlund (1980) developed a hybrid Finite Element (FEM) - CMM for the Neumarn boundary conditions. Bjørstad (1980) effectively applied an iterative CMM for solving the biharmonic problem on a rectangular region. The currently available program packages are described in Proskurowski (1978) for 2-D regions (included on the tape from the March 1978 Texas Conference on Mathematical Software) and in O'Leary-Widlund (1979) for 3-D regions and Dirichlet boundary conditions; see also Sweet (1979). The reported applications in which the CMM has been used include the measurement of biological shape change, see Bookstein (1978a, 1978b), simulation of trimmed film resistors, see Klebanoff (1979), and solving stationary Navier-Stokes equations, see Rapoport (1980).

For a detailed description of the CMM see survey by Proskurowski (1980) and references therein. Here we shall only mention the most essential facts. Using methods of the classical potential theory one can reduce the original problem (1.1) to solving a boundary integral equation with an operator, $I+K$, which is a compact perturbation of the identity. With the proper approach one can make the capacitance matrix C a close approximation of such an operator. The problem of solving the discretized equation (1.1) is reduced to:

(1) Solving the capacitance matrix system for s, the monopole or dipole density at the boundary of the region.

(2) Computing the final solution u of $Bu = b + Ws$, where B represents a Fast Helmholtz Solver, b includes the right hand side f and the boundary data, and W is a properly chosen rectangular matrix with columns that represent discrete monopoles or dipoles.

The conjugate gradient iterations applied to a linear equation with an

(I + K) - type operator converge superlinearly as shown by Hayes (1954); see also Winther (1980). On the other hand the capacitance matrix C can be factored into a product of sparse matrixes and a matrix representing a FHS. These two facts together allow for a computationally efficient way of calculating s in (1) at a cost proportional to $N(\log N)^2$. In addition, the capacitance matrix is not required in an explicit form and therefore the storage requirements are greatly reduced.

Extensive numerical experiments have been carried out on different computers ranging from CDC 7600 to DEC 10, as originally reported in O'Leary-Widlund (1979) for 3-D problems and in Proskurowski (1978, 1979) and Proskurowski-Widlund (1976, 1980) for 2-D problems. We limit the presentation to problems with no discretization error; however experiments with other smooth solutions and with solutions generated randomly show only minor differences in the rate of convergence. Denote by M the number of mesh points in the region Ω, by N the number of points in the rectangle in which Ω is imbedded and by P the number of irregular mesh points, which is also the order of the capacitance matrix. The conjugate gradient iterations are terminated when the norm of the residuals, $\|r\| = (r^T r/P)^{1/2}$, drops below a prescribed tolerance. For many problems a condition of the form $\|r_k\| < 10^{-3} \|u\|$ (estimated), provides a satisfactory accuracy, see Proskurowski-Widlund (1980). In Table 1 we have collected results on circular (spherical) regions with a radius of about 0.4 and for problems with c=0. It consists of three parts: 2-D and 3-D problems with the Dirichlet boundary conditions in part 1 and 2, 2-D problems with the Neumann boundary conditions solved with the FEM approach in part 3. In all these cases when the mesh is refined and P, the order of the capacitance matrix, grows correspondingly from 60 to 1500 (approximately) the number of conjugate gradient iterations does not grow faster than logN. Additional numerical evidence shows that the rate of convergence remains essentially the same for regions with sufficiently smooth boundaries, see Proskurowski-Widlund (1980), and for constants $c > 0$. The rate of convergence is slower if c is negative and lies in the vicinity of an eigenvalue of the discrete Laplacian, see O'Leary-Widlund (1979). Roughly 70 to 80 percent of the total time is spent using the fast solvers. Information about the form of the boundary of the region and the data at $\partial\Omega$ is stored in a few vectors of length P. This information is preprocessed once for all and the overhead is below 10% of the total computational expenses. For the conjugate gradient iterations only a storage of small multiple of P locations is needed if the Sparse Solver is used. For example, in the algorithm with the FEM approach a total of 15 vectors of length P was used. The FHS used here are comparable in efficiency to the best existing fast solvers, see Proskurowski (1978, 1979) for 2-D problems. It should be noted that a CMM program can be speeded up whenever a new and faster solver is available. In Table 1 the execution times in seconds are given: on a CDC 7600 computer in part 1, on a CDC 6600 in part 2 and on a DEC 10 in part 3. One can conclude that computational expenses are modest for this size of problems. If the capacitance matrix had been used in an explicit form then only the smallest problems, $P < 200$, could have been solved without using auxiliary storage. Here only core memory was used even for the largest problems. In the present form the CMM solvers are second order accurate in L^2, both for the Dirichlet and Neumann problems (the latter one with the FEM approach). The accuracy at the boundary $\partial\Omega$ is the same as in the interior of the region. The properties of the hybrid finite element-capacitance method are highly promising as the boundary conditions are easy to handle and a fast solver is used only once per conjugate gradient iteration. It is expected that one can develop higher order accurate schemes using more general triangulation at the boundary and extend the method to more complicated problems.

N	M	P	k		time (sec) Total	FHS		remarks
32x32	437	64	5		0.18	0.01	(1)	Dirichlet problem
64x64	1789	132	5		0.68	0.04		Run on CDC 7600
128x128	7209	268	6		2.7	0.16		
256x256	28,913	540	7		11.9	0.62		
512x512	115,777	1084	8	(*)	53.2	2.9		
650x650	212,201	1468	8	(*)	77.5	4.2		
8x8x9	93	66	5		0.95	0.05	(2)	Dirichlet problem
16x16x17	1357	438	7		9.0	0.43		Run on CDC 6600
32x32x24	7556	1522	8		81.1	2.76		
16x16	145	68	4		0.57	0.08	(3)	Neumann problem
32x32	497	132	5		1.9	0.20		Run on DEC 10
64x64	1861	268	4		6.0	0.61		FEM approach
128x128	7001	524	3		20.4	2.18		
200x200	16,541	820	3	(*)	31.6	n.a.		
400x400	66,541	1644	2	(*)	101.40	n.a.		

Table 1. The effect of enlarging the size of the problems on the number of conjugate gradient iterations k. The prescribed tolerance is 10^{-3} and the constant c=0.

(*) The solutions were not computed for all the mesh points.

ACKNOWLEDGEMENTS

I would like to thank Roland Sweet and Paul Swarztrauber for initiating the work on this survey. I am especially grateful to Olof Widlund for his helpful remarks.

REFERENCES

Banegas A. (1978), Math. Comp., vol. 32, 441-446.

Bjørstad P. (1980), Ph.D. Thesis, Stanford, to appear.

Bookstein F.L. (1978a), Comp. Biom. Res., vol. 11, 435-458.

Bookstein F.L. (1978b), The measurement of biological shape and shape change, Springer Lecture Notes in Biomath., vol. 24.

Buzbee B.L., Dorr, F.W., George, J.A. and Golub, G.H. (1971), SIAM J. Numer. Anal., vol. 8, 772-736.

George J.A. (1970), Report CS 159, Stanford University.

Hayes R.M. (1954), in Contributions to the solution of systems of linear equations and the determination of eigenvalues, e.d. by O. Taussky, Nat. Bur. of Standards, Appl. Math. Series, vol. 39, 71-103.

Hockney R.W. (1968), J. Appl. Phys., vol. 39, 4166-4170.

Klebanoff V. (1979), SIAM Fall Meeting, Nov. 12-14, Denever, Colorado.

O'Leary O.P. and Widlund O. (1979), Math. Comp., vol. 33, 849-879.

Proskurowski W. (1978), Report LBL-7516, Lawrence Berkeley Laboratory.

Proskurowski W. (1979), ACM Trans. Math. Software, vol. 5, 36-45.

Proskurowski W. (1980), 'An iterative variant of the capacitance matrix method - a survey', to appear.

Proskurowski W. and Widlund O. (1976), Math. Comp., vol. 30, 433-468.

Proskurowski W. and Widlund O. (1980), "A finite element - capacitance matrix method for the Neumann problem for Laplace's equation", to appear.

Rapoport D. (1980), J. Comp. Phys., vol. 35, 393-407.

Shieh A.S. (1978), Numer. Math., vol. 29, 307-327.

Shieh A.S. (1979), Numer. Math., vol. 31, 405-429.

Swarztrauber P. and Sweet R. (1979), Algorithm 541, ACM Trans. Math. Software, vol. 5, 352-364. Appeared also as NCAR-TN/1A-109 (1975).

Sweet R.A. (1979), Report Stan-CS-79-704, Stanford.

Temperton C. (1979), J. Comp. Phys., vol. 31, 1-20.

Widlund O. (1978), Proc. Oberwolfach meeting 1976, Springer Lecture Notes in Math., vol. 631, 209-219.

Wilhelmson R.B. and Eriksen J.H. (1977), J. Comp. Phys., vol. 25, 319-331.

Winther R. (1980), SIAM J. Numer. Anal., vol. 17, 14-17.

GEM SOLUTIONS OF ELLIPTIC AND MIXED PROBLEMS WITH NON-SEPARABLE 5- AND 9-POINT OPERATORS[1]

Patrick J. Roache
Ecodynamics Research Associates, Inc.
Albuquerque, New Mexico

INTRODUCTION

This paper describes the results of timing tests on the initial version of the GEM (General Elliptic Marching) code. The GEM code solves elliptic and mixed discretized two-dimensional partial differential equations by direct (non-iterative) spatial marching methods. Both 5-point and 9-point stencils may be solved, with no requirement that the coefficients be separable. For example, it solves the usual second-order accurate discretization of

$$aF_{xx} + bF_{yy} + cF_{x} + dF_{y} + eF_{xy} + fF = g \qquad (1)$$

where a,b,...g are all functions of x and y. The only other direct method available for solving such equations is some version of Gaussian elimination, which has a much higher operation count and storage requirement. The methods used in the GEM code are described in detail in (1). The current code is based on "simple marching" and depends on problem parameters to control the instability incurred in marching elliptic equations; for realistic physical problems, this primarily depends on a large cell aspect ratio $\Delta x/\Delta y$.

Operation counts θ are given in some detail in Roache (1978). For a simple Poisson equation (5-point operator) without making use of symmetry, in a square array, this gives

[1] Research sponsored by the U.S. Army Research Office.

ISBN 0-12-632620-7

$$\theta_{\text{initiation}} = 4M^3 + \frac{3}{2}M^2 , \quad \theta_{\text{repeat}} = 14M^2 \tag{2}$$

The initiation count is less than that required to establish a single solution by point SOR, and the repeat count is less than 2 point SOR iterations. Since operation counts like these neglect many overhead and subscripting operations, it is necessary to validate them with actual timing tests, especially since Eq. (2) indicates such remarkable efficiency.

When a 9-point operator is used, the marching solution proceeds a line at a time (like line SOR) and requires a tri-diagonal solution at j+1 at each step in the march. This of course increases the operation counts, but not their order (i.e. repeat solutions are still optimal, with $\theta \propto M^2$). For other aspects of the method, see (1).

PROBLEM DESCRIPTION IN THE GEM CODE

The discretization of the continuum partial differential equation is left to the user. The code is written in Fortran IV, and the subroutine GEM solves the stencil

$$\begin{bmatrix} C_1 & C_2 & C_3 \\ C_4 & C_5 & C_6 \\ C_7 & C_8 & C_9 \end{bmatrix} F_{ij} = C_{10} \tag{3}$$

All the coefficients $C_1, C_2 \ldots C_{10}$ are arrays stored in BLANK COMMON. The Subroutine Call is of the form

CALL GEM(INIT,F,IL,JL,ILD,N59,IPER,ICOR,NRCOND,RCOND) (4)

INIT = 0 initiates only, = 1 initiates and solves, >1 back-solves only. The solution is stored in F. The problem size is ILxJL, with the actual first DIMENSION of the arrays being ILD. N59 = 5 or 9 gives the 5-point or 9-point operator solution. (If N59 = 5, the corner coefficients C_1, C_3, C_7 and C_9 are ignored.) For IPER = 1, periodic boundary conditions are used in the x-direction (normal to the marching direction y). ICOR is the number of corrective clean-up iterations used to reduce round-off error accumulation; usually, ICOR = 0 is used, but in some cases of marginal stability, ICOR = 1 or 2 may be used. The LU decomposition and back-solve of the influence coefficient matrix is done through LINPACK subroutines (2) which are selected by the option indicator NRCOND. For NRCOND = 1, LINPACK routines are used which give an estimate of the inverse of the condition number RCOND. The time penalty is small, and in the author's experience, RCOND has been valuable as a debugging aid.

Boundary conditions are also specified by the coefficients $C_1 - C_{10}$, as indicated by the following stencils.

$C_5\ C_6$ C_8	$C_4\ C_5\ C_6$ C_8	$C_4\ C_5$ C_8
$C_5\ C_6$ C_8		$C_4\ C_5$ C_8
C_2 $C_5\ C_6$	C_2 $C_4\ C_5\ C_6$	C_2 $C_7\ C_8$

$$F = C_{10} \qquad (5)$$

For example, this stencil indicates that in the lower left-hand corner, at i = 1 and j = 1, the boundary condition is

$$\begin{aligned} &C2(1,1)*F(1,2) + C5(1,1)*F(1,1) \\ &+ C6(1,1)*F(2,1) = C10(1,1) \end{aligned} \qquad (6)$$

The general form of Eq. (5) allows for all linear combinations of boundary conditions such as Dirichlet, Neuman, mixed, ratio of derivatives ($af_x + bf_y = c$), etc. However, the requirement for separability of boundary conditions in the marching y-direction dictates that C2 cannot be used at the side boundaries. In x, the periodic option indicator NPER = 1 over-rides the matrix specification in Eq. (5). For the 9-point operator, the periodic tridiagonal solution is obtained by the method of Ref. (3). The unused coefficients such as C4(1,J), C2(JL,I) are used by the subroutine for work storage, and are not available to the user.

TEST PROBLEMS AND RESULTS

One set of test problems used pseudo-random number generation for all coefficients, which was useful in debugging all the options. A second set used a simple Poisson equation modified by a cross-derivative term $VC*f_{xy}$, formulated with centered second-order differences.

The test problems were run on a CDC 6600. A sampling of the results is presented in Table 1. Note that the 81x81 mesh problem for the 5-point operator initiates in ~ 1 ms/cell, the equivalent of 64 Point SOR iterations, and solves repeat solutions in the equivalent of 2 Point SOR iterations. For the simple Poisson equation with $\Delta x/\Delta y = 10$, the maximum residual error (which occurs only at the end of the march) is 3.9×10^{-6}.

The 9-point operator with non-periodic boundary condition requires about 67% more initiation time and about 31% more repeat time. With periodic boundary conditions, the 9-point operator requires about 3.2x as long for initiation and about 2.6x as long for repeat solutions.

Current work is progressing on a stabilized code, a 3D code using FFT in z, and "driver" codes to set up representative problems.

Table 1. Timing tests of the GEM code on a CDC 6600 with Level 2 Optimization of the FORTRAN IV code. "init" refers to initiation times, "rep" refers to repeat solution times, "SOR" refers to times for a single iteration of a Point SOR method including a convergence test but without boundary calculations, θ refers to predictions based on theoretical operation counts. Total times are in seconds, times/cell are in milliseconds, based on the minimum of three consecutive runs which included one initiation and five repeats.

Problem grid	31 x 31	51 x 51	81 x 81	81 x 81	periodic 81 x 81
(operator)	(5 pt.)	(5 pt.)	(5 pt.)	(9 pt.)	(9 pt.)
init time	0.42	1.74	6.61	15.75	30.24
init time/cell	0.47	0.70	1.03	2.46	4.73
init time/SOR	29.7	42.8	64.3	107.1	205.7
rep time	0.035	0.089	0.206	0.384	0.750
rep time/cell	0.039	0.036	0.032	0.060	0.117
rep time/SOR	2.48	2.19	2.00	2.61	5.10
init time/rep time	12.0	19.6	32.1	41.0	40.3
%error, θ_{rep}	-32	-22	-21	-28	-32
%error, $\theta_{init}/\theta_{rep}$	-22	-19	-15	-5	-3

REFERENCES

(1) Roache, P. J., "Marching Methods for Elliptic Problems: Part 1", Numerical Heat Transfer, Vol. 1, No. 1, 1978, pp. 1-25. "Part 2", Vol. 1, No. 2, pp. 163-181. "Part 3", Vol. 1, No. 2, pp. 183-201.

(2) Dongarra, J. J., Moler, C. B., Bunch, J. R., and Stewart, G. W., "LINPACK Users' Guide", SIAM, Philadelphia, 1979.

(3) Roache, P. J. and Temperton, C., "Algorithms for the Solution of Cyclic Tridiagonal Systems with General Coefficients", J. Comp. Phys., to appear.

A PARALLEL BLOCK STIEFEL METHOD FOR SOLVING POSITIVE DEFINITE SYSTEMS

Y. Saad[1]
A. Sameh[2]

Department of Computer Science
University of Illinois at Urbana-Champaign
Urbana, Illinois

I. INTRODUCTION

The main purpose of this short note is to compare the performance of block generalizations of the conjugate gradient (C.G.) algorithm and the Stiefel iterations in a multiprocessor environment. The C.G. algorithm, developed in 1952 by Hestenes and Stiefel [3] as a direct method, has been receiving wide attention as an effective iterative method. Following Golub and Underwood, [1], O'Leary gave a block generalization of the C.G. method (B-C.G.), [5]. Such a block generalization enhances the rate of convergence of the method and is more economical when one is faced with the problem of solving several systems with the same coefficient matrix. Stiefel iteration [7], an iterative scheme proposed by Stiefel as a modification of Richardson's method, is not an optimal procedure in the same sense as the C.G. method, and in general is not preferred to the C.G. algorithm on a sequential

---

[1]Supported by DOE grant DE-AC02-76ERO2383.A003. On leave from Laboratoire d'Informatique et de Mathematiques Appliquees de Grenoble.

[2]Supported by NSF grant MCS79-18394.

ISBN 0-12-632620-7

computer. In a multiprocessor environment, however, we show that a block generalization of Stiefel's iterations (B-St.) is superior to B-C.G.

Throughout this paper we adopt the notational conventions of Householder, [4].

II. THE BLOCK-STIEFEL ITERATION (B-St.)

Let the linear system under consideration be $Ax = f$, where A is positive-definite with eigenvalues $0<\lambda_1\leq\lambda_2\leq\lambda_3\leq\cdots\leq\lambda_n$ and eigenvectors v_i, $i = 1,n$. The basic idea of the Stiefel iterative scheme (or Chebyshev iterations) is to produce an approximation x_k such that the residual $r_k = f-Ax_k$ is given by

$$r_k = P_k(A)r_0 \tag{1}$$

where

$$P_k(\xi) = T_k(\beta-\alpha\xi)/T_k(\beta)$$

in which $T_k(\xi)$ is the Chebyshev polynomial of degree k, $\alpha = 2/(\mu-\nu)$, $\beta = (\mu+\nu)/(\mu-\nu)$, and $0<\nu<\lambda_1\leq\cdots\leq\lambda_n<\mu$. The iteration can be written as

$$x_k = x_{k-1} + d_{k-1} \quad , \quad k = 1,2,\ldots \tag{2}$$

where x_0 is chosen arbitrarily, $d_0 = (\alpha/\beta)r_0$, and

$$d_k = \alpha_k r_k + \beta_k d_{k-1} \quad , \tag{3}$$

in which $\alpha_k = 2\alpha T_k(\beta)/T_{k+1}(\beta)$, and $\beta_k = T_{k-1}(\beta)/T_{k+1}(\beta)$. If the iterative scheme is used with the optimal parameters $\nu = \lambda_1$ and $\mu = \lambda_n$, the contributions of the eigenvectors v_1, $v_2,\ldots,v_n$ to r_k are quickly damped out, i.e. $r_k \to 0$ as $k \to \infty$.

An algorithm that is suitable for parallel computation may be outlined as follows. Assuming that $\lambda_1\leq\cdots\leq\lambda_p<\lambda_{p+1}\leq\cdots\leq\lambda_n$, we create p linearly independent vectors $f^{(i)}$, $i = 1,2,\ldots,p$, where $f \equiv f^{(1)}$. Taking $\nu = \lambda_{p+1}$ and $\mu = \lambda_n$, we simultaneously start (2) with p different initial iterates $x_0^{(i)}$ to obtain p sequences of residuals $r_k^{(i)}$ whose components along the eigenvectors $v_1, v_2, \ldots, v_p$ are large, while those along $v_{p+1}, v_{p+2}, \ldots, v_n$ are damped out even faster then the classical iteration with the optimal parameters $\nu = \lambda_1$ and $\mu = \lambda_n$, see Figure 1.

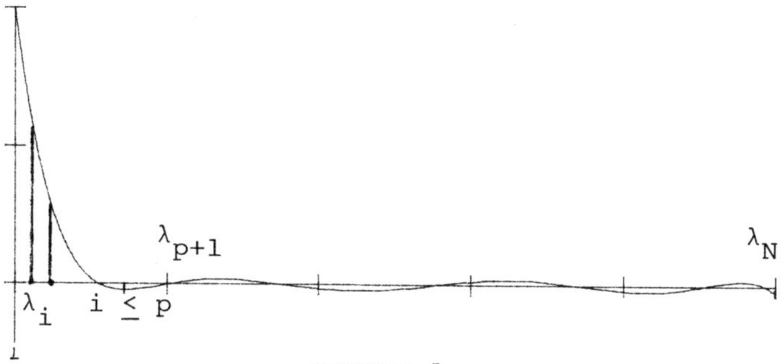

FIGURE 1.

If the eigenvectors $v_1, v_2, \ldots, v_p$ were known, however, their contributions to each $r_k^{(i)}$ can be eliminated via a projection process. This elimination can also be achieved by projecting on a subspace containing good approximations to $v_1, v_2, \ldots, v_p$. Such a subspace is provided by the span of $R_k = [r_k^{(1)}, r_k^{(2)}, \ldots, r_k^{(p)}]$. Assuming optimal parameters, $\nu = \lambda_{p+1}$ and $\mu = \lambda_n$, the above block-Stiefel algorithm may be summarized as:

(a) Simultaneously, for $i = 1,2,\ldots,p$, perform m steps of the Stiefel iterations

$$x_k^{(i)} = x_{k-1}^{(i)} + d_{k-1}^{(i)} \; ; \; r_k^{(i)} = f^{(i)} - Ax_k^{(i)}$$
$$d_k^{(i)} = \alpha_k r_k^{(i)} + \beta_k d_{k-1}^{(i)} \; , \; k = 1,2,\ldots,m \; . \tag{4}$$

(b) Projection process, (i) obtain the orthogonal factorization $R_m = Q_m L_m^T$ where Q_m is orthonormal and L_m is lower triangular; (ii) take as an improved solution

$$\tilde{x}_m^{(1)} = x_m^{(1)} + Q_m (Q_m^T A Q_m)^{-1} Q_m^T r_m^{(1)} \; . \tag{5}$$

In fact, it can be shown that

$$\|\tilde{x}_m^{(1)} - x\|_A \le \tau / T_m(\gamma_{p+1}) , \tag{6}$$

where

$$\gamma_{p+1} = (\lambda_n + \lambda_{p+1})/(\lambda_n - \lambda_{p+1}) ,$$

$$\tau = \|x - x_0^{(1)}\|_A + \sum_{j\le p} |v_j^T(x - x_0^{(1)}) \tan_A \theta(v_j, y_j)| \tag{7}$$

in which $y_j \in \text{span}[R_m]$ such that $y_j^T v_k = \delta_{jk}$, and $\|z\|_A^2 = z^T A z$.

In practice the eigenvalues of A are not known beforehand, and the parameters ν and μ are dynamically estimated. This is further enhanced by the fact that the eigenvalues of $Q_k^T A Q_k$ provide estimates of $\lambda_1, \lambda_2, \ldots, \lambda_p$ that improve with increasing k. Once the components of the residual $r_k^{(1)}$ along the eigenvectors $v_1, v_2, \ldots, v_\ell$ $(\ell \le p)$ are negligible we can shift ν further to the right resulting in marked improvement in the convergence to the solution. Another practical consideration is that, if the improvement after a projection step is not satisfactory it becomes necessary to restart the algorithm. By restarting we mean that stages (a) and (b) above are repeated with $x_0^{(1)} := \tilde{x}_j^{(1)}$ and $x_0^{(i)} := x_j^{(i)}$, for $i \ge 2$, rather than $\tilde{x}_j^{(i)}$ so as to preserve the good approximation of $v_1, v_2, \ldots, v_p$ in span$[R_k]$. In the case we are interested in solving the systems $Ax^{(i)} = f^{(i)}$, $i = 1,2,\ldots,q$, where $q \le p$, we update all $x_m^{(i)}$ according to (5). The process becomes more effective if we take $p \ge 2q$.

Assuming that we have p processors, one for each right-hand side $f^{(i)}$, and assuming that each arithmetic operation requires one time step, the time required by arithmetic operations per iteration of stage (a) is $n(3 + 2\sigma)$, in which σ is the maximum number of nonzero elements per row of A.

Here we have ignored the additional time, O(p), required by the adaptive procedure which takes place less frequently.

III. THE BLOCK CONJUGATE GRADIENT METHOD

The simplest form of the block C.G. algorithm may be described as follows, see [O'Leary, 1980]. Start with an initial iterate $X_0 = [x_0^{(1)}, x_0^{(2)}, \ldots, x_0^{(p)}]$, compute the residual matrix $R_0 = F - AX_0$, where $F = [f^{(1)}, f^{(2)}, \ldots, f^{(p)}]$, take $P_0 = R_0$, and for $k = 1,2,\ldots$, compute

$$X_k = X_{k-1} + P_{k-1}A_{k-1} ,$$

$$R_k = R_{k-1} - AP_{k-1}A_{k-1} , \text{ and}$$

$$P_k = R_k + P_{k-1}B_{k-1} ,$$

in which $A_j = (P_j^T AP_j)^{-1} R_j^T R_j$, and $B_j = (R_j^T R_j)^{-1} R_{j+1}^T R_{j+1}$. Note that $R_k^T R_j = P_k^T AP_j = 0$ for $k \neq j$. The rate of convergence of this B-C.G. algorithm is superior to that of the classical C.G. algorithm. In fact, we can show that

$$\|x^{(i)} - x_k^{(i)}\|_A \leq \eta^{(i)} / T_k(\gamma_p)$$

where $\eta^{(i)}$ is given by a simpler formula than that proposed in [O'Leary, 1980],

$$\eta^{(i)} = \|x^{(i)}\|_A + 2 \sum_{j=1}^{p-1} \frac{\lambda_n + \lambda_p - 2\lambda_j}{\lambda_n - \lambda_p} \cdot |v_j^T x^{(i)} \tan_A \theta(v_j, y_j)|$$

with y_j defined as in (7), see [6]. Using p processors, the arithmetic operations of one iteration require time $2n(6p + \sigma)$.

IV. SOME REMARKS AND NUMERICAL EXPERIMENTS

Based on arithmetic alone, the time required by one B-C.G. iteration is approximately $1 + \frac{6p}{\sigma}$ larger than that required by one B-St. iteration. If, in addition, we were to consider the time spent in interprocessor communication, the above ratio would be even larger. If A is very sparse and $p >> \sigma$, then for a given amount of work, the accuracy of the approximate solution of the B-St. algorithm will far exceed that of the B-C.G. algorithm in spite of the superior rate of convergence of the B-C.G. method. The above leads us also to believe that if one is interested in solving the systems $Ax^{(i)} = f^{(i)}$, $i = 1,2,\ldots,p$, using the conjugate gradient method on a multiprocessor, it will be more efficient to deal with each system independently, one system per processor, as any advantage gained by the B-C.G. algorithm will be negated by the time spent in interprocessor communication. Finally, we note that any preconditioning techniques that may be incorporated in the B-C.G. algorithm can also be added to the B-St. algorithm without difficulty.

The sequential B-C.G. is compared with the sequential B-St. method on two examples using the Cyber 175 with single precision (machine precision $\simeq 10^{-14}$). In the first example we took A to be Diag(1,2,20,21,...,217). Figure 2 compares the (classical) Stiefel iteration using optimal parameters (curve a), the B-St. iteration using an adaptive procedure for estimating the parameters (curve b: $p = 3$; curve c: $p = 5$), the C.G. method (curve d), and the B-C.G. method (curve e: $p = 2$; curve f: $p = 5$). We have observed that while the amount of work for B-C.G. increases drastically with the block dimension p, the rate of convergence does not improve accordingly (compare curve d (C.G. algorithm) with curve e (B-C.G. with $p = 2$)). It was stated in [5] that the B-C.G. is more effective than the C.G. algorithm in the case when a cluster of small eigenvalues is widely separated from the rest of the spectrum, as in our first example. The closeness of curves d and e in Figure 1 and the fact that curve e requires double the work seem to contradict the statement in [5].

The second example deals with the Laplacean 5-point difference operator of order 300. Figure 3 compares, for the value p = 5, the convergence of the B-C.G. (curve g) and the B-St. methods (curve h). We must mention that in order to obtain a residual norm of 10^{-11} the B-St. algorithm consumed ≈ 3 cpu seconds compared to 7.5 cpu seconds for the B-C.G. algorithm, despite the fact that the B-St. method requires roughly double the number of iterations.

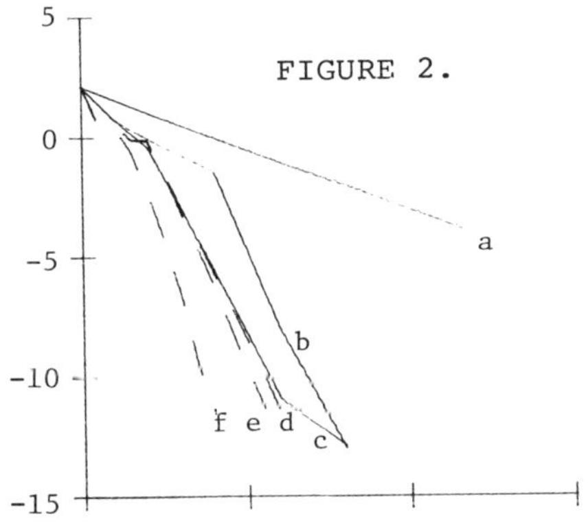

FIGURE 2.

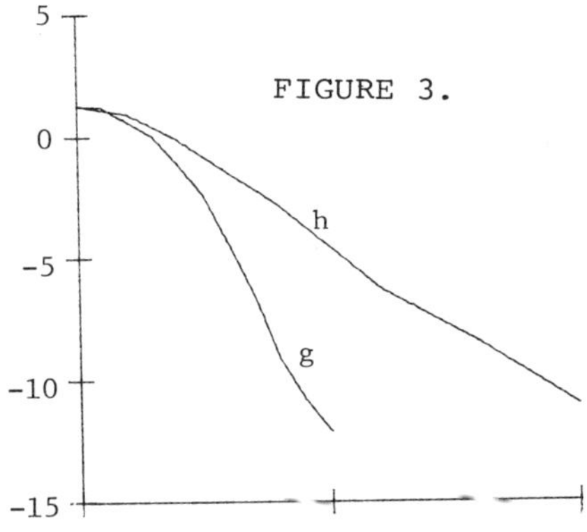

FIGURE 3.

REFERENCES

1. Golub, G., and Underwood, R., "The block Lanczos method for computing eigenvalues," in Mathematical Software III (J. Rice, ed.), pp. 361-377, Academic Press, New York, (1977).
2. Golub, G., and Varga, R., "Chebyshev semi-iterative methods, successive over relaxation iterative methods and second order Richardson iterative methods," Numer. Math. 3, 147-168 (1961).
3. Hestenes, M., and Stiefel, E., "Methods of conjugate gradients for solving linear systems," J. Res. Nat. Bur. Stand. 49, 409-436 (1952).
4. Householder, A., The Theory of Matrices in Numerical Analysis, Blaidell (1964).
5. O'Leary, D., "The block conjugate gradient algorithm and related methods," Lin. Alg. Appl. 29, 243-322 (1980).
6. Saad, Y., and Sameh, A., "On some iterative methods for parallel computers," to appear.
7. Stiefel, E., "Kernel polynomials in linear algebra and their applications," Nat. Bur. Stand. Appl. Math. Ser. 49, 1-22 (1958).

Numerical Solution of Coupled Systems of Partial Differential Equations in One Spatial Variable and Time

N. L. Schryer

Bell Laboratories
Murray Hill, New Jersey 07974

1. Introduction

There is a natural hierarchy in numerical analysis:

$\mathbf{Ax} = \mathbf{b}$ - linear equations
$\mathbf{f}(\mathbf{x}) = \mathbf{0}$ - nonlinear equations
$\mathbf{f}(t, \mathbf{x}, \mathbf{x}_t) = \mathbf{0}$ - ordinary differential equations

and the numerical solution of partial differential equations (PDE's) is a natural extension of it.

There are many possible extensions of the above hierarchy to PDE's. The classical, text-book extension has typically taken the form

$$\frac{\partial}{\partial x} a(t, x, u, u_x, u_t, u_{tx}) = f(t, x, u, u_x, u_t, u_{tx}) \quad \text{for } x \in [L, R] \tag{1.1}$$

with boundary conditions (BC's) of the form

$$b(t, u, u_x, u_t, u_{tx}) = 0 \tag{1.2}$$

at $x = L$ and R, where $u(x,t)$ is the unknown solution. This form takes advantage of the fact that all physical laws, which are second order in space, are in divergence form.

However, there is something seriously wrong with formulation (1.1)-(1.2). The PDE (1.1) is purely local in nature, as are the BC's (1.2). Thus, the user can only deal with conditions at every point x (the PDE) or at the end points $x = L$ or R. This inability to deal with a specific interior ($L < x < R$) point precludes the user from automatically and efficiently dealing with dynamic coordinate systems, shock waves, integro-differential equations, and much more.

The next section presents another extension of the above hierarchy to PDE's which doesn't suffer from these handicaps. The object is to make software for solving PDE's more widely applicable to and usable in the physical sciences.

It should be noted that there are many excellent schemes for spatially discretizing (1.1)-(1.2), reducing them to a system of ordinary differential equations (ODE's), and thus solving the problem. One of the most elegant, and most acceptable to people in the physical sciences, techniques is Galerkin's method [1]. That method has the property that the term $a(.)$ in (1.1) is forced to be continuous everywhere. Since $a(.)$ is a flux, physically, and physicists expect fluxes to be continuous, this means that Galerkin's method gives the solution the physicists want. Other spatial discretization methods such as finite-differences and collocation do not automatically make $a(.)$ continuous.

ISBN 0-12-632620-7

2. The Extension

The formulation will consider two classes of dependent variables, $\mathbf{u}(x,t)$ and $\mathbf{v}(t)$, those which depend upon both space (x) and time (t) and those which only depend upon time. The PDE's of (1.1) are then extended to the form

$$\frac{\partial}{\partial x}\mathbf{a}(t,\, x,\, \mathbf{u},\, \mathbf{u}_x,\, \mathbf{u}_t,\, \mathbf{u}_{tx},\, \mathbf{v},\, \mathbf{v}_t) = \tag{2.1}$$

$$\mathbf{f}(t,\, x,\, \mathbf{u},\, \mathbf{u}_x,\, \mathbf{u}_t,\, \mathbf{u}_{tx},\, \mathbf{v},\, \mathbf{v}_t),$$

the BC's of (1.2) are extended to

$$\mathbf{b}(t,\, \mathbf{u},\, \mathbf{u}_x,\, \mathbf{u}_t,\, \mathbf{u}_{tx},\, \mathbf{v},\, \mathbf{v}_t) = 0, \tag{2.2}$$

and ODE's

$$\mathbf{d}(t,\, \mathbf{U},\, \mathbf{U}_t,\, \mathbf{v},\, \mathbf{v}_t) = 0 \tag{2.3}$$

are used to determine $\mathbf{v}(t)$, where

$$\mathbf{u}(t,x) = \sum_i \mathbf{U}_i(t) B_i(x)$$

is the representation of the solution $\mathbf{u}$ in terms of the spatial discretization's basis functions $B_i(x)$. Since any statement about $\mathbf{u}(x,t)$ is equivalent to a statement about $\mathbf{U}(t)$, this last equation (2.3) gives a *functional* coupling between $\mathbf{u}(x,t)$ and $\mathbf{v}(t)$.

A software package exists for solving problems of the form (2.1)-(2.3) [2].

We now show how the formulation (2.1)-(2.3) can be used to solve physically interesting problems.

3. Examples

The author knows of no way to even state the following two problems in formulation (1.1)-(1.2), yet they are easily stated and solved in (2.1)-(2.3).

The first example considers a parabolic PDE with a moving interface. Let us study the PDE

$$u_t = (c(u)u_x)_x \qquad \text{for } x \in [0,1], \tag{3.1}$$

where $c(u)$ is a piecewise analytic function of u, depending on the sign of u, and at the position $x(t)$ where u changes sign there is an interface condition

$$c(u)u_x\Big|_{x_-(t)}^{x_+(t)} = \alpha\frac{dx}{dt} \tag{3.2}$$

First, rewrite the PDE (3.1) as

$$u_t = \Big\{c(u)u_x - H(x,x(t))\alpha x'(t)\Big\}_x \tag{3.3}$$

where

$$H(x,x(t)) \equiv \begin{cases} 0 & x < x(t) \\ 1 & x > x(t) \end{cases}.$$

In this latter form, when Galerkin's method is used, the interface condition (3.2) is forced to hold.

Then, all we need to do is solve (3.3) in a moving coordinate system which tracks $x(t)$. This is quite simple to accomplish. Let $x(\xi; v_1(t)) : [0,2] \rightarrow [0,1]$, with $x(1,v_1(t)) = v_1(t)$, be a change of independent spatial variable x to ξ. In such a mapping, we will want to choose $v_1(t)$ to be $x(t)$, which is unknown. Specifically, let

$$x(\xi; v_1(t)) \equiv \begin{cases} \xi v_1(t) & 0 \le \xi \le 1, \\ v_1(t) + (1-v_1(t))(\xi-1) & 1 \le \xi \le 2. \end{cases} \tag{3.4}$$

We can rewrite the PDE (3.3) in terms of $w(\xi,t) \equiv u(x(\xi; v_1(t)), t)$ and $v_1(t)$. To determine $v_1(t)$ we use

$$\mathbf{d}: \quad w(1,t) = 0.$$

Note that since the mapping (3.4) is piecewise analytic, $w(\xi,t)$ will have the same property. Thus, it is important for the spatial discretization of w to allow w_ξ to be discontinuous at $\xi = 1$. For example, if B-splines of order k are used then the mesh point $\xi = 1$ must be of multiplicity $k-1$.

The above spatial mapping scheme has been used to study the use of lasers in annealing semiconductor substrates [3] and the motion of magnetic domain walls [4].

The second example is the solution of an integro-differential equation:

$$u_t = u_{xx} + \int_0^1 c(x,y)\, u(y,t)dy \qquad \text{for } x \in [0,1], \tag{3.5}$$

where $c(x,y)$ is some given kernel. This is clearly beyond the scope of formulation (1.1)-(1.2). It is trivial to solve using (2.1)-(2.3). Simply introduce ODE variables $v_1(t),\ldots,v_N(t)$, where N is the number of basis functions to be used in the spatial discretization of the problem. Let

$$\mathbf{d}: \quad \mathbf{v} - \mathbf{U} = \mathbf{0}.$$

Then, $\mathbf{v} \equiv \mathbf{U}(t)$ and we can functionally couple u back into the PDE (3.5) via $\mathbf{v}$. To see this, note that

$$\int_0^1 c(x,y)u(y,t)dy \equiv \sum_i U_i \int_0^1 c(x,y)B_i(y)dy.$$

If we let $c_i(x) \equiv \int_0^1 c(x,y)B_i(y)dy$, this lets us rewrite (3.5) as

$$u_t = u_{xx} + \sum_{i=1}^{N} v_i(t)c_i(x). \tag{3.6}$$

which has the desired form (2.1)-(2.3). This is a very simple use of (2.1)-(2.3) to solve an integro-partial differential equation. If the integral operator were nonlinear, rather than linear, a suitable quadrature rule would have to be employed, rather than defining the $c_i(x)$.

4. Two Dimensions

It is clear that the formulation of (2.1)-(2.3) in one space dimension has implications for PDE's in two dimensions. However, it is not clear just what those implications are.

A reasonable first approximation to that extension would be to consider two-dimensional PDE's, one-dimensional PDE's and ODE's all "coupled" together. A very tentative step in that direction has been taken by L. O. Wilson [5]. In that time-varying free boundary problem, a one-dimensional PDE is used to parameterize the boundary of the domain and a two-dimensional PDE is solved on that domain. The one-dimensional PDE is solved using the software for (2.1)-(2.3) [2] and the two-dimensional static PDE is solved using the multi-grid code of Bank and Sherman [6].

From this example, it seems clear that the spatial variable in the one-dimensional problem (for example, arc-length along the boundary) need not have anything to do with the two-dimensional spatial variables.

A realistic extension of (2.1)-(2.3) to two space variables requires further experimentation and experience.

Bibliography

[1] G. Strang and G. Fix, **An Analysis of the Finite Element Method,** Prentice-Hall, New York, 1973.

[2] N.L. Schryer, "Numerical Solution of Time-Varying Partial Differential Equations in One Space Variable", Bell Laboratories Computing Science Technical Report #53, 1977.

[3] C. M. Surko, A. L. Simons, D. H. Auston, J. A. Golovchenko and R. E. Slusher, "Calculation of the Dynamics of Surface Melting during Laser Annealing", ***Appl. Phys. Lett.*** **34,** (10), 635-637(1979).

[4] N.L. Schryer and L.R. Walker, "The Motion of 180° Domain Walls in Uniform dc Magnetic Fields", ***J. Appl. Physics*** **45,** No. 12, 5406-5421(1974).

[5] L. O. Wilson, "Two-dimensional Simulation of Silicon Oxidation", to be submitted to ***J. Appl. Physics.***

[6] R. Bank and A. H. Sherman, "An Adaptive Multi-Level Method for Elliptic Boundary Value Problems", ***Computing,*** to appear.

ON THE CHOICE OF DISCRETIZATION FOR SOLVING P.D.E.'S ON A MULTI-PROCESSOR

L. Ridgway Scott

Department of Mathematics
University of Michigan
Ann Arbor, Michigan

There are several algorithmic questions associated with solving partial differential equations (p.d.e.'s) in a multi-processor architecture. To name three, there are 1) the type of discretization technique used (finite difference, finite element, boundary integral, etc.), 2) mapping an irregular (discretized) problem domain onto a (regular) grid of processors, and 3) solution of linear equations. The topic of this note will be the first of these three; here we shall simply summarize the contents of the report [6] because of spatial limitations. Topic number 2) above has been studied in the context of the so-called "finite element machine" architecture, to be described later, by Bokhari [2]. It reduces mathematically to a graph-theoretic mapping problem. Topic 3) has been discussed by several different authors at this conference, hence no further attention will be given here.

ISBN 0-12-632620-7

The main focus of our attention will be on data communications on a multi-processor, and on how different discretization techniques utilize the communications network more or less efficiently. Our model architecture is a multiple-instruction/multiple-data-storage device (MI/MD) with direct "nearest-neighbor" communication links between adjacent processors. A multi-processor using a single-instruction-stream architecture has been discussed at this conference by David Hunt. An example of an MI/MD architecture is the so-called "finite element machine" (FEM) being developed at the NASA Langley Research Center. It is presently configured as an $n \times n$ grid of micro-processors ($n \leq 32$) with each processor connected to eight nearest-neighbors, together with a global bus connecting all processors. For more information on the FEM, see the works of Jordan and Podsiadlo [4] and Shah [7], as well as the references therein.

The local, direct connections of a grid of micro-processors such as the FEM are particularly well suited to solving partial differential equations, since differential operators are themselves local. To evaluate (an approximate value of) the derivative of a function at one point, one needs only to know the function's values at nearby points. If the discretized problem domain is mapped onto the grid of processors in such a way that nearby spatial points go to nearby processors, then derivatives (differences) are easy to evaluate using the local, or nearest-neighbor, connections for data communication.

If the discrete problem domain is a regular grid that is isomorphic to (a subset of) the grid of processors, then it is easy to envision how the multi-processor is utilized. Each discrete unknown is assigned to a different processor (via the isomorphism), and that processor sees to it that the corresponding equation for its unknown is solved (say, via an iterative process). For such a problem, a single-instruction-stream multi-processor is probably sufficient. At any rate, it seems unrealistic to assume that the number of processors available will not be less than the number of discrete unknowns in the problem. Thus a natural question is how to <u>consolidate</u> several discrete unknowns, to be assigned as a group to one processor, in a simple way.

One natural consolidation comes via the use of high-order finite elements which have several nodal variables associated with each mesh point. All standard nodal variables are associated with either a vertex, edge, or triangle interior (we shall think from now on about triangular elements for two-dimensional problems). With a rather coarse mesh, one can achieve good accuracy by using high-order piecewise polynomials, even for problems having singular solutions (cf. Rossow <u>et</u> <u>al</u>. [5] and Babuška <u>et</u> <u>al</u>. [1]). Such elements lead to a small number of <u>nodes</u> (e.g., vertices and edges) having many unknowns associated with each node. Thus a natural consolidation consists of grouping all nodal variables associated with a particular vertex or edge (interior nodes may be eliminated via static condensation) and assigning each group to a different processor.

The exact communications requirements of finite element schemes, consolidated as above, have been described in [6]. The main conclusion is that vertex nodes require more than just nearest-neighbor connections, even on a regular grid.

This leads to two questions requiring further study: a) the use of next-nearest-neighbor connections (via software) on a multi-processor and b) the development of higher-order elements having no vertex nodes. Since the nodal interactions (i.e. required communication links) of vertex nodes are still local, a significant amount of information could be passed via nearest-neighbor links between processors that are not themselves nearest-neighbors, by which we mean a next-nearest-neighbor connection. Concerning elements with no vertex nodes, a piecewise-linear, nonconforming element exists (Crouzeix and Raviart [3]), but a higher-order generalization does not appear to have been developed yet. The difficulties of such a construction are discussed briefly in [6].

ACKNOWLEDGEMENT

The research reported here was carried out in part while the author was visiting ICASE, NASA Langley Research Center.

REFERENCES

[1] I. Babuška, B.A. Szabo, and I.N. Katz, "The P-version of the finite element method, "Report WU/CCM-79/1, Washington Univ. (1979).
[2] S. Bokhari, "On the mapping problem," Proc. 1979 International Conf. on Parallel Processing, Aug. 1979, pp. 239-248.
[3] M. Crouzeix and P.A. Raviart, "Conforming and nonconforming finite element methods for solving the stationary Stokes equations," R.A.I.R.O. vol. R-3, pp.33-76 (1973).
[4] H.F. Jordan and D.A. Podsiadlo, "A conjugate gradient program for the finite element machine," preprint, Computer Systems Design Group, Dept. of Elec. Eng., Univ. of Colorado, 1980.
[5] M.P. Rossow, K.C. Chen, and J.-C. Lee, "Computer implementation of the constraint method," Computers & Structures vol. 6, pp. 203-209 (1976).
[6] L.R. Scott,"On the choice of discretizations for solving p.d.e.'s on the finite element machine,"ICASE Report,to appear.
[7] A.K. Shah,"Group broadcast mode of interprocessor communication for the finite element machine," thesis, Computer Systems Design Group, Dept.of Elec.Eng.,Univ.of Col.,1980.

A SOFTWARE PACKAGE FOR ELLIPTIC PARTIAL DIFFERENTIAL EQUATIONS[1]

Vincent G. Sigillito

Applied Physics Laboratory
The Johns Hopkins University
Laurel, Maryland

1. INTRODUCTION

The FORTRAN software package described in this brief paper uses explicit a priori inequalities and Chevyshev basis functions to compute approximate solutions of the Dirichlet, Neumann, and Robin boundary value problems for the constant coefficient second-order linear elliptic operator

$$Lu = au_{xx} + bu_{xy} + cu_{yy} + du_x + eu_y - qu \tag{1}$$

on rectangular domains.

Providing one can find an appropriate a priori inequality for the boundary value problem of interest, these inequalities allow one to (i) compute approximate solutions using basis functions which do not need to satisfy either the differential equation or the boundary conditions, (ii) automatically give norm error bounds for the approximation computed, and (iii) allow computation of pointwise error bounds, providing a fundamental solution or paramatrix is available for the operator (1), (this feature was not built into the software package described here).

Space considerations do not permit the development here of the main ideas behind the use of explicit a priori inequalities; for this information the reader is referred to the

[1]This work was supported by the Department of the Navy, Naval Sea Command under Contract No. N00024-C-5384.

ISBN 0-12-632620-7

papers (1-5) and the book (6) where most of the work with explicit a priori inequalities is summarized. A more detailed discussion of the package will be given in (7). Suffice it to say the method leads to the linear system $CA = D$ where C is a matrix which depends only on the operator L, the boundary operator B, and the basis function ϕ_i; the vector $A = (a_1, a_2, \cdots, a_n)^T$ is the vector of unknown coefficients of the approximate solution $u = \sum a_i \phi_i(x,y)$ and the vector D depends on the prescribed data and the basis functions.

2. ORGANIZATION OF THE PACKAGE

For our purposes the package is best considered to be composed of a main program (DRIVER) which calls a subroutine (APPROX) in which all the action occurs. Actually, this subroutine is itself organized in a modular fashion and calls a number of subroutines and functions; its organization is shown in Figure 1.

The principal use of DRIVER is to set aside the proper storage for the matrices generated by the approximation method and allow the user to read in the coefficients of the elliptic operator, the ordered pairs which define the rectangle, and a logical variable which indicates the type of boundary value problem to be solved.

In subroutine APPROX the first thing which is done is to compute the coefficients of the elliptic operator transformed to the standard domain $-1 \leq x \leq 1$, $-1 \leq y \leq 1$ and then determine the explicit constants for the appropriate a priori inequality to be used (the determination of these constants

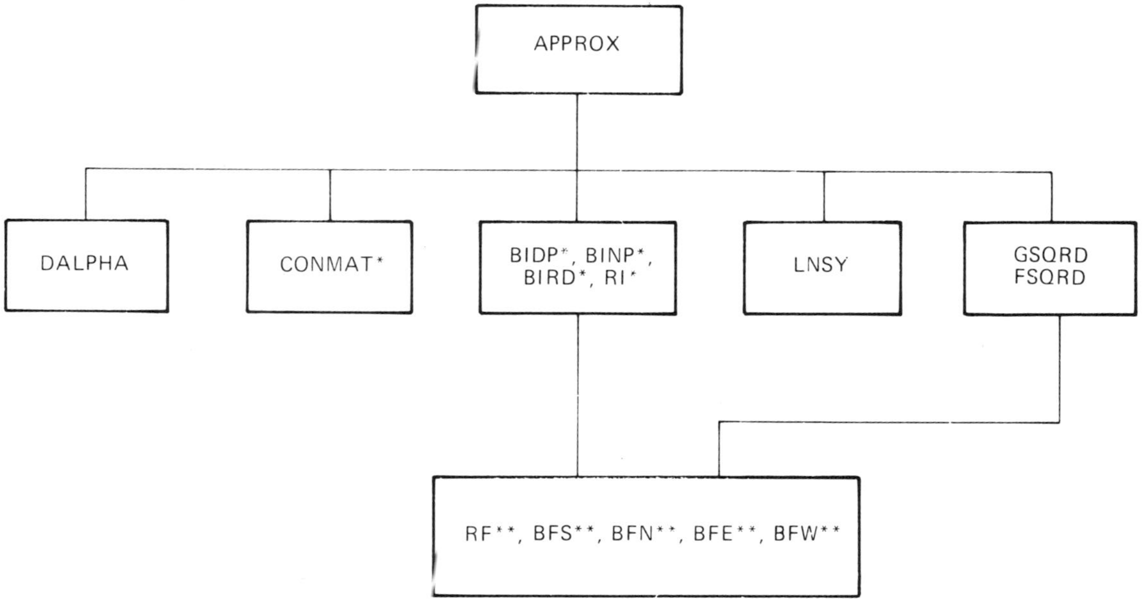

* Denotes routines or functions which call user supplied functions.
** Denotes functions which the user must supply.

for the case of the Dirichlet problem is sufficiently involved that these calculations have been relegated to a separate subroutine DALPHA). Now the main matrix C is constructed (in CONMAT). It should be emphasized that this subroutine is independent of the problem treated but does depend on the choice of basis functions (at this point we have used only products of Chebyshev polynomials, later work will involve writing CONMATs which will be based on other types of basis functions including bicubic Hermite polynomials).

The computation of the D vector involves the prescribed boundary functions and homogenous right-hand side of the differential equation. These, of course, must be supplied by the user and need be only single-line FORTRAN functions. In Figure 1 they are represented by the names BFS, BFN, BFE, BFW, and RF. The functions starting with B denote prescribed boundary functions and the last letter S, N, E, or W denotes the portion of the boundary, for instance, S stands for south, N for north, etc. Thus, the boundary function BFS gives the prescribed boundary function on the lower horizontal boundary of the standard region. The functions BIDP (= boundary integrals for Dirichlet Problem), BINP, BIRD, and RI use these functions in computing the appropriate components of the D vector. Once the system CA = D has been constructed, the routine LNSY is used to solve the system. The method uses Gaussian elimination with equilibration, partial pivoting, and backward substitution. The functions GSQRD and FSQRD are used to compute the norm errors on the boundary and over the region, respectively, which are then combined into a single norm error bound. As indicated in Figure 1, these functions also use the five user supplied functions RF, BFS, BFN, BFE, and BFW.

Space does not permit a discussion of typical results obtained using this package. This topic will be dealt with in detail in (7).

REFERENCES

1. Payne, L. E. and Weinberger, H. F., New bounds for solutions of second order elliptic partial differential equations, Pac. J. Math. 8, 551 (1958).
2. Bramble, J. H. and Payne, L. E., Bounds in the Neumann problem for second order uniformly elliptic operators, Pac. J. Math. 12, 823 (1962).
3. Bramble, J. H. and Payne, L. E., Some integral inequalities for uniformly elliptic operators, Contributions to Differential Equations 1, 129 (1963).
4. Sigillito, V. G., On a continuous method of approximating solutions of the heat equation, J. Assoc. Comp. Mach. 14, 732 (1967).
5. Sigillito, V. G., A priori inequalities and approximate solutions of the first boundary value problem for $\Delta^2 u = f$, SIAM J. Numer. Anal. 13, 251 (1976).
6. Sigillito, V. G., "A Priori Inequalities with Applications to Boundary Value Problems," Pitman Publishing Limited, London (1977).
7. Sigillito, V. G., Software for elliptic PDEs, (in preparation).

AN EMPIRICAL INVESTIGATION OF METHODS FOR NONSYMMETRIC LINEAR SYSTEMS

Andrew H. Sherman[1]

Department of Computer Sciences
University of Texas at Austin
Austin, Texas

This paper compares six techniques for the solution of nonsymmetric linear systems arising from finite difference discretizations of the convection-diffusion equation.

I. INTRODUCTION

In this paper we compare methods for solving linear algebraic systems that arise from finite difference discretizations of the elliptic convection-diffusion equation

$$-(a_1u_x)_x - (a_2u_y)_y + b_1u_x + b_2u_y + cu = f \qquad (1)$$

in a planar region Ω with Dirichlet boundary conditions. Such

[1]Work supported in part by the University of Toronto, by NASA under grant NSG-1632, and by the Office of Naval Research under grant N00014-80-C-0645. Much of this work was performed while the author was on leave at the University of Toronto.

ISBN 0-12-632620-7

linear systems are typically of the form

$$Ax = b \tag{2}$$

where A is an NxN sparse nonsymmetric matrix. We are especially interested in nonseparable equations (1) for which the coefficient functions are general functions of the spatial variables. Hence we consider only general purpose methods for solving (2), excluding methods that work well for less general problems.

II. DISCRETIZATIONS

For purposes of this section, we assume that a regular rectilinear mesh of width h has been imposed on Ω. We consider three discretizations: central differences (CD), upstream differences (UD), and modified upstream differences (MD). All of these use a standard five-point difference approximation to the self-adjoint terms in (1): $-(a_1u_x)_x - (a_2u_y)_y + cu$ (see [1]). They differ only in their treatments of the convection terms.

In CD, second-order centered differences are used to approximate convection terms. This gives an approximation that is $O(h^2)$ accurate at each mesh point. The difficulty is that numerical stability is only guaranteed when the local Peclet number

$$\kappa = \max\{|b_1(x,y)|/h|a_1(x,y)|,|b_2(x,y)|/h|a_2(x,y)|\} \tag{3}$$

satisfies $\kappa \le 2$ (see [2]).

To avoid this problem, upstream differences are often used. Here forward or backward differences are used for the convection terms, with the choice at each point made so that the contribution to the diagonal term of A is positive. Now stability is guaranteed since A is a diagonally-dominant M-matrix having a positive inverse. However, the approximation is only $O(h)$ accurate.

Modified upstream differences are a compromise suggested in [2-4]. The idea is to include the first term of the truncation error expansion of upstream differences (corresponding to a second partial derivative) in the self-adjoint portion of (1) at each point. Then, based on the sign of the convection coefficient, the self-adjoint terms are rewritten to guarantee a diagonally-dominant matrix. The overall scheme is $0(h^2)$ accurate at each point, but the constant for the leading term of the error is a factor of κ larger than those for CD or UD. For large κ, this is a disadvantage, even though the method remains stable.

As an example of MD, consider the operator

$$- (a(x)u_x)_x + b(x)u_x \tag{4}$$

and assume $b(P) < 0$ at mesh point P. The leading term of the error in the upstream approximation of $b(P)u_x$ is $(b(P)/2)hu_{xx}(P)$. Adding this into the self-adjoint portion of (4) gives the operator $- [(a(x) + \frac{b(P)h}{2})u_x]_x$, which will not lead to a positive diagonal matrix entry for large negative $b(P)$. However, a second-order approximation to this operator is $(u_x/(a(x) - b(P)h/2))_x$, and this modified operator does have the desired properties. An analagous modification is used when $b(P) \geq 0$.

III. SOLUTION METHODS

We consider six methods for solving (2). Three variants of Gaussian elimination have been chosen as representatives of state-of-the-art software for direct methods under different assumptions about pivoting. In addition, we have included three iterative methods that appeared particularly promising.

The three direct methods are the compressed storage code from the Yale Sparse Matrix Package (YSMP) [5], NSPFAC (similar to [6]), and MA28 [7]. All are extensively documented elsewhere, so we provide little information here. We only note that YSMP does no pivoting for stability, while NSPFAC and MA28 do threshold column and complete pivoting, respectively. (We used threshold values of .1.) In addition, we used the YSMP ordering package as a pre-ordering for NSPFAC.

Two of the iterative methods are versions of the pre-conditioned conjugate residual iteration [8]. Both methods use a pre-conditioning corresponding to an incomplete LU factorization of A. In CR1, the sparsity pattern of L+U was restricted to that of A (see [8]), while in CR2, L and U were obtained by applying NSPFAC to A without allowing fill-in.

The third iterative method is a variant of the generalized conjugate gradient procedure (GCG) of [9-10]. In this method A is pre-conditioned by its symmetric part $M = (A+A^T)/2$, and we used the symmetric codes in YSMP to solve systems involving M. (For general problems, the fast direct methods suggested in [9-10] do not apply.)

IV. COMPARISONS AND CONCLUSIONS

We have compared the six methods described in the last section on several moderately large equations (1). In this section we present the results for one series of problems which are typical of all our tests. The equation (see[11]) is

$$- u_{xx} - ((1+xy)u_y)_y - \beta(\cos(x)u_x + (\exp(-x)+x)u_y) + 3u = f \quad (5)$$

on the unit square with f and the boundary conditions chosen to make the true solution

$$u(x,y) = \exp(-y^2) + \sin(x-y+1/2) \quad (6)$$

β took on the values 0,1,10,100,1000, and 10000. All tests were run in double precision on an IBM 370/165 using the Fortran IV H-extended compiler with OPT=2. N=900 in our examples.

Figures 1-3 present plots of time versus β for all solution methods with each discretization, and Table 1 presents the approximate amount of array storage required for each solution method. The iterative methods were stopped when the L_2 error fell below 10^{-5}. Not all methods succeeded in all cases. No firm conclusions can be drawn from these results, but it appears that YSMP and NSPFAC do quite well throughout and that CR1 and CR2 are especially effective for large β. We note that the time increase for YSMP and NSPFAC for large β is due entirely to the system overhead time in processing underflow exceptions. (This might serve as a warning to those who compare codes -- underflow may have a significant effect on statistics, if not numerical results.)

REFERENCES

1. Varga, R.,"Matrix Iterative Analysis". Prentice-Hall, Englewood Cliffs, NJ, 1965.
2. Axelsson, O. and Gustafsson, I., *JIMA 23*, 321 (1979).
3. Barrett, K., *J.Mech.Appl.Math. 27*, 57 (1974).
4. Ilin, A., *Mat.Zametki 6*, 237, *Math.Notes 6*, 986 (1969).
5. Eisenstat, S., Gursky, M., Schultz, M., and Sherman, A., "The Yale Sparse Matrix Package", I and II. Reports 112 and 114, Dept. of Computer Science, Yale University (1977).
6. Sherman, A., *Trans.Math.Soft. 4*, 391 (1978).
7. Duff, I., "MA28-A set of Fortran Subroutines for Sparse Unsymmetric Linear Equations," Report AERE-R8730, HMSO, London (1977).
8. Eisenstat, S., Elman, H., Schultz, M. and Sherman, A., "Fifth SPE Symposium on Reservoir Simulation," 127 (1979).
9. Concus, P. and Golub, G., in "Lecture Notes in Economics and Mathematical Systems," vol. 134, Springer-Verlag, Berlin, 1976.
10. Widlund, O., *SIAM J.Num.Anal.15*, 801 (1978).
11. Houstis, E., Lynch, R., Papatheodorou, T., and Rice, J. *J.Comp.Phys. 27*, 323 (1978).

TABLE 1. Storage Requirements (Double Precision Entries)

Method	YSMP	NSPFAC	MA28	CR1	CR2	GCG
Storage	35000	37000	52000	14000	19000	35000

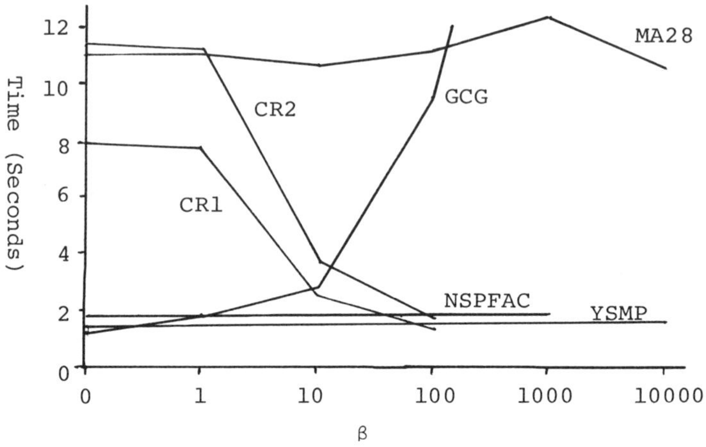

FIGURE 1.

Results for CD

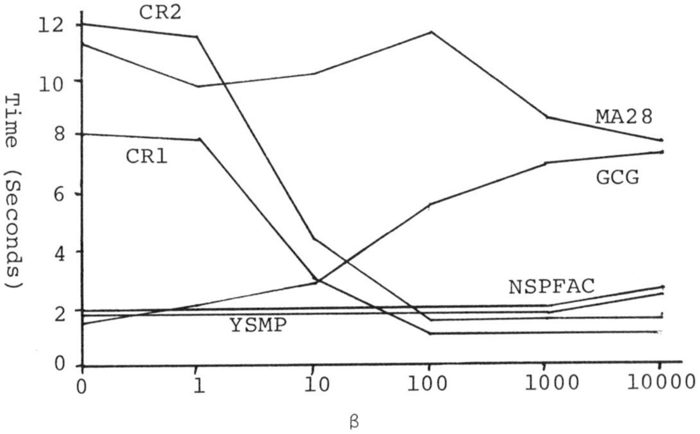

FIGURE 2.

Results for UD

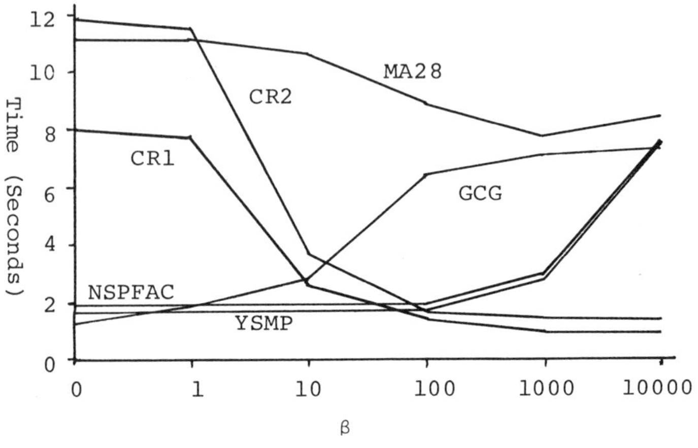

FIGURE 3.

Results for MD

SEMICONDUCTOR DEVICE SIMULATION

C. L. Wilson
Electron Devices Division

J. L. Blue
Scientific Computing Division
National Bureau of Standards
Washington, DC

The basic semiconductor device equations [1] can be normalized [2] to a coupled nonlinear elliptic system:

$$-\nabla\cdot(\varepsilon\nabla\Psi) = N + \exp(\phi_p-\Psi) - \exp(\Psi-\phi_n) \quad (1)$$

$$\nabla\cdot[\mu_n \exp(\Psi-\phi_n)\nabla\phi_n] = R \quad (2)$$

$$-\nabla\cdot[\mu_p \exp(\phi_p-\Psi)\nabla\phi_p] = R \quad (3)$$

$$R \equiv \frac{\exp(\phi_p-\phi_n)-1}{\tau_{po}\ [\exp(\Psi-\phi_n)+1] + \tau_{no}\ [\exp(\phi_p-\Psi)+1]} \quad (4)$$

All quantities are dimensionless; ε is the permittivity of the material, N is the net doping density, Ψ is the electrostatic potential, ϕ_n and ϕ_p are the hole and electron quasi-Fermi levels, μ_p and μ_n are hole and electron mobilities, and τ_{po} and τ_{no} are the asymptotic hole and electron life times. The independent variables in this system are Ψ, ϕ_n and ϕ_p.

The known solutions to this system of equations have very large dynamic ranges, $\pm 10^4$ for Ψ, -10^4 to $+20$ for $(\Psi-\phi_n)$ and $(\phi_p-\Psi)$ and $\pm 10^{10}$ for N. The solutions contain abrupt transitions from one boundary value extreme to another at locations which depend sensitively on the problem boundary conditions, net doping density, and geometry.

Some of the unique properties of semiconductor device simulations can be seen by studying the simulation of a

ISBN 0-12-632620-7

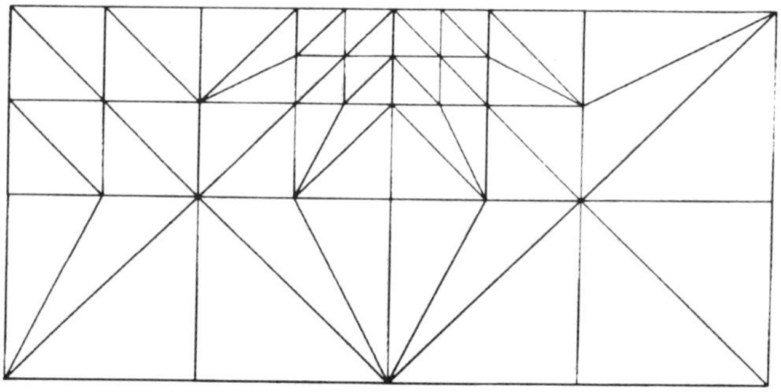

Figure 1a. Schottky Barrier - Initial Mesh

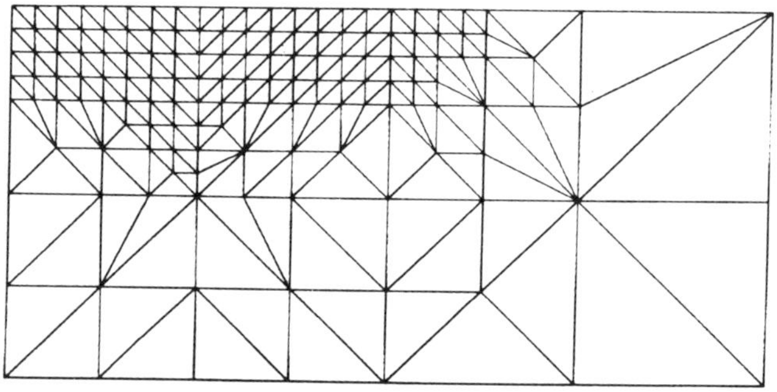

Figure 1b. Schottky Barrier - First Refinement

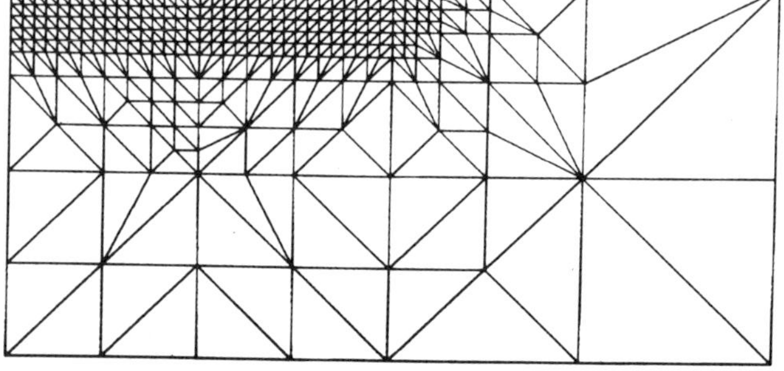

Figure 1c. Schottky Barrier - Second Refinement

Schottky barrier diode and the charge-sheet model of a MOS transistor. A good approximation to the Schottky barrier diode on the region shown in Fig. 1 can be obtained from

$$\nabla^2 u = N - e^u \tag{5}$$

The value of the solution is specified on the Schottky barrier on the left half of the top boundary as u = -50 and on the bottom boundary as u = log (N). All other boundaries have zero normal derivatives. Ignoring n and p requires that $V_{bias} \leq 0$ if N>0; for simplicity, we take N to be constant, $N=10^5$. The problem is solved using linear finite elements on triangles with adaptive triangulation as described elsewhere [3]. On each level of triangulation an initial guess of the solution is made and the non-linear equation is solved using a damped Newton method. Using the solution to the nonlinear equations, the error in each triangle is estimated, and the triangles with the largest errors are subdivided to produce a finer mesh. The sequence of meshes generated by this process is shown in Fig. 1a, 1b and 1c. This is evident from the contour plot of the equipotential of the solution shown in Fig. 2. This problem was run on a DEC-10 computer using 10,355 words of data storage and 32 sec. of execution time.

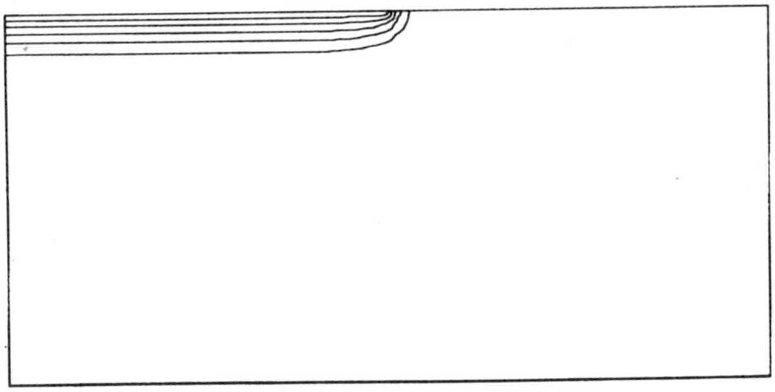

Figure 2. Schottky Barrier Equipotentials

The charge sheet model of a MOS transistor is obtained by approximating the hole quasi-Fermi level by log(N) + Ψ_{sub}, where Ψ_{sub} is the boundary condition on the bottom edge of the irregular region shown in Fig. 3. The two straight ends have

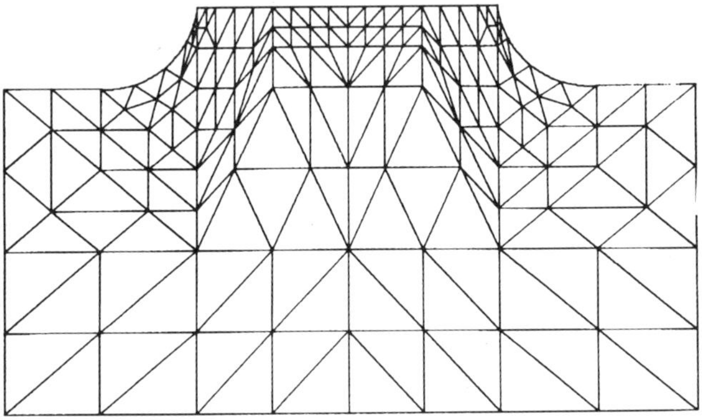

Figure 3. MOS Transistor Mesh

zero normal derivative boundary conditions. The boundary conditions on the two partially curved upper corner regions are Ψ_S and Ψ_D. The upper edge of the region is subject to

$$\varepsilon_{ox} \frac{\partial \Psi}{\partial n}\Big|_{0-} - \varepsilon_s \frac{\partial \Psi}{\partial n}\Big|_{0+} = N_s(x) = \int_0^x \exp(\Psi - \phi_n)\,dy \qquad (6)$$

This is computed by a variant of the charge sheet method discussed in [4]. In the MOS transistor a thin oxide layer covers the top of the transistor. The normal derivative, $\partial\Psi/\partial n|_{0-}$, is obtained in the oxide using fast Poisson methods [5]. A typical mesh is shown in Fig. 3 and a typical potential solution is shown in Fig. 4. This solution was obtained on a DEC-10 computer using 5645 words of data storage and 37 sec. of execution time.

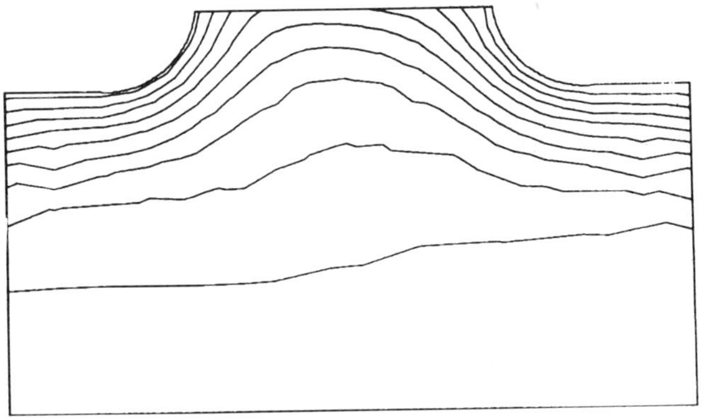

Figure 4. MOS Transistor Equipotentials

For the semiconductor device equations, using adaptive meshes can provide significant savings in computer time.

Acknowlegments

The authors would like to acknowledge the computer support obtained from G. W. Brown and C. W. Gwyn of Sandia National Laboratory, Albuquerque, N.M.

References

1. Sze, S. M., "Physics of Semiconductor Devices", p. 65-72, Wiley, New York, (1969).
2. Gummel, H. K., IEEE Trans. Electron Devices, ED-11, 455 (1964).
3. Bank, R. E., and Sherman, A. H., in "Sparse Matrix Proceedings" (I.S. Dugg and G. W. Stewart, ed.), SIAM, (1979).
4. Brews, J. R., Solid State Elect., 21, 345, (1978).
5. Swarztrauber, P., and Sweet, R., NCAR note, NCAR-TN/lA-109, (1975).

Index

N

O

P

Q

R

S